AF443331

Wind Power Politics and Policy

WIND POWER POLITICS AND POLICY

Scott Victor Valentine

OXFORD
UNIVERSITY PRESS

OXFORD
UNIVERSITY PRESS

Oxford University Press is a department of the University of
Oxford. It furthers the University's objective of excellence in research,
scholarship, and education by publishing worldwide.

Oxford   New York
Auckland   Cape Town   Dar es Salaam   Hong Kong   Karachi
Kuala Lumpur   Madrid   Melbourne   Mexico City   Nairobi
New Delhi   Shanghai   Taipei   Toronto

With offices in
Argentina   Austria   Brazil   Chile   Czech Republic   France   Greece
Guatemala   Hungary   Italy   Japan   Poland   Portugal   Singapore
South Korea   Switzerland   Thailand   Turkey   Ukraine   Vietnam

Oxford is a registered trademark of Oxford University Press
in the UK and certain other countries.

Published in the United States of America by
Oxford University Press
198 Madison Avenue, New York, NY 10016

Library of Congress Cataloging-in-Publication Data
Valentine, Scott V.
Wind power politics and policy / Scott Victor Valentine.
pages cm
Includes bibliographical references and index.
ISBN 978–0–19–986272–6 (alk. paper)
1. Wind power industry. 2. Wind power—Economic aspects. 3. Wind power—Government policy.
I. Title.
HD9502.5.W552V35 2015
333.9'2—dc23
2014014224

9 8 7 6 5 4 3 2 1
Printed in the United States of America
on acid-free paper

CONTENTS

ACKNOWLEDGMENTS

This book represents a synthesis of case study research that I have undertaken over the past decade. As my work has progressed in this field, I cannot help but be fascinated by the contextually infused energy sector dynamics that are evident in each nation. The disparate influences on wind power diffusion that were exhibited in each nation inspired me to try and seek a way to conflate the insights that emerged. I hope that I have done this quest justice, but if I have fallen short or erred in the process, the responsibility is mine alone to bear.

As I struggled to relate what I was observing in wind power policy back to a broader policy context, I found myself supported by new and old friends—some of whom I spoke with in person and others who spoke to me through their insightful prose. I am extremely grateful to the authors of all the works that have been cited in the chapter notes. Without these academic foundations, the observations from my research would lack substance and support.

In the energy policy and the policy theory fields, I have been blessed by being able to hone my craft under a number of skilled academics who have provided me with valuable insights. They are too numerous to list here without running the risk of forgetting someone, but you know who you are, and I truly thank you for helping me to hone my craft.

During the course of my research, I received financial support from a number organizations to which I am extremely grateful. Thanks are owed to the Lee Kuan Yew School of Public Policy at the National University of Singapore for hosting the early stages of this research. Thanks also to the Alliance for Global Sustainability at the University of Tokyo for funding some of the case study research. Finally, I owe a debt of gratitude to the Department of Public Policy and the School of Energy and Environment at the City University of Hong Kong for providing me with support to complete this project.

At Oxford University Press, Jeremy Lewis, Augustine Leo and the rest of the team have been highly responsive and supportive throughout the process. My warm thanks go out to the entire team at OUP and to the anonymous reviewers who helped refine the book. Thanks also to Lolita Ko for help on the *Political SET* graphic.

Last but certainly not least, a special thank you is extended to my wife, Rebecca, and my daughter, Elle, for providing me with both the support and motivation needed to complete this book. To you both and to the memory of my first teachers—Ellery and Victor, I dedicate this work.

Scott Victor Valentine
July 14, 2014
Hong Kong

Wind Power Politics and Policy

CHAPTER 1

Introduction: The Global Imperative

The climate centres around the world, which are the equivalent of the pathology lab of a hospital, have reported the Earth's physical condition, and the climate specialists see it as seriously ill, and soon to pass into a morbid fever that may last as long as 100,000 years. I have to tell you, as members of the Earth's family and an intimate part of it, that you and especially civilisation are in grave danger.

—James Lovelock, 2006[1]

Climate change presents a unique challenge for economics: it is the greatest and widest-ranging market failure ever seen. . . . Our actions over the coming few decades could create risks of major disruption to economic and social activity, later in this century and in the next, on a scale similar to those associated with the great wars and the economic depression of the first half of the 20th century. And it will be difficult or impossible to reverse these changes.

—Sir Nicholas Stern, 2006[2]

This is a global problem, and it will require a global coalition to solve it. Our climate knows no boundaries; the decisions of any nation will affect every nation. . . . It is estimated that if we fully pursue our potential for wind energy on land and offshore, wind can generate as much as 20 percent of our electricity by 2030, creating as many as 250,000 jobs in the process. As with so many clean energy investments, it's win-win: good for the environment and great for our economy.

—US President Barack Obama, 2009[3]

1.1 THE GLOBAL IMPERATIVE

The years 2006–2007 represented an intellectual tipping point for climate change advocacy. Over this short period of time, there was ample evidence of a general convergence of understanding between many environmentalists and economists on the perilous threat posed by climate change.

In the summer of 2006, the release of Al Gore's *An Inconvenient Truth* turned climate change into an issue of public concern in the United States. The domestic debate that the film helped inspire escalated over the next year to a point where energy policy suddenly became a vote swaying issue in American politics. This development became a topic of interest for the rest of the world because signs of a weakening in American reticence toward climate change mitigation would have significant repercussions for the 128 nations that were struggling to keep the Kyoto Protocol from falling apart.[4]

In October 2006, a comprehensive independent study called the *Stern Review*, commissioned by the Chancellor of the Exchequer in the United Kingdom, presented an assessment of the anticipated impacts of climate change. As a foreboding sign of the content which would follow, the report began by describing climate change as "the greatest and widest ranging market failure ever seen" (p. i).[5] The report concluded that the long-term costs of climate change were expected to be so great that early action to abate global warming was the most cost-effective alternative. It estimated that the net benefits (benefits less costs) from reducing greenhouse gas (GHG) emissions to achieve a stabilization level of 550 parts per million (ppm) by 2050 would be in the neighborhood of US$2.5 trillion.

In February 2007, the first of four reports that comprise the Fourth Assessment Report of the United Nation's Intergovernmental Panel on Climate Change (IPCC) was released. The goal of this first report was to "describe progress in understanding of the human and natural drivers of climate change, observed climate change, climate processes and attribution, and estimates of projected future climate change."[6] Overall, the report upgraded international agreement on the likelihood of human activities being responsible for global warming from *likely* (66% or greater probability) to *very likely* (90% or greater probability). The data presented in the report was unexceptional in the sense that it mirrored data already available in the public domain; however, the report was significant in that it represented a consensus view of UN member nations. Symbolically, it represented humanity's formal acceptance of culpability for causing climate change, because it was endorsed by all United Nation (UN) member nations.

In April 2007, the second of four reports that comprise the Fourth Assessment Report of the IPCC was released. This second report focused on "current scientific understanding of impacts of climate change on natural, managed and human systems, the capacity of these systems to adapt and their vulnerability."[7] Comparatively, the report was less comprehensive than the Stern Review in its assessment of the current and anticipated economic impacts of climate change on humanity and global ecosystems. However, it did serve to solidify the emergent consensus that climate change was significantly harming hydrological, terrestrial, and biological systems.

Given the emergent international consensus that climate change was an immediate threat to both the social and economic well-being of humanity, the intuitive international response should have been to cast vested national interests aside, hoist the sails of initiative, and embark on a journey of rigorous greenhouse gas abatement. However, such departures have not materialized. In fact, one is tempted to glibly question whether members of the international policy community have misconstrued the Stern Review's admonition "delay in taking action on climate change would make it necessary to accept both more climate change and, eventually, higher mitigation costs" (p. xv) as a policy recommendation.[8]

1.2 ENERGY AND THE GLOBAL IMPERATIVE

Of the six greenhouse gases covered under the Kyoto Protocol—carbon dioxide (CO_2), methane (CH_4), nitrous oxide (N_2O), and three fluorine gases (HFCs, PFCs, and SF_6)—CO_2 emissions represent by far the largest anthropogenic contribution to elevated greenhouse gas concentrations due to the sheer volume of annual CO_2 emissions. Figure 1.1 provides an indication of how substantial CO_2 emissions are in comparison to other key gases.[9] As the chart illustrates, CO_2 emissions from all sources accounted for approximately 75% of all GHG emissions in 2004, while energy-related CO_2 emissions alone accounted for over 57% of all GHG emissions. In terms of the

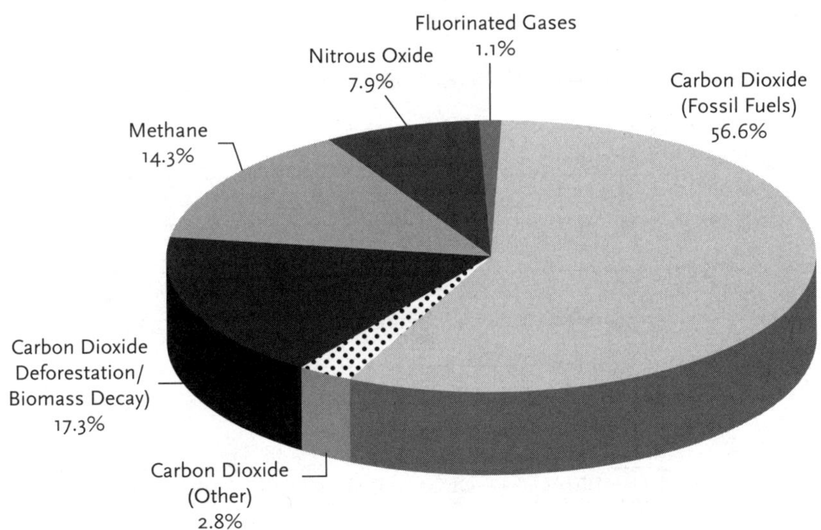

Figure 1.1. Global Greenhouse Gas Emissions, 2004
Source of data: IPCC AR4 (2007).

other key greenhouse gases, methane emissions accounted for approximately 14% of total GHG emissions and nitrous oxide emissions accounted for approximately 8% of the total. The remaining three fluorinated gases represent a very small proportion of GHG emissions.

The main hurdle stymieing international efforts to reduce CO_2 emissions appears to be the difficulty, that all countries are having, breaking free from a dependence on fossil fuel energy. As UN Secretary General Ban Ki Moon pointed out in his 2008 World Environment Day Message:

> Addiction is a terrible thing. It consumes and controls us, makes us deny important truths and blinds us to the consequences of our actions. Our world is in the grip of a dangerous carbon habit. . . . The environmental, economic and political implications of global warming are profound. Ecosystems—from mountain to ocean, from the poles to the tropics—are undergoing rapid change. Low-lying cities face inundation, fertile lands are turning to desert, and weather patterns are becoming ever more unpredictable.[10]

As Figure 1.1 indicates, CO_2 emissions from fossil fuel *combustion* accounted for approximately 57% of all GHG emissions. Clearly, if humanity is to avoid the worst effects of global warming alluded to by the Stern Review and the IPCC 4th Assessment Report, progress in terms of reducing emissions related to fossil fuel combustion is essential. Unfortunately, data points to increasing—not decreasing—trends in fossil fuel-related CO_2 emissions. As Figure 1.2 indicates, total combustion-related

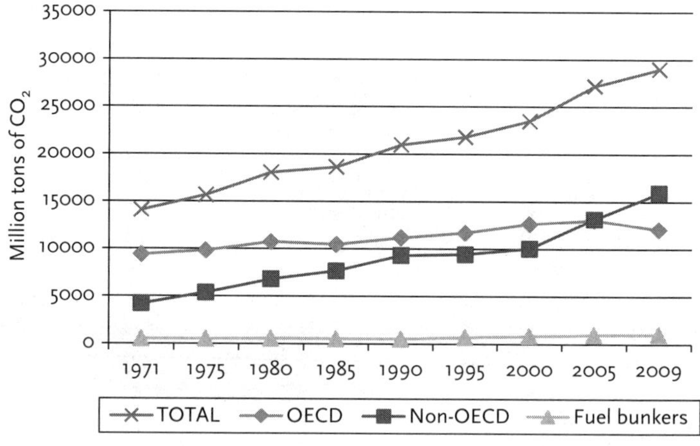

Figure 1.2. Global CO_2 Emission Trends
Source: IEA, CO2 Emissions from Fuel Combustion (2011).

CO_2 emissions doubled between 1971 and 2009.[11] In 2005, for the first time, CO_2 emissions in non-OECD nations in aggregate exceeded CO_2 emissions from OECD nations. Since then, CO_2 emissions from non-OECD nations have risen to the point where, in 2009, they exceeded aggregate CO_2 emissions from OECD nations by 32%. This is troubling given that most of the non-OECD nations are not Annex I nations under the Kyoto Protocol, and as such have no formal international obligations to reduce GHG emissions to meet agreed upon targets.

Looking forward, according to the International Energy Agency's (IEA) *World Energy Outlook 2010*, global CO_2 emissions associated with fossil fuel combustion are expected to reach 35.4 billion tons by 2035.[12] This represents a 22% increase over 2009 levels and produces an overall GHG emissions portfolio, which equates with emissions projected under the IPCC AR4's worst-case scenario that projected an average global temperature rise of between 2.4°C and 6.4°C by 2100. Thus, despite indications that CO_2 emission reductions of up to 80% are needed to abate the worst impacts of global warming,[13] CO_2 emission trends indicate that emissions will increase rather than decrease.

It is notable that a great deal of global interest has arisen regarding the prospects of carbon capture and sequestration technology (CCS technology). The premise behind CCS technology is to capture CO_2 emissions from a point source (i.e., a coal-fired power plant) and then store the emissions either aquatically (deep sea injection), chemically (biological absorption), or geologically (in natural geological storage chambers)—thereby preventing CO_2 from dispersing directly into the atmosphere. Unfortunately, the volume of CO_2 that must be sequestered each year to abate global warming is of such magnitude that the management of captured CO_2 presents economic, ecological, and logistical hurdles, thereby rendering discussions about how to safely sequester such volumes to be moot.

Geological CCS, as it stands today, requires underground injection of CO_2, often in a supercritical state.[14] This generates a lot of effluent. Just how much? Stanford University's David Victor estimates that if the CO_2 generated from all the coal-fired power plants in the United States were stored geologically at pressures typical for injection, approximately 50 million barrels per day of CO_2 infused fluid would be generated.[15] This volume is over three times greater than the daily oil production in the United States.[16] Globally, over 85 million barrels of oil per day are distributed by a network that has taken decades to create.[17] Accordingly, not only would enormous distribution networks be required to transport the effluent associated with CCS technology, the potential for environmental disaster caused by injecting so much effluent into geological or aquatic storage sites is almost unfathomable.[18] In short, CCS technology may be somewhat viable as part of a short-term solution to abate the worst effects of global warming, but in its current technological manifestation it is far from a responsible solution to the global GHG emissions management challenge.

1.3 ELECTRICITY AND THE GLOBAL IMPERATIVE

1.3.1 Electricity Generation Technologies

In 2007, electricity generation constituted 37.9% of all primary energy consumed. As Figure 1.3 illustrates, the role that power generation plays in total energy demand is expected to progressively increase, eventually reaching nearly 42% of all energy demand by 2030. In short, electricity generation represents the big piece of the pie when it comes to identifying energy intensive activities.

Table 1.1 tells a bleak tale about the trajectory of CO_2 emissions associated with electricity generation: it is the Energy Information Administration's (USEIA) 2030 global electricity use forecast from 2010 broken down by fuel source. Coal-fired electricity, which is the most CO_2-intensive fossil fuel resource, is expected to exhibit the strongest growth rate of 2.7% per year. Meanwhile, the role of renewable energy technologies in global electricity generation is expected to continue to be minor. If the IEA projections hold true, by 2030 electricity generated by wind, geothermal, solar, and tide and wave technologies will still only constitute

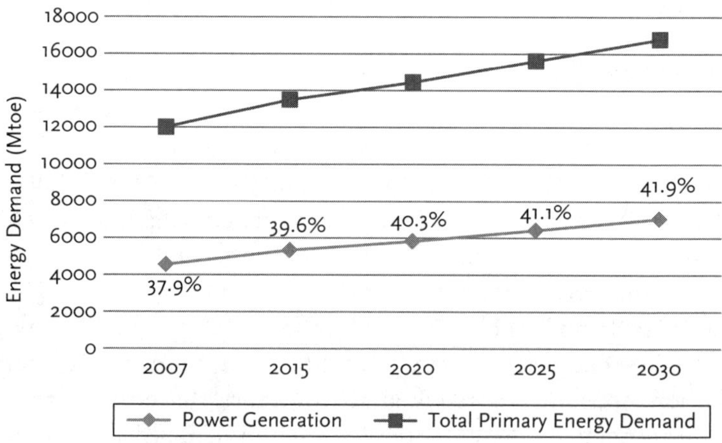

Figure 1.3. Proportion of Total Primary Energy Demand Attributed to Power Generation
Source: IEA, World Energy Outlook (2010).

Table 1.1 GLOBAL ELECTRICITY PRODUCTION BY SOURCE

(Data in Terawatt Hours)	2007	2030	Avg. Annual Growth Rate
Coal	8216	15,259	2.7
Oil	1117	665	-2.2
Gas	4126	7058	2.4
Nuclear	2719	3667	1.3
Hydro	3078	4680	1.8
Biomass and waste	259	839	5.2
Wind	173	1535	9.9
Geothermal	62	173	4.6
Solar	5	402	21.2
Tide and wave	1	13	14.6
TOTAL	**19,756**	**34,291**	**2.4**

Source: EIA, World Energy Outlook (2008).

6% of total electricity generation. On the other hand, coal-fired electricity will deliver 44.5% of the total electricity produced in 2030, up from 41.5% in 2007. These projected trends are in spite of a consensus that climate change presents an immediate, perilous threat to humanity,[19] and in spite of expectations that costs of fossil fuels will rise while the costs of wind power and other renewable power will continue to decline.[20] Although it should be emphasized that these projections were done prior to the nuclear disaster in Fukushima, Japan in March 2011, there is no guarantee that a diminished rate of nuclear power expansion will result in gains for renewable technologies over fossil fuel technologies. In fact, there are strong indications that a decline in nuclear power capacity will be of a greater benefit to fossil fuel technologies.

These trends do not bode well for prospects of energy-related CO_2 emission mitigation. As Figure 1.4 demonstrates, the IEA expects total energy-related CO_2 emissions to double, from 20,941 million tons in 1990 to 40,225 million tons in 2030. This increase will be driven by a 140% increase in CO_2 emissions associated with electricity generation, with massive increases in emissions from coal-fired and gas-fired electricity generation (Figure 1.4).

At this rate of increase in CO_2 emissions, the IEA asserts that humanity is on path that will see GHG concentrations in excess of 1,000 parts per million (ppm) of CO_2 equivalent (eq) GHG by 2150[21]—a level that would result in global warming in excess of 6°C, sea level rise in excess of 3 meters and unpredictable ecological disaster.[22] The IEA and IPCC both acknowledge that even holding atmospheric CO_2 concentrations between 450–490

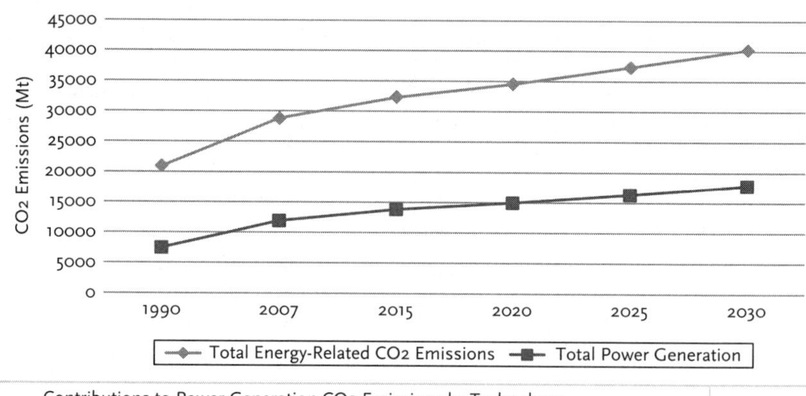

Contributions to Power Generation CO2 Emissions by Technology							
in million tons (Mt) of CO2	1990	2007	2015	2020	2025	2030	2007–2030 Increase
Coal-fired Power Generation	4929	8681	10556	11504	12680	13873	160%
Oil-fired Power Generation	1196	900	701	633	579	530	−59%
Gas-fired Power Generation	1346	2315	2562	2817	3085	3421	148%

Figure 1.4. CO_2 Emissions Associated with Primary Energy Usage and Power Generation
Source: IEA, World Energy Outlook (2010).

ppm will require CO_2 emission reductions in the range of 50 to 85% below 2000 levels by 2050; even if these targets are achieved, there is still no guarantee that global warming can be kept below 2°C. In other words, at the risk of stating the obvious, a revolution in how humanity generates electricity is needed almost immediately to abate the worst impacts attributed to climate change. The popular belief is that facilitating such a transition requires nurturing a willingness to pay higher energy prices—but as the next section details, in terms of electricity generation, this is not necessarily true.

1.3.2 The Dynamics of Electricity Prices

Historically, the sluggish diffusion of renewable energy has been rationalized in economic terms. Until recently, the cost disparity between fossil fuel power options (specifically coal and natural gas) and renewable energy alternatives has been capacious enough to discourage transition to alternative energy. However, fossil fuel prices have edged significantly higher in recent years, substantially eroding this historical competitive cost advantage.

High-grade US Northern Appalachian Coal exemplifies the volatility of fossil fuel prices. As Figure 1.5 illustrates, between December 2000 and December 2003 the trading range of this commodity was between US$25–35

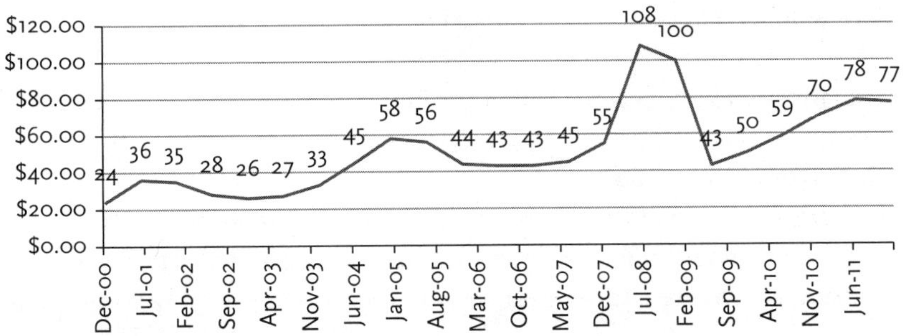

Figure 1.5. An Indication of Coal Spot Price Trends

US Energy Information Administration website: www.eia.doe.gov/coal/page/coalnews/coalmar.html. This data is based on the market prices for one short ton of Northern Appalachian coal (13,000 Btu; less than 3.0 lbs SO2 per mmBtu) as of October 2013.

Source: US Energy Information Administration.

per short ton. A late winter price spike in 2003–2004 was superseded by a ratcheting up of the trading range to US$40–45 per short ton between December 2005 and December 2007. In 2008, the commodity spiked once again, bursting through the US$100 per short ton barrier. In the aftermath, the market for US Northern Appalachian Coal has become increasingly volatile; however, the current trading range of US$70–80 per short ton appears to represent a new trading plateau. The bottom line is that coal is now trading at three times what it was at the start of 2000.[23]

Estimating the kilowatt hour (kWh) cost of energy generated by coal depends significantly on the grade of coal used and the generation technology employed; however, broadly speaking, the cost of the feedstock for generating 1 kWh can be estimated to be approximately US 3.25¢, assuming that i) Northern Appalachian coal has a thermal energy content of approximately 6,150 kWh/ton; ii) the coal sells for US$80 per short ton; and iii) the combustion technology employed exhibits a moderate 40% electricity conversion ratio. When the price was US$150 per short ton in September of 2008, the cost of feedstock to generate 1 kWh of electricity would have been approximately US 6¢. Note that this is just a cost estimate for the feedstock. It does not include capitalization, operation costs, or decommissioning costs, which inflate generation costs by a factor of two or more (total costs are summarized in Figure 1.8).

The case for renewable technologies is strengthened when upward price pressure on fossil fuel feedstocks are factored into the decision. For example, the IEA estimates that global coal consumption will increase by 53% between 2007 and 2030.[24] Many analysts believe that such levels of consumption will dangerously deplete already dwindling coal reserves. In a study for the European Commission, Kavalov and Peteves provide a succinct overview of trends in the coal industry:

- (Due mostly to accelerated consumption), from 2000 to 2005, the world's proven reserves-to-production ratio of coal in fact dropped by almost a third, from 277 to 155 years.
- Coal production costs are steadily rising all over the world due to the need to develop new fields, increasingly difficult geological conditions, and additional infrastructure costs associated with the exploitation of new fields.
- The USA and China—former large net exporters—are gradually turning into large net importers with an enormous potential demand, together with India.
- These trends suggest a likely significant increase of world coal prices in the coming decades.[25]

Recently, the costs of other fossil fuel stocks have not fared much better than coal. Throughout the twentieth century, the price of oil averaged US$24.98 per barrel with major price fluctuations occurring only during times of major global economic disruption.[26] However, as Figure 1.6 illustrates, since mid-1990 oil prices have sharply escalated. Given that the price plunge in 2008 can be attributed to a massive global economic downturn, it appears that an equilibrium price exceeding US$100 per barrel is likely.

Figure 1.6.   Europe Brent Spot Price FOB (US$ Dollars per Barrel in nominal prices)
Source of Data: US Energy Information Administration website (2011).

It may be tempting to try and equate the recent inflation of oil prices to the sudden price increases in oil during the 1970s. After all, if the circumstances are analogous, the world can expect oil prices to fall back to preinflationary levels as they did between 1985 and 1998. Unfortunately the circumstances are not analogous. The escalation of oil prices in the 1970s was due to a supply shock. Specifically, oil-producing nations in the Middle East curtailed supplies. The current trend of escalating oil prices is caused by demand-side pressure. The emergence of new economic powerhouses such as China and India, along with unabated increases in oil consumption in established industrialized countries, are taxing the ability of oil-producing nations to meet demand.[27] Not only are there concerns that oil capacity expansion efforts will continue to lag demand growth over the next few decades, there are a growing number of experts within the oil industry who acknowledge that the global supply of oil may have peaked.[28] For example, the Japanese government, which is a major importer of oil, estimates that commercially recoverable reserves of oil will be exhausted in 40 years.[29] If oil has indeed peaked, it will become increasingly scarce and more costly to procure as rampant demand continues to deplete available supplies. It is for these reasons that the IEA anticipates that the price of oil will remain above US$100 per barrel (in 2008 constant dollars) for the next three decades (see Figure 1.7).

For over 50 years, major oil-producing countries have been in the driver's seat in terms of controlling the price of oil. The Saudis in particular, who still boast over one quarter of the world's proven oil reserves, have played an active role in ensuring stable oil prices by controlling supply and pressuring other OPEC nations to follow their lead. Leaders in Saudi Arabia have astutely recognized that high oil prices provide incentives for nations to consider adopting other energy technologies.[30] The fallout from the oil crises of the 1970s taught this lesson. In response to high

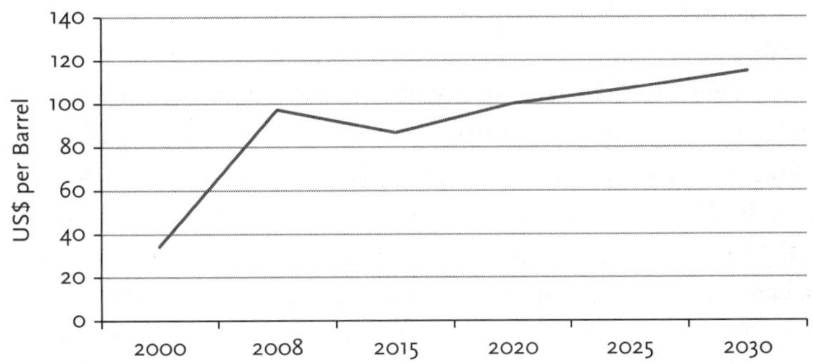

Figure 1.7. IEA Estimates for Oil Prices 2008–2030
Source: IEA, World Energy Outlook (2010).

oil prices, nations such as the United States and Germany adopted more aggressive renewable energy promotion policies.[31] On the other hand, if oil prices are too low, oil producers squander profit opportunities because the demand for oil is relatively inelastic between the \$30–\$60 per barrel range.[32] Consequently, the oil producing nations have sought to maintain a balance that optimizes profitability without precipitating a shift to alternative energy forms. However, the demand for oil has escalated over the past decade to the point where oil producers have lost control of the market.[33] Opening the supply taps in order to maintain low enough oil prices to discourage adoption of alternative energy sources has simply accelerated depletion of oil reserves.[34]

Robert Hefner, the founder of the GHK Company which specializes in the development of natural gas projects sums up the coal and oil situation thusly:

> Unfortunately, our existing energy infrastructure and its principal fuels of coal and oil are basically 18th, 19th and 20th century technologies that have not changed that much and can no longer meet our 21st century needs.[35]

Natural gas is increasingly viewed as an attractive substitute for oil in many energy applications due to superior combustion efficiency and lower CO_2 emissions. On average, in comparison to electricity generated from coal, natural gas emits less than half the CO_2 for every kilowatt hour generated.[36] The IEA anticipates that between 2007 and 2030 global demand for natural gas demand will increase by 41%, with the power generation sector accounting for 45% of this increase.

Unfortunately, the supply of natural gas exhibits the same undesirable characteristics as the supply of oil does. For starters, the nations that have rich reserves of natural gas are almost as unstable as the oil-producing nations. In fact, in many cases, they are one and the same in that natural gas and oil are frequently found in combination with one another.[37] Russia, for example, which is the number one producer of oil in the world, is also the number one producer of natural gas. It possesses 26% of global natural gas reserves and has demonstrated a propensity to use this resource for political gain and to exploit periods of high demand to gouge consumers.[38] For example, a week prior to the conclusion of negotiations over the Black Sea Fleet in 1993, Russia cut natural gas supplies to the Ukraine by 25%. In 1998, it threatened to curtail natural gas provisions to Moldova unless Russia was permitted to retain troops in a breakaway region of the country. Then, in 2006 and 2008, Russia again cut off gas supplies to the Ukraine in the middle of winter when the Ukraine refused to renegotiate a favorable contract that they had in place for Russian natural gas. Russia exhibited similar tactics in January 2007 by curtailing delivery of oil to Belarus amidst purchase price negotiations.[39]

Like oil and coal, natural gas is also a finite resource. Currently, the global reserves-to-production ratio of natural gas is estimated at 63 to 66 years.[40] Although history has demonstrated that fossil fuel reserves tend to grow as exploration activities expand, it is becoming more evident that the projected demand boom for natural gas will significantly outpace the expansion of supply.[41] In short, like the prices of coal and oil, an upward escalation in the price of natural gas is likely. It is for this reason that the IEA predicts that natural gas prices will rise to US$11.36 per million British thermal units (Mbtu) in the United States by 2030. Compared to the 2000 price level of US$4.79, the IEA forecast represents a 137% increase (in constant 2008 US$).[42]

While the costs of fossil fuels are on a decidedly upward trajectory, the costs of most mainstream alternative energy technologies continue to decline as higher volumes of installed capacity lead to improved economies of scale and technological innovations improve generation efficiency. The end result is a convergence of generation costs for the various technologies. Figure 1.8, taken from a joint study sponsored by the US Department of Energy, the US Environmental Protection Agency, the World Resources Institute, SEPA Green Power Partnership, and the Center for Resource Solutions, provides levelized cost estimates for electricity generation by various technologies.[43] Levelized costs refer to aggregated construction, fuel, operation, and maintenance costs that have been spread out over the lifetime of the technology.

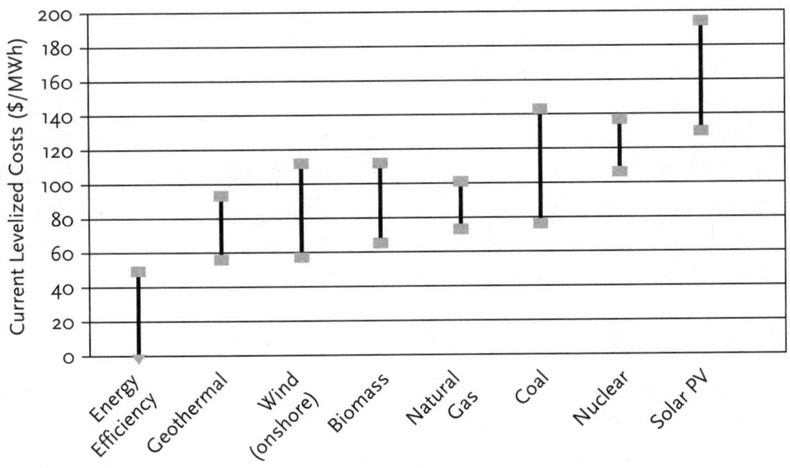

Figure 1.8. Levelized Cost of New Power Generation Technologies in 2008
Source: US Department of Energy (2010).

The wide variation in cost estimates for each of the technologies in Figure 1.8 is worth highlighting, because it touches on a key aspect of the economic debate over whether or not renewable energy is truly cost competitive with fossil fuel sources. The cost of power generation using any of the major generation technologies depends on, among other things, the technology used, the quality and prevailing costs of fuel feedstock, geographic influences, component costs, construction costs, operating costs, maintenance costs, decommissioning costs, connectivity costs, project-specific costs, and equipment performance. Predictably, each of these factors is influenced by prevailing conditions in the locale where the technology is being installed. In short, aggregate estimates of the type presented in Figure 1.8 do not shed much light on whether the cost of generating electricity through a particular technology in a given country will be at the upper end or lower end of the cost spectrum.

In fact, in some nations, certain technologies have an absolute advantage over others. For example, the cost of generating power through geothermal resources may be comparatively higher in Australia than in a nation like Japan, because Australia`s geothermal reserves are located far away from electricity grids. Therefore, grid connection costs in Australia substantially increase the cost of geothermal energy. Meanwhile, in other nations—such as Thailand, for example—geothermal reserves would require deep drilling technologies that could send the cost of geothermal energy well beyond the range outlined in Figure 1.8. For some countries, certain renewable technologies simply are not feasible for large-scale application.

Some technologies also exhibit more variable economic profiles in certain nations. For example, proponents of coal-fired power could make the argument that minimizing the cost of coal combustion is largely dependent on the choice of technology; as such, producing electricity at the lower-cost range for coal-fired power (i.e., US7.9¢/kWh) is simply a matter of technology selection. Conversely, producing electricity at the lower cost range for geothermal, biomass, and wind power is largely dependent on geographic attributes, which are not controllable. In other words, some countries can produce wind power for US6¢/kWh, but other countries can only produce wind power at the higher range (i.e., US11¢/kWh). This perspective suggests that although wind power may be more economical than coal-fired power in some nations, it is not more economical in all nations.

Unfortunately for coal-fired power advocates, this perspective is invalidated when the full costs associated with electricity generation are incorporated into levelized cost estimates. Energy expert Benjamin Sovacool refers to this process as "adjusted levelized costs." In addition to amortizing the aggregate construction, fuel, operation, and maintenance costs over the

Technology	Adjusted LCOE, US ¢/kWh ($2007)
Onshore wind	6.0
Geothermal	7.1
Hydroelectric	7.8
Landfill gas	10.8
Biomass (combustion)	13.6
Advanced nuclear	16.0
Advanced gas and oil combined cycle (AGOCC)	20.2
AGOCC with carbon capture	24.8
Integrated gasification combined cycle (IGCC)	25.9
Scrubbed coal	26.3
IGCC with carbon capture	27.9
Solar photovoltaic	39.9

Table 1.2 ADJUSTED LEVELIZED COSTS FOR THE UNITED STATES

Source: Sovacool (2008).

operational lifetime of a given electricity technology, Sovacool argues that all environmental costs and subsidies associated with a given technology should be also added to the cost estimate, because these reflect real costs that taxpayers bear. Table 1.2 reproduces Sovacool's estimates on the impact that internalizing these external costs has on electricity source cost profiles.[44]

As Table 1.2 illustrates, based on Sovacool's estimates for electricity costs in the United States, wind power, geothermal power, and hydroelectric power emerge as decisively the most economical when all of the external costs are internalized. It should be noted that any such comparison of electricity costs comes with inherent biases that influence the results. For example, the data presented in Table 1.2 stems from a levelized estimate that is contingent on a host of assumptions made regarding costs associated with infrastructure, fuel, operation and maintenance, downtime costs, financing, and decommissioning. Furthermore, the adjustments made to the levelized data is appurtenant to assumptions made regarding the cost of dominant negative externalities such as CO_2 emissions or health costs associated with pollutants coming from electricity generation.

Accordingly, for the purposes of this chapter, the data presented in Figure 1.8 or Table 1.2 is not intended to support definitive quantitative proclamations regarding the comparative cost of electricity technologies; rather, it is intended to lend general support to the assertion that commercially viable alternative electricity generation technology is available today.

A bounty of studies investigating the cost of externalities associated with fossil fuel electricity generation have all arrived at the same conclusion that even if conservative estimates regarding the cost of externalities (i.e., using the current price of carbon credits as a proxy for all external costs) are employed, fossil fuel electricity sources become more expensive than hydropower and wind power.[45] While the specific cost data may vary with the source, the general conclusion that many alternative energy technologies are viable competitors on economic grounds to fossil fuel technologies is becoming less of a point of contention.

Looking forward, the trends of amplified fossil fuel generation costs and declining alternative energy costs are expected to continue. A 2010 report by Mott and MacDonald for the UK Department of Energy and Climate Change also looked at the levelized costs of major electricity generation technologies.[46] As Figure 1.9 indicates, onshore wind power and nuclear power are considered to be the most economically attractive generation technologies for the United Kingdom over the next decade. Although the Mott MacDonald estimates clash somewhat with the US Department of Energy estimates (Figure 1.8), this should not be misinterpreted as one study being more accurate than the other, but rather as evidence that generation cost estimates are dependent on a lot of local factors and modeling assumptions that render comparison between countries to be fraught with ambiguity. Again, the point of this discussion is not to establish a definitive estimate for electricity costs broken down by technology, but to highlight the commercial viability of wind power.

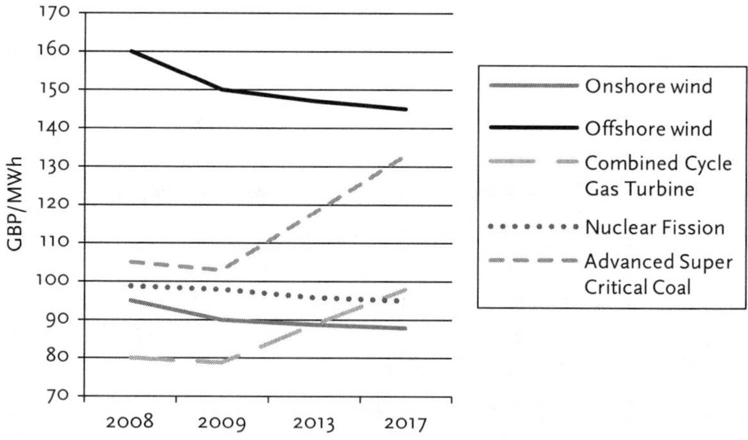

Figure 1.9. Current and Projected Levelized Costs for Electricity Technologies in the United Kingdom
Source: Mott MacDonald (2010).

An added economic inducement that is strengthening the appeal of wind power pertains to job creation. A number of studies have determined that the wind power provides far more jobs than conventional energy technologies do, when compared on an equitable basis. The World Resources Institute summarized comprehensive research by Wei, Patadia, and Kammen to demonstrate the job creation benefits attributed to wind power.[47] As Figure 1.10 illustrates, median estimates indicate that wind power generates 21% more jobs than nuclear power and 55% more jobs than both natural gas-fired power and coal-fired power.

The most important insight that the trends convey is that electricity sector market dynamics are changing due to international concerns over global warming, the progressive narrowing of the cost differential between fossil fuel electricity generation and alternative generation sources, and the enhanced job creation prospects attributed to wind power. From a policy perspective, a transition away from fossil fuel electricity generation technologies presents new opportunities and new threats. Accordingly, the next two sections examine the potential impact of such a transition on national interests. Section 1.4 examines opportunities and threats from the perspective of industrialized nations, while Section 1.5 takes a developing nation perspective. As will be demonstrated, after weighing the opportunities and threats associated with such a transition, there is a strong argument to be made for adopting aggressive policies to expedite such a transition.

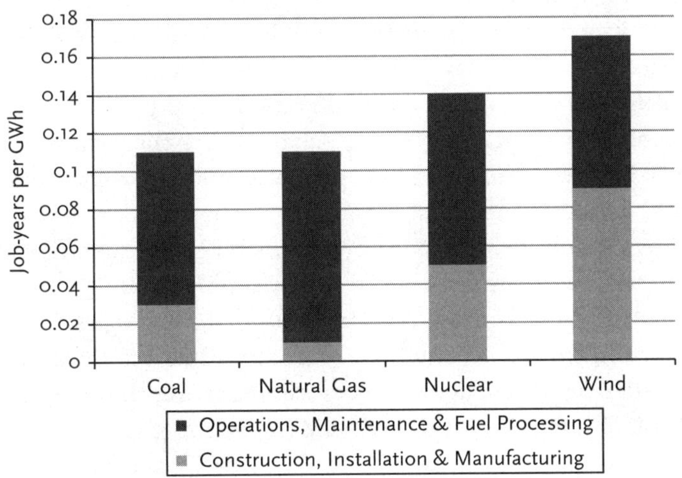

Figure 1.10. Job Benefits from Wind Power
Source: World Resources Institute (2010).

1.4 ENERGY MARKET CHANGE AND INDUSTRIALIZED NATIONS

In industrialized nations, energy has played a largely unheralded role in wealth creation and the cultivation of military might. Energy drives the high-tech production processes that provide industrialized nations with technological advantage over developing nations. It also fuels machines of war and supports military production processes that provide industrialized countries with international clout and domestic defense capabilities. Accordingly, any changes in energy market dynamics that alter the comparative cost structure of the nation's energy mix can potentially undermine national competitive advantage and destabilize national security. Overall, there is an ineluctable connection between energy policy, environmental policy, economic policy, national security policy, and foreign policy.[48] As the allure of fossil fuel energy technology continues to diminish, the once disparate objectives within these policy realms are exhibiting convergence.[49]

1.4.1 Convergence and Alternative Energy

It can be argued that a common created competency exists for most industrialized nations—effective strategic management of energy resources for the purposes of supporting industrial mechanization.[50] The top economies have learned how to create core competencies at different stages in the energy value chain. Canada (in oil and natural gas) and Australia (in coal) have exploited abundant reserves of fossil fuels to become global suppliers. The United Kingdom (British Petroleum), Holland (Shell), and the United States (Exxon) created national competitive advantages in wholesaling by nurturing the development of multinational energy firms.[51] Singapore established a core competency as an Asian hub for the refinement of fossil fuels. Japan leads the world in energy utilization efficiency and nuclear technology development.[52] In short, many countries that have achieved economic prosperity have been aided by establishing and exploiting strategic strengths in one or more links of the energy supply chain.

As a global transition to alternative energy technologies materializes, new opportunities will emerge for nations to establish entrenched positions of leadership throughout the stages associated with these new energy value chains. Nations that are successful in assuming leadership roles will develop core competencies that will facilitate national competitive advantage. Viewed from a defensive perspective, industrialized nations that fail to make the transition in a strategic manner may find their historical advantages usurped by developing nations. This threat is increasingly evident in recent times, as the technological advantages that have been enjoyed by firms in industrialized nations are increasingly eroded. As wind advocate Tom Wizelius summarizes, "Even if the economic subsidies for wind power during its early stage of development are relatively expensive for the economy, politicians have calculated that in the longer run it will generate economic benefits."[53]

In national defense, the strategic disadvantages of fossil fuels are becoming increasingly evident. Fossil fuels are largely imported (using tankers, barges, trucks, or pipelines that make easy military targets), scarce (thus, increasingly expensive) and subject to high degrees of international competition.[54] As energy expert Daniel Yergin points out, domestic energy supply limitations restrict a nation's capabilities to sustain lengthy military operations. Historically, insufficient access to oil at strategic stages of warfare contributed significantly to the downfall of both the German and the Japanese armies during the 1940s.[55] In recent times, the world witnessed the perils associated with foreign energy dependency when Russia curtailed delivery of liquid natural gas supplies to the Ukraine.[56] Clearly, establishing a national energy portfolio that focuses on encouraging the cultivation of domestically derived sources of energy represents a prudent initiative in the context of national security. Although very few countries can boast fossil fuel production that exceeds annual demand, all countries can bolster domestic energy security to some extent by harnessing alternative energy sources (geothermal, wind, hydro, solar PV, biofuels, etc.).

This should not be misconstrued to imply that complete independence in energy is a goal that all nations should strive to achieve.[57] Clearly, for many nations there will be resource barriers which inhibit such a goal.[58] Moreover, the economic theory of comparative advantage suggests that complete energy independence may in fact be economically suboptimal.[59] However, it is clear that for many nations, the current reliance on fossil fuel supplies imported from unstable foreign countries subverts national security.

The influence that energy has on other aspects of global stability was summed up succinctly by authors Kurt Campbell and Jonathon Price in the context of US national security:

Every major issue confronting the United States today—including climate change, the rise of China and India, jihadist financing, an increasingly bellicose Russia, worrisome trends in Latin America, and endemic hostilities in the Middle East—is either inextricably linked to or exacerbated by decisions associated with energy policy.[60]

1.4.2 The Need for Speed

There is strategic value in policies that encourage expedience in regard to a transition to domestically cultivated alternative energy supplies. First and foremost, as just discussed, some alternative energy technologies will represent the prudent economic choice in decades to come. Therefore, industrialized nations that outperform others in nurturing alternative electricity generation technology lay the foundation for preserving a competitive edge in this important factor endowment. This is because policies that most effectively anticipate and support superior technology expedite competitive shakeout, whereby the most efficient competitors leverage market expansion opportunities to attain competitive advantage through economies of scale. Eventually the market consolidates to a pool of highly proficient market leaders,[61] providing consumers with economically optimized products—in this context, cleaner and cheaper energy.

Firms that establish advanced competencies in critical stages of the renewable supply chain can use this competitive edge to establish unassailable market positions in foreign energy markets. The mechanisms that link domestic market conditions to international competitiveness are fairly straightforward. In order to achieve a dominant position in a given market, a firm must develop the core competencies that allow it to produce and deliver goods and services that meet market requirements in a competitively superior manner.[62] Many of these core competencies can only be honed through experience. In short, market pioneers can gain a competitive advantage over slow market entrants by learning from experiences in highly competitive markets and adopting better practices.[63] Firms that succeed in highly competitive domestic markets often find that lessons learned domestically are often transferable to competitive forays into foreign alternative energy markets. Moreover, first-movers can establish defensive beachheads in markets to more effectively deter market entry attempts by competing firms.[64] They can establish early brand recognition, early market share leads, and economies of scale that make it difficult for competitors to usurp.[65] The Dutch firm Vestas, which is the world's largest wind turbine manufacturer, is a testament to the effectiveness of domestic energy policy in nurturing firms that are capable of competing successfully in global markets.

As all this unfolds, governments that have helped nurture the development of alternative energy industry leaders begin to benefit through enhanced tax revenues and job creation as the firms grow first domestically and then internationally. Employment is a particularly salient issue in regard to alternative energy development because studies suggest that in comparison to fossil fuel technologies, alternative technologies such as wind and solar employ well over double the number of people on a per kilowatt hour basis. In a 2009 study, Rutovitz and Atherton found that a transition to clean fuel would result in an increase in the number of people employed in the energy sector by 2 million in 2020 and 2.7 million in 2030 (compared to business-as-usual).[66] Moreover, evidence from countries such as Germany, Denmark, and Spain indicates that alternative energy technologies encourage the proliferation of high value-added small businesses, which manufacture component parts and provide support services to alternative energy facilities.[67]

It is ironic how reluctant many leaders of industrialized nations have been to provide leadership in facilitating a transition away from fossil fuel dependence, given the increased threats that fossil fuel reliance poses to economic well-being and national security. Islamic extremism; unrest in the Middle East; the rise of nationalism in countries such as Russia, Venezuela, and Iran; global warming; the international drug trade; and global financial instability all have roots stemming from this global addiction to fossil fuel energy.[68] The often heard laments espoused by leaders of industrialized countries that moving away from fossil fuel energy will increase the cost of doing business for domestic firms and impinge upon economic growth prospects is a false belief predicated on a misperception that fossil fuel energy technology is actually cheaper than other forms of energy. As outlined earlier, even excluding external costs, wind energy, for example, is now cost competitive with fossil fuels (see Figure 1.8). Including external costs, fossil fuel energy is economically inferior to any alternative energy form (Table 1.2).

1.5 ENERGY MARKET CHANGE AND DEVELOPING NATIONS

Unsurprisingly, strategic energy mix planning also has extensive economic, security, and social repercussions in developing countries.

1.5.1 Economic Considerations

For firms from developing nations that compete in international markets, a key competitive advantage is the ability to tap into a cost base that is significantly lower than that found in industrialized nations.[69] Accordingly, if the energy trends outlined earlier continue and alternative energy becomes less expensive than fossil fuel energy, exporting firms from developing countries will be at a strategic disadvantage if they must continue to pay higher prices for electricity produced by fossil fuel sources.

1.5.2 Economic Security Considerations

Volatile electricity costs are of particular concern in developing countries. This is because developing countries are frequently characterized by both low per capita rates of saving and low levels of government fiscal resilience.[70] Consequently, unanticipated increases in the cost of a resource, that is as important to social and economic activity as energy is, can significantly influence the economic well-being of firms and citizens. Clearly, anything that can be done by policymakers in developing countries to encourage price stability should be done.

Alternative energy technology represents an avenue for enhancing electricity price stability. As demonstrated earlier, fossil fuel prices have fluctuated considerably while inching higher over the past few years and are expected to surge higher in the decades to come.[71] On the other hand, the costs of many alternative sources of energy have been declining consistently over the past decade. The only degree of volatility that exists for many alternative energy technologies lies in uncertainty over the timing and the degree to which costs will decline.[72] In short, renewable energy represents an opportunity to inject a degree of cost stability into a nation's energy mix.

1.5.3 Economic Empowerment

The technological diversity of alternative energy options allows policymakers in developing nations to target and support technologies which mesh most effectively with existing economic infrastructure and national competencies. Governments in developing nations that attempt to fast-track economic development by importing advanced technology often experience suboptimal results, because existing economic and social infrastructures fail to support the technology.[73] Development experts Todaro and Smith contend that a more effective national economic development strategy is to identify strategies to encourage the development of forward and backward linkages associated with existing industries.[74] In the alternative energy industry there are biofuel options that can be integrated with agricultural activities, there are solar options that can provide electricity to areas where electricity grid infrastructure is insufficient, and there are biomass energy options that can add value to industries that produce biomass as waste byproducts. Clearly, the diversity of technical options in alternative energy allows developing countries to match strategic energy mixes with national conditions.

1.5.4 Social Considerations

In developing countries, abatement of climate change is just one benefit associated with a transition away from fossil fuel energy. Economic growth supersedes environmental governance in most developing countries. Consequently, lax environmental regulations governing electricity generation and transportation emissions give rise to significant environmental and social problems. To put this into perspective, according to a 2007 World Health Organization (WHO) report, air pollution was responsible for diseases that cause over 650,000 pre-mature deaths in China each year and over 2 million premature deaths worldwide.[75] Air pollution is now the leading cause of death in China.[76] Similarly, in India air pollution accounts for 527,000 fatalities each year. When compared to the number of fatalities related to air pollution in the United States each year (approx. 41,400), one begins to understand the scale and scope of the damage that pollutive forms of energy cause. If a transition to cleaner forms of energy could be facilitated in an economically effective manner, citizens in developing countries such as China could enjoy the benefits of enhanced affluence without also having to suffer the negative externalities associated with economic growth.

1.5.5 The Need for Speed

Previously, an argument was put forth that industrialized nations, which embrace more proactive policies for expediting a transition to alternative energy, can nurture the development of internationally competitive, domestic alternative energy forms. A similar case for expedience applies in developing nations. The China case study presented in this book provides a good illustration of how government support for alternative energy in developing nations can also sire domestic alternative energy firms that are capable of competing successfully internationally. Due to China's aggressive wind power expansion program, four of the world's ten largest wind power firms in 2010 were Chinese firms—Sinovel (second-largest), Goldwind (fourth), Dongfang (seventh), and United Power (tenth) (see Table 1.3).

There is another benefit to proactive alternative energy development policies that applies solely to developing nations. Currently there are a

___

Table 1.3 WORLD`S LARGEST WIND TURBINE
MANUFACTURERS BY MARKET SHARE IN 2010

Vestas	Denmark	14.8%
Sinovel	China	11.1%
GE Wind Energy	US	9.6%
Goldwind	China	9.5%
Enercon	Germany	7.2%
Suzulon	India	6.9%
Dongfang	China	6.7%
Gamesa	Spain	6.6%
Siemens Wind	Germany	5.9%
United Power	China	4.2%

Source: Bayar, T., 2011. World Wind Market: Record Installations, But Growth Rates
Still Falling, Renewable Energy World Online Edition.

number of financial mechanisms—the Clean Development Mechanism
(CDM), the Global Environmental Facility, the World Bank Clean Energy
Fund, and a number of other overseas development assistance funds—
that developing nations can tap into to help finance a transition away from
fossil fuel energy. However, these financial support funds will not last
forever. As more nations expand alternative energy capacity, competition
for these funds will heat up and donor agencies will be faced with difficult
choices in regard to allocation. If history is any indicator, this in turn will
result in more conditions being placed on the funds.[77] Furthermore, a stage
will inevitably be reached where international willingness to finance such
energy projects will wane. Forebodingly, a number of CDM projects are
already being rejected for not meeting the CDM condition of additionality
(that the project would not have been carried out without support from
the CDM program).[78] It appears that the market for funds is already tightening. Developing nations that move quickly to take advantage of these
financial mechanisms will gain a leg up on their developing country rivals
by procuring alternative energy generation capacity at subsidized rates.

1.6 WHEN FORCES FOR SPEED MEET THE NEED FOR SPEED

The analysis presented to this point indicates that energy market dynamics are gradually shifting in favor of alternative energy technologies; and
indeed, for industrialized and developing countries alike, there are strong
emergent incentives for political leaders to embrace aggressive policies to
facilitate expedient transition. Fortuitously, the benefits associated with

such a transition mesh seamlessly with the need to respond assertively to abate global warming.

In the Stern Review—the oft-quoted economic impact assessment of climate change—climate change was called "the greatest and widest-ranging market failure ever seen." The review concluded that "the benefits of strong, early action (to abate global warming) considerably outweigh the costs." In emphasizing the importance of expedience in facilitating a transition away from fossil fuel dependence, the report declared:

> The effects of our actions now on future changes in the climate have long lead times. What we do now can have only a limited effect on the climate over the next 40 or 50 years. On the other hand, what we do in the next 10 or 20 years can have a profound effect on the climate in the second half of this century and in the next.[79]

The IPCC's Fourth Assessment Report on Climate Change also echoed the appeal made in the Stern Review that expediency in developing and implementing mitigation measures is of utmost importance. The report stated:

> Many impacts can be reduced, delayed or avoided by mitigation. Mitigation efforts and investments over the next two to three decades will have a large impact on opportunities to achieve lower stabilisation levels. Delayed emission reductions significantly constrain the opportunities to achieve lower stabilisation levels and increase the risk of more severe climate change impacts.[80]

In fact, in November 2011, the IEA released a statement in the lead up to the COP17 Climate Change talks in Durban, South Africa, which warned that the window for taking action was rapidly closing and that failure to make effective decisions to alter the trajectory of CO_2 emissions within the next five years will lead to a technological lock-in, whereby the chance to avoid dangerous climate change will be lost forever.[81] Therefore, it is promising that the forces which justify an expedient transition to alternative energy are amassing during a period of time when just such expediency is required.

1.7 THE DICHOTOMY OF ALTERNATIVE ENERGY

Yet the pace of alternative energy development has been phlegmatic *despite* emergent levelized cost data such as the data presented earlier (Figure 1.8), which indicate that wind power, hydro power, geothermal power, and biomass combustion are all economically competitive with all forms of fossil fuel power (with or without carbon capture and sequestration); *despite* the potential benefits to nations (both industrialized and developing) that undertake a transition to these alternative energy forms in an expedient manner; and *despite* the global warming mitigation imperative that demands national commitments to reduce CO_2 emissions.

Most certainly the growth rates attributed to some of the more commercially attractive alternative energy technologies are impressive when considered in isolation. For example, the World Wind Energy Association reports that installed wind power capacity has grown more than tenfold since 2000.[82] However, in absolute terms, the inroads that wind power has made into the electricity generation sector have been minor. Total global installed wind power capacity at the end of 2012 amounted to approximately 282,275 MW,[83] enough to satisfy only about 3% of global electricity consumption—nowhere near the level of penetration necessary to make significant contributions to global warming abatement.

This, then, is the emergent dichotomy involving wind energy; although strong environmental, economic and political justifications exist for nations to adopt aggressive programs for supporting a transition to economically viable forms of alternative energy such as wind power, the nations of the world remain highly committed to fossil fuel electricity generation. In the lead-up to the 15th Conference of the Parties in Copenhagen (COP15), there were indications that the commitments to be undertaken by developed countries would be in the neighbourhood of 8 to 12% below 1990 levels by 2020 after accounting for forestry credits.[84] This lies in stark contrast to the 25 to 40% reduction level described as necessary by the IPCC. Yet even these modest targets failed to gain acceptance at COP15. Meanwhile, developing countries lag far behind the industrialized countries in terms of curtailing GHG emissions with most still resisting the concept of making formal commitments.[85]

Clearly a degree of dynamic tension exists within electricity policy regimes in all nations. On the one hand, all of the 191 nations that have ratified the Kyoto Protocol (as of June, 2012) have introduced initiatives to support the development of renewable energy. Many of the top economies of the world now have specific renewable energy targets supported by policy instruments such as feed-in tariffs, production tax credits, mandatory renewable energy quotas, and production subsidies, all intended to encourage a greater uptake of renewable energy. Best practices in policy development have been internationally disseminated, and policymakers who are intent on pursuing aggressive renewable energy development strategies can access numerous accounts of how countries such as Germany, Spain, and Denmark have recorded successes in encouraging development of renewable energy through various policy instruments.[86] However, with the exception of a few smaller nations such as Denmark, Norway, Portugal, and the Netherlands, the contribution of nonhydro, renewable energy sources to the national electricity mix is well below 5%. Given the justifications outlined to this point for supporting wind energy development and indications of the exigency of policies for catalyzing such a transition, the phlegmatic results are perplexing. Simply put, what is preventing a more robust diffusion of wind energy in the electricity sector?

1.8 OBJECTIVES AND OUTLINE OF THIS BOOK

This book seeks to identify and explain forces that catalyze wind power development and forces that prevent a more robust diffusion of wind power in the electricity sector.

The choice to focus on wind power diffusion (as opposed to other renewable energy technologies) will serve the interests of a number of stakeholders. First, wind power is a commercially competitive energy source. Therefore, understanding catalysts and barriers to wind power diffusion arguably represents the most important challenge that wind power developers face in terms of optimizing industry growth prospects. Second, wind power is an egalitarian technology. Aside from geographic constraints that limit capacity, wind power can be adopted by any nation, and in doing so, yields many of the benefits outlined earlier in this chapter. Therefore, energy policymakers, who possess an understanding of the dos and don'ts of effective wind power development policy, stand to add value to their respective nations. Third, as was pointed out earlier, the need for an expedient transition to carbon-free forms of electricity generation is imperative if humanity is to have any chance of avoiding severe ecological impacts associated with global warming. As will be detailed in Chapter 2, wind power could conceivably provide at least 20% of humanity's electricity demand within months if the political will and economic commitment existed. In short, understanding how to optimize wind power development can make a strategic contribution to meeting the challenge of facilitating change immediately, so as to stave off the need for more substantial economic sacrifices in the future. In short, this book should be of interest to wind power developers, energy policymakers, and members of the general public who are interested in exploring ways to avoid saddling future generations with economic and environmental costs that no generation should have to bear.

This book has been organized in three sections. The first section includes this introduction and two other chapters. Chapter 2 provides the reader with an overview of wind power technology in order to advance understanding of the strengths and weaknesses of wind power. The reader will learn that by and large, wind power technology is riddled with misconceptions that engender a number of false barriers to development. Chapter 3 provides an introduction to the complexities influencing the evolution of energy and the formulation of energy policy. It also provides an introduction to the *Political SET* framework that is used for analyzing the six case studies that comprise the core of this book. The main thesis of this chapter also establishes the rationale for writing this book. Succinctly put, energy policy is influenced a seamless web of political, social, economic, and technological forces that complicate analysis. If we are to succeed in our quest to understand how to more effectively influence wind power development policy, we must document the diverse yet influential variables that impact energy policy, highlight and understand the nature of interdependencies between these variables, and attempt to apply this newfound knowledge to guide the development of more effective policy.

The next section presents the case studies which are analyzed using the *Political SET* framework. Six nations have been selected as target nations for studying and documenting the forces which influence wind power development. In selecting the nations to study, a conscious decision was made to choose two nations (Denmark, Chapter 4 and Germany, Chapter 5) that have enjoyed historic success in encouraging wind power development; two critical nations (China, Chapter 6 and the United States, Chapter 7) that can be considered breakout nations in terms of wind power diffusion; and two nations (Canada, Chapter 8 and Japan, Chapter 9) that can be considered as laggard nations in respect to wind power diffusion. By investigating wind power diffusion in nations that have exhibited varying degrees of wind power success, we will gain a clearer picture of what conditions support wind power, what conditions tend to inhibit wind power, and what conditions, if altered, can initiate improved wind power diffusion. It is also felt that by choosing two nations from North America, two from Europe, and two from Asia that the study will be able to partially address the external validity of the findings by highlighting any sociocultural anomalies that might confound generalizations.

The final section consists of two chapters that draw together the insights from the case studies to document the forces that impact wind power diffusion, highlight interdependencies between these forces, and interpret these findings from a policy perspective. These concluding chapters aim to equip policymakers with a clear understanding of which variables must be proactively managed in order to ensure wind power development policy encourages prolonged and sufficient diffusion of the technology.

NOTES

1. Lovelock, James. 2006. "The Earth is About to Catch a Morbid Fever that May Last as Long as 100,000 Years." *The Independent*. /www.independent.co.uk/voices/commentators/james-lovelock-the-earth-is-about-to-catch-a-morbid-fever-that-may-last-as-long-as-100000-years-523161.html.

2. Stern, Nicholas. 2006. *The Stern Review: Report on the Economics of Climate Change*. London: Cabinet Office, HM Treasury.

3. Excerpt from Barack Obama`s Speech on Energy, April 22, 2009, Newton Iowa. www.whitehouse.gov/the-press-office/remarks-president-newton-ia.

4. The Kyoto Protocol is the main international agreement that government efforts to reduce greenhouse gas emissions. When the Kyoto Protocol entered into force on February 16, 2005, there were 128 nations that had ratified the agreement. As of June 2013, there were 191 nations that have ratified the Kyoto Protocol.

5. Stern, Nicholas. 2006. *The Stern Review: Report on the Economics of Climate Change*. London: Cabinet Office, HM Treasury.

6. Intergovernmental Panel on Climate Change (IPCC). 2007. *Contribution of Working Group I to the Fourth Assessment Report of the Intergovernmental Panel on Climate Change: Summary for Policy Makers*. Geneva: Intergovernmental Panel on Climate Change.

7. IPCC. 2007. *Contribution of Working Group II to the Fourth Assessment Report of the Intergovernmental Panel on Climate Change: Summary for Policy Makers*. Geneva: Intergovernmental Panel on Climate Change.

8. Stern, Nicholas. 2006. *The Stern Review: Report on the Economics of Climate Change*. London: Cabinet Office, HM Treasury. See executive summary pages i–vi for more on this perspective.

9. IPCC. 2007. *Climate Change 2007: Synthesis Report*. Geneva: Intergovernmental Panel on Climate Change.

10. Ban, Ki-Moon. 2008. *Message for World Environment Day 2008*. New York: United Nations.

11. International Energy Agency (IEA). 2011. *CO_2 Emissions from Fuel Combustion: Highlights*. Paris: International Energy Agency.

12. IEA. 2010. *World Energy Outlook 2010*. Paris: International Energy Agency.

13. According to the Stern Review. Stern, Nicholas. 2006. *The Stern Review: Report on the Economics of Climate Change*. London: Cabinet Office, HM Treasury.

14. Hefner, Robert A. 2008. "The Age of Energy Gases: The Importance of Natural Gas Energy Policy." In *The Global Politics of Energy*, edited by K. M. Campbell and J. Price, pp. 149–178. Washington: The Aspen Institute.

15. Victor, David G. 2008. "Sources of Alternative Energy and Energy Market Innovations." In *The Global Politics of Energy*, edited by K. M. Campbell and J. Price, pp. 135–147. Washington: The Aspen Institute.

16. IEA. 2010. *World Energy Outlook 2010*. Paris: International Energy Agency.

17. Ibid.

18. Victor, David G. 2008. "Sources of Alternative Energy and Energy Market Innovations." In *The Global Politics of Energy*, edited by K. M. Campbell and J. Price, pp. 135–147. Washington: The Aspen Institute.

19. In the words of the Stern Review.

20. Cost of fossil fuels will rise according to the IEA; IIEA. 2008. *World Energy Outlook 2007*, Paris: International Energy Agency. Regarding wind power and other renewable sources, this reflects conclusions found in the following studies: Celik, A. N.,

T. Muneer, and P. Clarke. 2007. *An Investigation into Micro Wind Energy Systems for their Utilization in Urban Areas and their Life Cycle Assessment*, Proceedings of the Institution of Mechanical Engineers, Part A: *Journal of Power and Energy* 221 (8): 1107–1117; and Brown, Brit T., and Benjamin A Escobar. 2007. "Wind Power: Generating Electricity and Lawsuits." *Energy Law Journal* 28 (2): 489–516; and DeCarolis, Joseph F., and David W. Keith. 2006. "The Economics of Large-Scale Wind Power in a Carbon Constrained World." *Energy Policy* 34 (4): 395–410.

21. IEA. 2010. *World Energy Outlook 2010*. Paris: International Energy Agency.
22. IPCC. 2007. *Climate Change 2007: Synthesis Report*. Geneva: Intergovernmental Panel on Climate Change.
23. Source: The Energy Information Administration, www.eia.doe.gov/cneaf/coal/page/coalnews/coalmar.html.
24. IEA. 2010. *World Energy Outlook 2010*. Paris: International Energy Agency.
25. Found on pages 4–5 of Kavalov, Boyan, and S. D. Peteves. 2007. *The Future of Coal*. Luxembourg: Institute for Energy, European Commission.
26. Source: WTRG Economics website: "Oil Price History and Analysis," www.wtrg.com/prices.htm.
27. Yergin, Daniel. 2008. "Energy Under Stress." In *The Global Politics of Energy*, edited by K. M. Campbell and J. Price, pp. 26–43. Washington: The Aspen Institute.
28. For an insightful review of peak oil see: Deffeyes, Kenneth S. 2005. *Beyond Oil: The View from Hubbert's Peak*. New York: Hill and Wang.
29. Agency for Natural Resrouces and Energy (ANRE). 2006. *Energy in Japan 2006: Status and Policies*, pp. 1–28. Tokyo: Agency for Natural Resources and Energy, Ministry of Economy, Trade and Industry.
30. Ross, Dennis. 2008. "Arab and Gulf Perspectives on Energy." In *The Global Politics of Energy*, edited by K. M. Campbell and J. Price. pp. 46–75. Washington: The Aspen Institute.
31. Sovacool, Benjamin K. 2008. *The Dirty Energy Dilemma: What's Blocking Clean Power in the United States*. Westport, CT: Praeger Publishers.
32. Deffeyes, Kenneth S. 2005. *Beyond Oil: The View from Hubbert's Peak*. New York: Hill and Wang.
33. Yergin, Daniel. 2008. "Energy Under Stress." In *The Global Politics of Energy*, edited by K. M. Campbell and J. Price, pp. 26–43. Washington: The Aspen Institute.
34. Deffeyes, Kenneth S. 2005. *Beyond Oil: The View from Hubbert's Peak*. New York: Hill and Wang.
35. Hefner, Robert A. 2008. "The Age of Energy Gases: The Importance of Natural Gas Energy Policy." In *The Global Politics of Energy*, edited by K. M. Campbell and J. Price, pp. 152. Washington: The Aspen Institute.
36. Ibid.
37. Deffeyes, Kenneth S. 2005. *Beyond Oil: The View from Hubbert's Peak*. New York: Hill and Wang.
38. Stent, Angela. 2008. "An Energy Superpower? Russia and Europe." In *The Global Politics of Energy*, edited by K. M. Campbell and J. Price, pp. 76–95. Washington: The Aspen Institute.
39. Ibid.
40. Supported by the following two studies: ANRE. 2006. *Energy in Japan 2006: Status and Policies*. Tokyo: Agency for Natural Resources and Energy, Ministry of Economy, Trade and Industry, pp. 1–28; and IEA. 2008. *World Energy Outlook 2007*. Paris: International Energy Agency.
41. Deffeyes, Kenneth S. 2005. *Beyond Oil: The View from Hubbert's Peak*. New York: Hill and Wang.

42. IEA. 2010. *World Energy Outlook 2010*. Paris: International Energy Agency.

43. This insightful report is: US Department of Energy. 2010. *Guide to Purchasing Green Power: Renewable Electricity, Renewable Energy Certificates, and On-Site Renewable Generation*. Washington, DC: US Department of Energy.

44. Sovacool, B.K. 2008. *The Dirty Energy Dilemma: What's Blocking Clean Power in the United States*. Westport, CT: Praeger Publishers.

45. As representative studies, please consult: ATSE. 2009. *The Hidden Costs of Electricity: Externalities of Power Generation in Australia*. Parkville, Victoria: Australian Academy of Technological Sciences and Engineering, pp. 1–90; S. Krohn, S., Morthorst, P.E., and Awerbuch, S. (eds.). 2009. *The Economics of Wind Energy*. Brussels: European Wind Energy Association; Tester, J. W., Drake, E. M., Driscoll, M. J., Golay, M. W., and Peters, W. A.. 2005. *Sustainable Energy: Choosing Among Options*. Cambridge, MA: MIT Press; and Wizelius, T.. 2007. *Developing Wind Power Projects: Theory and Practice*. Oxford: Earthscan.

46. Mott MacDonald. 2010. *UK Electricity Generation Costs Update*. Brighton, UK: Mott MacDonald.

47. The World Resources Institute presents median estimates from the study, whereas the metastudy by Wei et al. presented a range. The original study can be referenced from: Wei, Max, Shana Patadia, and Daniel M. Kammen. 2010. "Putting Renewables and Energy Efficiency to Work: How Many Jobs Can the Clean Energy Industry Generate in the US?" *Energy Policy* no. 38 (2): 919–931, http://dx.doi.org/10.1016/j.enpol.2009.10.044.

48. Rothkopf, D. 2008. "New Energy Paradigm, New Foreign Policy Paradigm." In *The Global Politics of Energy*, edited by K. M. Campbell and J. Price, pp. 186–213. Washington: The Aspen Institute.

49. Biegan, S. 2008. "The Global AMERICAN Politics of Energy." In *The Global Politics of Energy*, edited by K. M. Campbell and J. Price, pp. 214–223. Washington: The Aspen Institute.

50. Yergin, D. 1993. *The Prize: The Epic Quest for Oil, Money & Power*. New York: The Free Press.

51. Ibid.

52. Campbell, K. M., and Price, J. (eds.). 2008. *The Global Politics of Energy*. Washington: The Aspen Institute.

53. Wizelius, T. 2007. *Developing Wind Power Projects: Theory and Practice*. Oxford: Earthscan, p. 133.

54. Campbell, K. M., and Price, J. 2008. "The Global Politics of Energy: An Aspen Strategy Group Workshop." In *The Global Politics of Energy*, edited by K. M. Campbell and J. Price, pp. 11–23. Washington: The Aspen Institute.

55. Yergin, D. 1993. *The Prize: The Epic Quest for Oil, Money & Power*. New York: The Free Press. This book is a tremendous reference for learning about the historical connections between national security and oil.

56. Campbell, K. M., and Price, J. 2008. "The Global Politics of Energy: An Aspen Strategy Group Workshop." In *The Global Politics of Energy*, edited by K. M. Campbell and J. Price, pp. 11–23. Washington: The Aspen Institute.

57. Yergin, D. 2008. "Energy Under Stress." In *The Global Politics of Energy*, edited by K. M. Campbell and J. Price, pp. 27–43. Washington: The Aspen Institute.

58. Farrell, D., and Bozon, I. 2008. "Demand-Side Economics: The Case for a New US Energy Policy Direction." In *The Global Politics of Energy*, edited by K. M. Campbell and J. Price, pp. 45–61. Washington: The Aspen Institute.

59. This ties in with the concept of comparative advantage. A theoretical explanation can be found in: Mankiw, N.G.. 1997. *Principles of Economics*. New York: Harcourt.

60. Campbell, K.M., and Price, J. 2008. "The Global Politics of Energy: An Aspen Strategy Group Workshop." In *The Global Politics of Energy*, edited by K. M. Campbell and J. Price, p. 11. Washington: The Aspen Institute.

61. Competitiveness theory is eloquently explained in Porter, M. E. 1998. *On Competition*. Boston: Harvard Business School Press.

62. For a theoretical review of core competency theory, see Porter, M. E. 1985. *Competitive Advantage: Creating and Sustaining Superior Performance*. New York: The Free Press.

63. A discussion on the benefits of first mover advantage can be found in Grant, R. M. 2005. *Contemporary Strategy Analysis*, 5th ed. London: Blackwell Publishing.

64. The essence of international strategy is discussed in Bartlett, C., Ghoshal, S., and Birkinshaw, J. 2003. Transnational Management, 4th edition ed. Chicago: McGraw-Hill/Irwin.

65. An insightful review of the economies of scale is provided in Doyle, P. 1998. *Marketing Management and Strategy*, 2nd ed. Harlow, UK: Prentice Hall Publishing.

66. Rutovitz, J., and Atherton, A. 2009. *Energy Sector Jobs to 2030: A Global Analysis*. Sydney: Institute for Sustainable Futures.

67. Valentine, S.V. 2011. Towards the Sino-American Trade Organization for the Prevention of Climate Change (STOP-CC). *Chinese Journal of International Politics* 4 (4): 447–474.

68. Rothkopf, D. 2008. "New Energy Paradigm, New Foreign Policy Paradigm." In *The Global Politics of Energy*, edited by K. M. Campbell and J. Price, pp. 186–213. Washington: The Aspen Institute.

69. Bartlett, C., Ghoshal, S., and Birkinshaw, J. 2003. *Transnational Management*, 4th ed. Chicago: McGraw-Hill/Irwin.

70. A theoretical review of this is provided in Perkins, D., Radelet, S., and Lindauer, D. 2006. *Economics of Development*, 6th ed. New York: Norton.

71. EIA. 2008. *International Energy Outlook 2008* Washington, DC: US Department of Energy, Energy Information Administration, pp. 1–260.

72. Neuhoff, K. 2005. "Large-Scale Deployment of Renewables for Electricity Generation." *Oxford Review of Economic Policy: The New Energy Paradigm* 21 (1): 88–112.

73. Perkins, D., Radelet, S., and Lindauer, D. 2006. *Economics of Development*, 6th ed. New York: Norton.

74. Todaro, M., and Smith, S. 2003. *Economic Development*, 8th ed. New York: Pearson Addison-Wesley.

75. Platt, K. H. 2007. "Chinese Air Pollution Deadliest in World, Report Says." National Geographic News. Online edition. http://news.nationalgeographic.com/news/200 7/07/070709-china-pollution.html.

76. Fairley, P. 2007. "China's Coal Future." Technology Review. www.technologyreview. com/featuredstory/407092/chinas-coal-future/.

77. Perkins, D., Radelet, S., and Lindauer, D. 2006. *Economics of Development*, 6th ed. New York: Norton.

78. Castro, P., and Michaelowa, A. 2008. *Climate Strategies Report: Empirical Analysis of Performance of CDM Projects*. Zurich: Institute of Political Science, University of Zurich. www.climatestrategies.org/component/reports/category/39/138.html.

79. Stern, N. 2006. *The Stern Review: Report on the Economics of Climate Change*, pp. i–ii. London: Cabinet Office, HM Treasury.

80. IPCC. 2007. *Climate Change 2007: Synthesis Report*. Geneva: Intergovernmental Panel on Climate Change.

81. This IEA Press release is available at www.iea.org/weo/docs/weo2011/pressrelease.pdf.

82. World Wind Energy Report (WWEA). 2011. *World Wind Energy Report 2010*. Bonn: World Wind Energy Association.

83. Capacity estimate taken from the *World Wind Energy Report 2012* published by the World Wind Energy Association.

84. Climate Tracker web-site: www.climateactiontracker.org. This site is created to allow members of the general public to keep track of political commitments made by nations.

85. Valentine, S.V. 2012. "Enhancing Climate Change Mitigation Efforts through Sino-American Collaboration." *Chinese Journal of International Politics*, 6 (2): 159–182.

86. For just a sampling of such accounts see Komor, P. 2004. *Renewable Energy Policy*. City: iUniverse; Wizelius, T. 2007. *Developing Wind Power Projects: Theory and Practice*. Oxford: Earthscan; Mallon, K. (ed.). 2006. *Renewable Energy Policy and Politics: A Handbook for Decision Making*. Oxford: Earthscan; and IRENA. 2012. *30 Years of Policies for Wind Energy: Lessons from 12 Wind Energy Markets*. Abu Dhabi: International Renewable Energy Agency.

CHAPTER 2

Understanding Wind Power Systems

2.1 THE IMPORTANCE OF UNDERSTANDING WIND POWER SYSTEMS

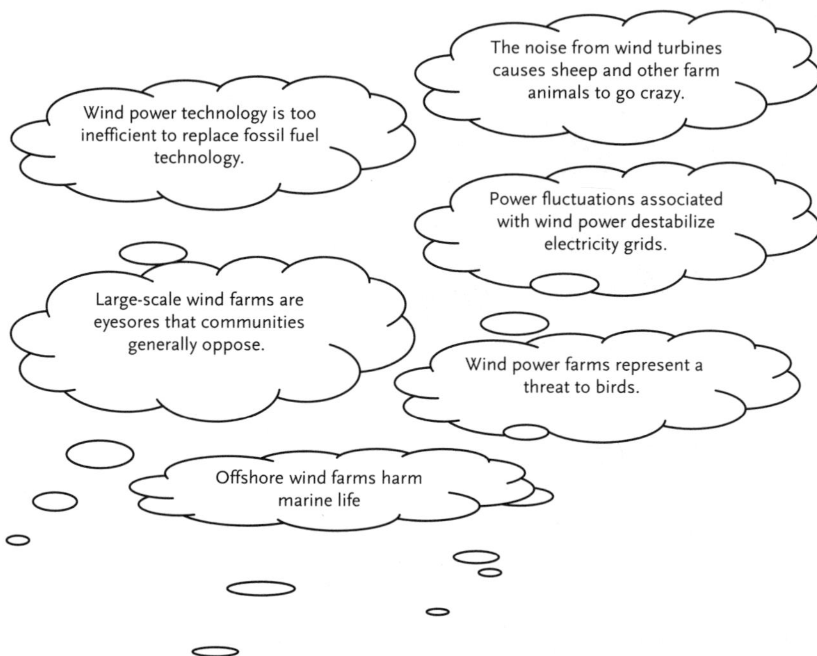

All of the above statements represent prominent objections to wind power development. For the most part, these statements are premised upon small truths that have been exaggerated by wind power opponents in order to generate public opposition. The intent of this chapter is to try and separate fact from fiction in order to give the reader a better technical understanding of the true hurdles faced by nations that embark on ambitious wind power development programs. Although a technical understanding of wind power systems is not necessary to understand the case studies presented in this book, enhanced technical understanding will help the reader better understand the possibilities and limitations of the technology.

This chapter begins by describing the basic components of a wind power system before exploring how technical choices made in regard to system components and site location influence generation costs. From this technical foundation, the discussion will shift to the stochastic (fluctuating) nature of wind power and examine existing solutions for smoothing power fluctuations. This will provide the reader with a better understanding of the potential of wind power systems to replace fossil fuel electricity generation technologies. In concluding sections of this chapter, an attempt will be made to separate truth from fiction in regard to community and environmental impacts commonly attributed to wind power systems. Hopefully, by the end of this chapter, the pros and cons associated with wind power development will be better understood.

2.2 FEATURES OF WIND POWER SYSTEMS

2.2.1 Main Components of a Wind System

There are basically two main wind turbine designs—vertical axis and horizontal axis. Vertical axis wind turbines (VAWT), which can resemble egg beaters placed on towers, are not widely used for electricity generation, so this section will focus on the main components of horizontal axis wind turbines (HAWT).

The main components of a wind turbine includes the rotor blade; the nacelle (which houses the gearbox, generator, and yaw motor); the tower upon which the wind turbine is placed; the foundation which anchors the tower to the ground; the control system and transformer (usually located at the base of the tower), which transforms the collected energy into electric current that can be delivered to the electricity grid; and the electrical conduits that connect the wind turbine to the electricity grid.[1] Various designs are available for each of these component parts in order to optimize wind power performance at a given site.

Most commercial wind turbines currently use a three-blade rotor, which tends to be more resilient than two-blade rotors. When the rotor blades turn, they cover a circular area known as the *swept area*. The power generated by any given turbine is proportional to the cube of the wind speed.

However, the energy collected from the wind turbine is also dependent on the size of the swept area. Therefore, the engineering challenge is to manufacture increasingly light yet resilient rotor blades to permit the manufacture of longer rotor blades, thereby increasing the swept area without proportionately increasing rotational friction.

Over the past 25 years, significant progress has been made in regard to increasing the swept area of wind power systems. As Figure 2.1 illustrates, in 1985 a typical 500 kW wind power turbines had a rotor diameter of about 15 m. In 2008, it was envisaged that sometime in the near future, a 8–10 MW wind system would become a reality. That day has arrived. As of June 2013, Vestas's largest wind power system—the V164–8.0 MW wind turbine—boasts a rotor diameter of 164 m, approximately twice the length of a 747. Its swept area is the size of three football pitches.

In order to expand the swept area of a wind system, the turbines must be mounted on enormous towers. For example, the Vestas V164–8.0 MW wind turbine described in the previous paragraph can sit on a tubular steel tower that is as high as 190 m, approximately the same height as Seattle's space needle and twice the height of the clock tower which houses Big Ben in London. One of the added advantages of mounting wind turbines on higher towers is that airflow at higher altitudes is less turbulent. Whereas many of the earlier wind power systems were mounted on trellis-like towers, almost all modern utility-scale wind power systems are mounted on tubular steel or concrete towers.

The component parts that are found in the nacelle—the gearbox, the generator, and the yaw motor—play vital roles in optimizing the capture of wind power. The rotor of a wind turbine is mounted on a main drive shaft

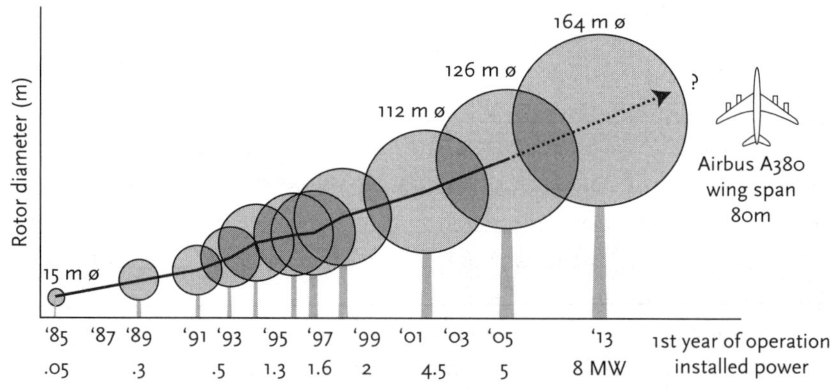

Figure 2.1. The Progressively Improving State Of Wind Turbine Technology

Source: DONG Energy (2008) *Final Report on Offshore Wind Technology*. Fredericia, Denmark: Risø National Sustainable Energy Laboratory, New Energy Externalities Developments for Sustainability Consortium.

which runs into the gearbox. The function of the gearbox is to transform the relatively low RPMs from the main driveshaft to a speed that maximizes electricity generation. Some turbines, known as direct drive units, have a generator that is directly connected to the rotor, thereby bypassing the need for a gearbox. Regardless of whether or not a gearbox is used, eventually the power generated by the turning rotor blades is transferred to a generator that turns the kinetic energy of the wind into electrical power. Generators come in an array of design features such as fixed speed, soft start, double generators, variable speed, direct drive, and slip controls. A review of these features is beyond the scope of this chapter; however, the rationale behind multifaceted design features is to optimize the generation of electricity given site-specific wind quality characteristics. This is akin to choosing computer specifications and software that best suit a user's requirements. The last major component found in the nacelle is the yaw motor, which turns the nacelle so that the rotors are consistently aligned perpendicular to the prevailing wind direction.[2]

The voltage produced by most of the large wind power systems is 690 volts (V). Although this can be connected to a factory for use, in order to feed the power into an electricity grid, the voltage must be stepped up by a transformer, normally to 10,000 or 20,000 kilovolts (kV). In larger wind systems, the transformer is often located at the base of the tower. A computerized control system is also typically installed at the base of the tower. These computer systems serve multiple functions, such as operation control (monitoring wind speeds, coordinating the yaw motor function, etc.); power management (dampening harmonics which have a negative impact on power quality); and system monitoring (temperature control, quality control, etc.).

2.2.2 Innovation and Cost

As outlined in Chapter 1, the cost of wind has declined from US28¢ to US6–11¢ per kWh over the past 30 years,[3] and both the US Department of Energy and UK government authorities project that costs will likely continue to decline over the next 30 years.[4] There are three clusters of forces that portend a declining cost profile for wind power systems.

First, advances in wind system technology have led to leaner, more durable components with improved wind capture capabilities. Rotor blades made of composite materials are much lighter but more durable than previous designs. Moreover, over the past 30 years a diverse array of wind system innovations have emerged to improve performance under varied conditions.[5] For example, many nacelles now house a small motor which automatically adjusts the pitch of the blades in response to wind speed variance.[6] In a location that is characterized by inconsistent winds, adjustable rotor blades will, provided other conditions remain the same, profitably improve power output and generation consistency.[7] As another example, the development of better tower technology makes it possible for higher towers that enable turbines to capture high quality wind which is less influenced by ground friction.[8]

Second, technical economies of scale have emerged as bigger units produce more energy for less cost. The rated power generation capacities of wind turbines have increased significantly since the early 1980s. State of the art 20 kW wind turbines of the 1980s now seem like school science projects in comparison to the 6000 kW turbines that are being erected today. All of the main components of wind systems—the nacelle components (blade, gears, and generator), the tower, the foundation and the balance of plant components (including transformer and transmission cables)—have declining cost profiles in relation to increased scale. For example, a turbine with twice the power capacity of another does not require twice the tower height, nor does it require twice the foundation materials or twice the resource inputs for balance of plant components. However, there are diminishing returns in regard to technical economies of scale. One wind expert contends that in the near future, technical economies of scale may be obviated by amplified increases in the cost of larger wind turbines. Simply put, building larger turbines higher off the ground eventually requires excessively expensive structural reinforcement. Furthermore, the weight of the rotor blade inevitably limits the extent to which the size of the swept area can be expanded;[9] however, engineers disagree over how much progress is still achievable. For the time being, technical economies of scale continue to be realized. Advances in turbine technology, lighter component materials, and improvements in wind capture engineering continue to drive down generation costs.

Third, there are production economies of scale that have contributed to the overall reduction of wind power generation costs in the same manner that Henry Ford's production line contributed to lower automobile assembly costs. Wind experts Ackerman and Soder estimated that in the 1990s, wind power cost declined by 20% every time the aggregate amount of global installed wind power capacity doubled.[10] Although experience from other industries suggests that further market growth will likely provide diminishing returns regarding production economies of scale, the consensus appears to be that over the next few decades, as the wind power market expands, production economies of scale will continue to contribute to lower generation costs and this will fuel further market expansion because market growth and cost enjoy a symbiotic relationship. Danish developer DONG

Energy provides a forward perspective on this phenomenon, estimating that the cost of wind power can be expected to decrease by 4 to 10% each time the aggregate market capacity doubles.[11]

Overall, progressive technological advances, technological economies of scale, and production economies of scale helps explain why many experts believe wind power costs will decrease substantially in coming decades.[12] Some estimate that the cost of wind power will fall to approximately US2¢ per kWh within the next three decades.[13] Others counter that rising steel costs will impede wind system cost improvement. Even if costs do fall further to a US2–5¢ per kWh level, this does not guarantee that a *specific* wind project will generate power in this range. Minimizing cost depends on a number of other factors that will be touched upon as this chapter progresses.

2.2.3 Delivering Wind Power to the Grid

Inauspiciously, wind farms are often sited in remote areas to take advantage of land availability and obviate social opposition.[14] The distance from the site to the electricity grid influences connection costs in two ways. First, spatial separation from power grids means that longer transmission networks need to be built. In the United States, this can add as much as US$80 per meter to the cost of a wind power project.[15] Second, energy dissipates as it travels along transmission lines. Power leakage increases as distance to the electricity grid increases. It has been estimated that transmission and distribution (T&D) leakage can be 10% or more of energy produced.[16] Two factors have the most influence on leakage—distance and the type of electricity conduit used—and knowledge of these two factors allow engineers to estimate leakage with a degree of accuracy.[17]

One other grid connection factor that influences generation cost is the voltage capacity of power lines installed to carry power to the grid. The voltage level limits the amount of power that can be delivered to the grid.[18] Consequently, power line capacity can influence the size of wind farms or necessitate investment in storage systems or substations to regulate voltage.

2.3 WIND POWER MANAGEMENT CHALLENGES

2.3.1 The Challenge of Matching Supply and Demand

Even in ideal situations, electricity load management is complicated by demand-side variances. Demand for electricity varies by season, by week, by time of day, and even by the second. Godfrey Boyle colorfully documents how a mass rush to brew tea at half time during the 1990 World Cup Semi-final between West Germany and England caused a demand spike of 2 GW over a 2–3 minute period.[19] An electricity supply system must be able to respond promptly to all these demand variations.[20]

This challenge is complicated by the fact that all power generation systems face hurdles in producing consistent energy flows and adjusting power output to meet demand. This is important to note, because one of the chief criticisms of wind power is that inconsistent power flows create costly power management challenges. This can be true, but it is not a weakness that only wind systems exhibit. Both nuclear and coal-fired plants, for example, can break down, suddenly removing a major source of electricity supply. The most prominent recent example pertains to Japan's nuclear power program where the 50 remaining plants were taken offline for safety checks after the Fukushima disaster.

Electricity generation technologies also differ in terms of ability to respond to sudden demand fluctuations. On one end of the spectrum are base-load technologies such as coal-fired power and nuclear power. These technologies cannot respond quickly to increased demand need, nor can they power down quickly. Therefore, they are most economically utilized to provide fixed power output—base-load. At the other end of the spectrum are peak-load technologies such as natural gas and hydropower. Both of these electricity generation technologies can adjust to demand fluctuations within minutes (for gas) or seconds (for hydropower). Wind power is most effectively used as a base-load technology because it allows the turbines to function at maximum capacity.

2.3.2 Stochastic Flows

With that said, the act of generating electricity from wind does create some arduous supply management challenges stemming from the stochastic nature of wind. The amount of energy generated from a wind turbine can vary significantly from minute to minute, hour to hour, day to day, week to week and year to year. However, as this section will demonstrate, insinuations that significant contributions from wind power will unfailingly destabilize electricity grids are largely exaggerated, because there are operational strategies and technological fixes that can be adopted to minimize the severity of the problem.

Strategic site planning, geographic dispersal of wind power facilities, and technical decisions made when selecting turbines (i.e., adjustable rotors, variable speed gearboxes, computerized yaw controls, etc.) can significantly attenuate wind power fluctuations.[21] In fact at low levels of contribution to an electricity grid, the stochastic nature of wind power poses limited threat to grid resilience, because spare capacity already embedded in the average electricity grid can smooth the flows.[22]

Currently, research indicates that depending on the electricity grid base-load profile, 10 to 40% wind energy can be integrated into an electricity grid without having to add storage or additional spare capacity. For grids that are dominated by coal-fired power stations, a 10 to 20% contribution from wind power represents the current benchmark beyond which additional storage or capacity additions become necessary.[23] For grids that are dominated by hydropower or interconnected to other national grids, a 30 to 40% contribution from wind power may be achievable without adding back-up systems or electricity storage. There are already examples of nations which rely on wind energy for up to 40% of total electricity demand.[24] Denmark has set a goal of producing 50% of its electrical power through wind energy by 2020 thanks to interconnections that Denmark has with the EU grid.[25] Indeed, some studies go as far as to conclude that even in systems with inflexible base-load energy sources (such as nuclear power), the potential contribution from wind energy (without adding reserve capacity) may cost effectively reach as high as 50% in coming decades through better dispersion of wind resources, improved generation technologies,[26] and new energy storage technologies.[27]

In short, it is clear that criticisms of the stochastic nature of wind power are largely invalid at current installed capacity levels for most nations. A high level of development can still be undertaken in most nations before the management of stochastic flows begins to have a significant economic ramification.[28]

To accommodate higher levels of installed wind power capacity, there are two technological approaches to improving technological resilience—but both increase the cost of electricity generated. The first approach is to increase generation capacity of peak-load support systems such as hydropower or natural gas-fired power plants. Ensuring higher capacity in highly responsive (peak-load) electricity generation technologies allows load engineers to compensate for fluctuations in wind power by adjusting output of the reserve generators. In nations where peak-load technologies such as gas-fired power plants are replacing coal-fired power plants, the resilience of the grid is naturally reinforced, meaning that higher levels of wind power can be added with few additional costs. However, in nations where peak-load capacity must be added specifically to accommodate wind power, this solution becomes costly because the investment is not fully exploited due to the downtime (and combustion inefficiencies) that is common to reserve generators.[29]

The other approach is to store wind energy that has been generated but not yet utilized. For utility-scale storage, prominent current options include advanced battery storage, pumped hydro, and compressed air energy storage.[30] Compressed air energy storage systems are purportedly the most efficient of the universally mobile storage technologies; however, the systems are still expensive to construct, require fuel to drive the compressor and leak energy (only a fraction of the energy gets stored).[31] In short, although storage is a feasible solution, it adds to the cost of electricity generated.

Although the added cost of adding storage or reserve back-up is frequently raised by wind critics as a reason to avoid amplified levels of wind power capacity, many of these cost concerns are exaggerated. For example, a report by the Australia Institute contends that adding approximately 5% wind power to the existing grid would only cost consumers AU$15–$25 per year extra.[32] Another study, indicates that the additional cost of backup generation (i.e., gas-fired generators) necessary to allow wind power to reach high contribution levels (i.e., 40%) in Australia would increase the cost of wind power by approximately 25%.[33] This amounts to roughly US1–2¢ per kWh.

In summary, the intermittent property of wind can indeed pose logistical problems for managing regional electricity grids but not at the current levels that exist in most countries. At low levels of wind power integration (i.e., 5–10%) existing generation capacity may be able to support additional wind power contribution without any additional costs. At higher levels (i.e., 20%+) adding spare capacity or energy storage systems will increase the cost of wind power, but not to the cataclysmic degree espoused by wind power critics.

2.4 ENVIRONMENTAL CONCERNS AND REALITIES

Aside from technological challenges, which as the previous section has demonstrated can be largely attenuated or eliminated altogether through strategic and technological fixes, there are a number of environmental threats that have been attributed to wind power. As this section will demonstrate, some of these concerns are valid and some are not. However, as the next two sections will demonstrate, even for valid issues, there are strategies which can be employed to significantly diminish the concern.

Many environmental concerns stand to be exacerbated as the size and the number of wind power projects increase. Therefore, there is understandable apprehension that the expansion of wind power capacity may begin to exhibit diminished growth potential due to social resistance if these concerns cannot be adequately addressed. In an underdeveloped market, wind power project developers have relatively unfettered rein in terms of site selection. Typically, this results in developers pursuing a strategy of prioritizing development of sites that possess the three attributes of technical attractiveness (overall wind quality and development costs, proximity to the electric grid and community acceptance). As the most attractive sites are developed, developers will be increasingly forced to consider development of sites that may be more socially contentious.[34] Social opposition can turn an economically beneficial wind power project into a political hot potato that can undermine completion of the project in question and influence prospects for future developments.

Research indicates that social impediments to wind power development fall under two broad themes—concerns over impairment of existing community endowments and concerns over impairment of existing ecosystems. Consequently, this chapter proceeds by sequentially addressing both of these themes and translating existing knowledge to applied policy insight.

2.5 IMPAIRMENT OF EXISTING COMMUNITY ENDOWMENTS

Community opposition to wind power projects is widely known by the acronym NIMBY ("not in my back yard"). Reasons for opposing wind projects are varied. For example, in a survey related to a proposed wind energy project in Cape Cod in the United States, eight justifications for opposition were uncovered. Responding stakeholders were concerned about adverse impacts on aesthetics, community harmony, the local fishing industry, pleasure boating, property values, bird life, marine life, and tourism.[35] Accordingly, developing effective community opposition mitigation strategies requires awareness of the varied motivations for opposition to a given project.[36] A starting point for enhanced understanding of community concerns is through public outreach initiatives (surveys, town hall meetings, etc.) to identify the nature of community concerns.

2.5.1 Separating Perception from Fact

A significant amount of research exists that suggests community concerns regarding wind energy projects are based more on perception than fact.[37] Concerns over turbine noise, shadow flicker, and threats to birdlife are not conclusively supported by actual data.[38] Moreover, in scenic areas there is a perception that the erection of wind turbines will adversely affect tourism. Yet surveys conducted in tourist areas in Germany, Belgium, and Scotland indicate that such concerns are unfounded.[39]

The trouble is that the general public rarely has ready access to information necessary to assess the pros and cons of wind power projects. Media reports tend to emphasize storylines that have popular appeal (i.e., famous figures who are opposed to a development, accusations of scandalous behavior, etc.).[40] Consequently, media coverage often fails to provide the full information that the public needs to effectively evaluate the merits of a project.[41] Moreover, as one wind expert points out, a great deal of misinformation about wind power has purportedly been propagated by fossil fuel and nuclear power special interest groups.[42]

The lesson for policymakers is that some forms of opposition can be mitigated by providing community members with more comprehensive information on a given project. In fact, not only will a more proactive media management strategy help mitigate opposition, it may actually engender enhanced support.

2.5.2 Perceptions Improve

Research also indicates that public perceptions generally improve after wind projects become operational.[43] Polls conducted with residents from communities that host wind energy developments in the United Kingdom, Scotland, France, the United States, and Finland have all demonstrated that wind farms that are properly planned and sited can engender positive project perceptions.[44] In fact, wind energy projects that have been planned to minimize adverse social and environmental impacts can even positively influence perceptions of wind energy once completed.[45] Even in Denmark, where there is a higher degree of wind power saturation, respondents to a survey who were living closer to wind turbines were not found to be more negatively disposed toward wind power, and overall attitudes toward additional on-land turbines were found to be positive.[46] From a project management perspective, it is noteworthy that positive perceptions are particularly strengthened when community members are offered opportunities to invest in the development.[47]

2.5.3 Aesthetic Concerns Overshadow All Others

Research indicates that local concerns trump global concerns and aesthetic impacts trump ecological impacts when community members evaluate the pros and cons of a wind energy project. Research by Robert Thompson of the University of Rhode Island found that a wind energy project's contribution to greenhouse gas (GHG) emission abatement will fail to mitigate project resistance associated with concerns that the project will degrade the aesthetics of a community.[48] The lesson is clear: arguing that wind farms are a necessary evil in light of climate change will simply not overcome opposition to poorly planned wind farms. A number of studies have also found that the cause of public disenchantment over a given wind project is frequently centered on concerns over erosion of aesthetic values rather than concerns over degradation of ecosystems.[49] A caveat with many of these studies is that they were done in the United States, and may not be representative of other advanced nations. However, there is enough anecdotal evidence to indicate that these general principles apply in other nations (i.e., Denmark, Sweden) as well.[50] Overall, it is highly likely that in any community, the perception that wind turbines represent aesthetic eyesores must be addressed either through technical solutions (improved siting, camouflaging turbine towers, etc.) or through better marketing of the community benefits associated with such projects.

2.5.4 Beyond NIMBY Opposition

Research shows that NIMBY resistance to wind energy projects is not the only type of resistance. In attempting to understand opposing factions in greater depth, Maarten Wolsink of the University of Amsterdam identified four types of resistance:[51]

- **Type A:** Individuals who support wind energy but are opposed to developing a specific site (this is the classic NIMBY group).
- **Type B:** Individuals who are generally opposed to all wind power developments (NIABY—not in *any* back yard).
- **Type C:** Individuals who were initially positive toward a specific project, but develop negative feelings as a project develops.
- **Type D:** Individuals who are opposed to a specific project due to poor planning or other technical reasons.

The importance of delineating opposition across the four typologies rests with the observation that each type of opposition demands a different strategic mitigation approach. As mentioned earlier, mitigating opposition from NIMBY opponents (Type A) involves a process that begins by seeking to understand the often diverse drivers behind such opposition. Once the sources of concern are identified, strategies can be developed to i) correct misperceptions, ii) design and implement solutions, or iii) attempt to dilute opposition by highlighting project benefits that offset the areas of concern.

On the other hand, it is unlikely that Type B opposition can be fully eliminated because such opposition frequently stems from misperceptions caused by entrenched and opposing ideologies. Although NIABY factions are typically small,[52] opposition by such factions can fuel opposition from other groups (such as Type C groups described in the next paragraph). Fortunately, as opposed to nuclear energy, NIABY opposition to wind power is rarely substantial enough to stimulate large-scale public protest.[53] With that said, in some countries there are well-organized, vocal groups that oppose wind power, such as the Country Guardians in the United Kingdom, the Association for Protection of the Landscape in Sweden, and Windkraftgegner in Germany.[54] Negotiation is typically the only way to allay opposition from such groups.

Opposition from Type C factions occurs when new information emerges which alters perceptions of a project. In some cases, negative perceptions are based on misinformation which can be intentional (i.e., opponents to a wind project spreading false information) or unintentional (i.e., media reports which focus on only one or two aspects of a wind project). It can be argued that the logical approach for diminishing opposition based on mis-information is to try and ensure that all stakeholders receive comprehensive information on the pros and cons of a given project prior to the project being approved. However, many wind power developers anecdotally note that in some circumstances presenting such a comprehensive analysis prior to project initiation could also unnecessarily exacerbate opposition if communications are not effectively handled. The lesson appears to be that more communication is best, but the effectiveness of communication cannot be ignored. When such opposition emerges, improved information dissemination may restore positive support. However, entrenched perspectives—even based on false information—can be hard to alter, especially if the source of the misinformation continues to perpetuate such perspectives.

In other cases, negative perceptions are based on justifiable concerns that have emerged. In these cases, revisions to the project may appease concerned parties. In yet other cases, the source of negative perception may be both well-founded and irresolvable. In these cases, mediation efforts may at least help to dilute the strength of opposition. A starting point for restoring support from Type C opponents is to first identify why negative perceptions have emerged and craft solutions accordingly.

Opposition from Type D factions can stem from either real or perceived problems. Therefore, mitigating opposition from Type D factions can be approached in a similar manner to mitigating opposition from Type C factions: i) correct misperceptions that fuel opposition, ii) amend real problems that can be viably resolved, and iii) employ mediation to defuse emotions when full resolutions are not possible.

2.5.5 Overall Lessons in Regard to Community Opposition

Overall, research tells us that a great deal of opposition can be avoided simply by correcting misperceptions. Improved communication can temper emotions and attenuate local opposition.[55] Public forums, project websites, community mailings, media management strategies, and opinion surveys present opportunities for creative dialogues to take place. Often interaction with stakeholders generates creative solutions.[56] Even when full resolutions are not possible, public interaction allows citizens to vent and express their opposition. While this may not fully appease dissatisfied factions, research indicates that allowing dissenters to voice concerns diminishes the heated emotional response that often underlies public protest.[57] There are also more advanced strategies to correct misperceptions. For example, taking civic leaders and key stakeholders to visit established sites and discuss wind power with citizens who have installations in their community can positively influence perceptions.

In addition to ongoing discourse with community stakeholders, five other principles have been proposed to mitigate opposition to wind energy projects. First, sufficient distance between the project site and residential areas should be preserved in order to minimize disruption caused by noise and shadow flicker. Second, in inhabited areas, turbines with noise dampening devices should be mandated. Third, a project that financially benefits the local community garners improved support. Therefore, initiatives to encourage project participation from local firms and to entice community ownership over wind power projects can help endear projects within communities.[58] Fourth, if participants in the development have social ties to the area, the propensity for community opposition diminishes. Therefore, project developers should endeavor to enlist the participation of local firms to the greatest extent possible. Last, providing avenues for ongoing community feedback purportedly diminishes extreme forms of resistance that can cause project delays.[59] This can take the form of project websites and hotlines, on-site administrative offices (for larger projects), or even just regular community meetings.

2.5.6 Government Agency Opposition

One final form of opposition that can derail a proposed wind project comes from government agencies. Two illustrative areas of conflict are concerns over disruptions to military installations and airport communications.[60] Military agencies, airport authorities, and telecommunication authorities may block wind power projects due to concerns that wind turbines can adversely influence radar surveillance and communication systems. Although studies show that interference is negligible, misperceptions can pose intractable barriers for project developers because often, these bodies have veto power over neighboring developments or at the very least can muster up a high degree of regulatory woe. With adequate buffer zones, such threats can be entirely negated. For planning reference, guidelines regulating minimum distance and maximum heights of wind turbines are often available through national civil aviation authorities.[61]

Mitigating opposition from government agencies shares the same basic precepts as mitigating public opposition. The threat of opposition can be minimized by seeking to understand concerns, rectifying misperceptions, working with stakeholders to develop agreeable mitigation measures when necessary, and engaging proactively with officials from government agencies who may be concerned about the impact of a proposed wind energy project.

In conclusion, as wind power projects expand in scale and scope, managing public perception will become increasingly important.[62] Wind power projects will increasingly encroach upon locations that are valued for aesthetic or environmental reasons. A degree of public resistance is unavoidable because scenic spots such as hilltops, ocean bluffs, and wide sweeping plains are often ideal locations for wind power projects.[63] Accordingly, a degree of reeducation may also be required in many communities in order to entrench understanding that a transition away from carbon-based electricity generation requires a degree of community commitment to accepting necessary trade-offs. As Dismukes and colleagues point out, "success of radical innovation (such as wide scale wind adoption) requires much of the community it affects: resolution of technical debates about approach, write-down of existing investments, unlearning and relearning of organisational behaviors and practices, creation of new businesses or even industries, perhaps even cultural change. These processes can take years."[64]

2.6 IMPAIRMENT OF EXISTING ECOSYSTEMS

Although the concerns over impairment of existing community endowments outlined to this point in the chapter are typically the strongest impediment to wind power development, there are documented cases where concerns over impairment of existing ecosystems have also catalyzed social opposition to wind power developments.

Many attractive wind power sites are located in ecologically sensitive areas.[65] Rural or offshore sites that are often richer in biodiversity appeal to wind project developers thanks to lower land costs and lower risks of public opposition.[66] Although coastal areas, mountain ridges, and mountain passes all present attractive siting options due to superior wind quality,[67] they are often among the most ecologically precious. In many countries, coastal areas are extensively developed and few undeveloped sites remain. Erecting wind farms in such areas can close off important migration corridors for keystone species that bridge coastal and inland habitats.[68] Similarly, mountain passes are often attractive wind power sites due to wind channels inherent to such passes; unfortunately, wind channels also serve as avian flight paths.[69]

2.6.1 Bird Mortality

Bird mortality is perhaps the most notorious of the ecological threats that wind farms pose. It is not uncommon for wind project developers to be confronted with public concern or even active protest over threats to the avian population.[70]

Statistically, as Table 2.1 illustrates, pollution, electrocution, collisions with vehicles or buildings, and collisions with electricity infrastructure associated with conventional power grid operations cause far more bird deaths than do collisions with wind turbines.[71] A study in 2001 conducted by the US National Wind Coordinating Committee estimated that there were 6,400 bird fatalities associated with 3,500 wind turbines investigated by the study.[72] Generally, research indicates that it is not the absolute number of bird kills, but rather, the rarity or ecological sensitivity of specific avian species that fuels the staunchest opposition to wind energy projects.

Despite low avian mortality rates, misperceptions fuelled by planning flaws associated with wind farms of the 1970s and 1980s can still fan the flames of protest. Early turbine models were erected on lattice towers which provided a nesting ground for birds.[73] However, newer turbine models are mounted on pylon-style towers which are not conducive to nesting.[74] Unfortunately, although the primary causes of avian fatalities have been significantly mitigated by improved tower construction, larger rotor blades that spin at a slower pace and improved siting strategies, the stigma remains that wind turbines pose unacceptably large threats to avian populations. Community engagement supported by avian impact assessments can help diffuse such dissonance.

Table 2.1 BIRD MORTALITY FROM
ANTHROPOCENTRIC CAUSES
IN THE UNITED STATES

Object	Mortality (Birds Per Year)
Power Grid	130–174 million
Cars and Trucks	60–80 million
Buildings	100–1000 million
Telecom Towers	40–50 million
Pesticides	67 million
Wind Turbines	6400

Source: Wizelius, Tore. (2007). *Developing Wind Power Projects: Theory and Practice*. Oxford: Earthscan.

2.6.2 The Challenge of Estimating Bird Mortality

One common method for assessing the impact of a wind energy project on the avian population is to estimate bird mortality, which is often expressed as the number of birds killed in a given area (i.e., bird kills per square kilometer per year). Separate bird mortality estimates are often calculated for any endangered species inhabiting an area.

Unfortunately, data can be misleading or altogether inaccurate due to a number of confounding factors. Many bird mortality estimates use data from other proxy wind power sites to generate rough estimates of bird kills. However, species, migration, and scavenging behavior of birds as well as the characteristics of each wind farm differ. Accordingly, estimates that are based on proxy data from other sites will never be directly transferrable.

Even studies done post-project construction can be inaccurate due to enumeration challenges. For example, bird mortality is usually calculated by counting the number of bird carcasses found within the site area. However, the number of carcasses found is dependent on the number of birds migrating through an area. Studies that fail to account for seasonal migration variations are unrepresentative. Moreover, counting bird carcasses found within a site boundary produces underestimates of true mortality. Birds that are injured by wind turbines can fly off to other areas where they perish. Furthermore, bird carcasses that fall to the ground are frequently carried off by scavengers.

Even when bird mortality estimates are relatively representative, absolute mortality numbers tell only part of the story. A thousand birds killed per year within the boundary of a wind site represents a significant mortality rate if ten thousand birds pass through the site each year. However, if ten million birds pass through the site each year, the mortality rate is less significant. In order to evaluate the bigger picture, a statistic known as *bird risk* is commonly used. Bird risk is defined as the number of bird fatalities as a percentage of the total number of birds observed in the area.[75]

Unfortunately this metric exhibits all of the potential problems associated with estimating bird mortality plus a host of other confounding threats associated with estimating the number of birds passing through an area. First, birds migrate into and out of habitats. Accordingly, bird numbers are rarely consistent throughout the day, month, or year. Second, some birds are nocturnal. This poses obvious enumeration challenges. Third, different birds fly at different heights and so the threat posed is not the same for all species. Finally, as implied earlier, avoiding fatalities of endangered birds should take priority. Therefore, enumeration efforts should ideally endeavor to separate endangered species from commonly found species. In practice, this is hard to accomplish.[76]

There are two useful lessons to draw from these observations. First, mortality studies can provide insight into the potential for public opposition from groups that are concerned about avian welfare. However, such studies are only useful if they avoid the threats to validity outlined earlier. Second, policymakers who reference avian mortality studies in order to gain insight into the threat that a wind energy project poses to the avian population should do so with a critical mindset. The methodology supporting the data should be clearly understood in order to ascertain the limitations associated with the study's conclusions. Absolute declarations of safety based on subjective assumptions can inadvertently inflame public opposition. Similarly, estimates of harm can be overstated.

2.6.3 Degradation of Animal Habitat

Disruptions to animal habitats associated with construction and operation of a wind facility can significantly influence foraging patterns and undermine the continued viability of the area to support resident species.[77] More effective planning can significantly mitigate threats to ecosystem integrity at the site preparation, construction, and operation stages.

Ecosystem-friendly site design requires a reassessment of the traditional approach to site development, which typically begins by clearing all vegetation from a site and leveling the site with bulldozers. Clearing a site in this manner creates ecologically barren wastelands that uproot animal habitats, disrupt foraging patterns and fragment animal populations.[78] This is true even if new vegetation is planted once construction is completed. The level of comfort that an animal has with its habitat is dependent on the familiarity it establishes with its environs. Changes to physical features of the environment or even to scent patterns attached to flora can severely disrupt foraging patterns.[79] A better way of developing sites is to clear only those areas of land that will be built upon. This will leave some of the original flora in place and provide a level of familiarity that will induce animals to return to the area after construction is completed. Moreover, careful attention should be given to selection of any vegetation to be replanted. Efforts should be made to ensure that new vegetation mirrors the type of vegetation lost.[80] Furthermore, the ecological intrusiveness of wind tower foundations can be significantly reduced by recovering foundations with soil and vegetation.[81]

Another flaw with traditional site development concerns fencing which is often erected around a site, often in adherence to public safety regulations. Utilizing traditional chain-link fencing prevents larger species from returning to the site. Construction standards that require access holes to be installed at various intervals along the fence to facilitate animal migration can attenuate this problem.[82]

It is worth noting that ecologically sensitive site design should not stop at site boundaries. One of the greatest threats associated with wind system developments stems from the clearing of pristine lands for access roads and transmission line towers. Not only do access roads potentially hinder animal migration, they also facilitate human access to ecologically sensitive areas. Again, the process of designing the project with these threats in mind can produce cost-effective solutions. Migration corridors can facilitate improved animal migration and lockable entry gates at the mouth of service roads can help regulate unauthorized access.

Different species of animals respond differently to external commotion. During the construction stage, noise and commotion from construction activities can either scare off predators or prey; in doing so, they can unintentionally upset the ecological balance.[83] Identifying the types of animals native to a site—resident animal profiles—and developing impact assessment and mitigation strategies for the identified species can help minimize the disruptive impact of construction activities. It is particularly important in the development of resident animal profiles that endangered species and keystone species are prioritized to ensure a given project does not cause irreparable ecological damage.[84]

Unfortunately, ecological disruption caused by wind farms does not entirely disappear upon completion of construction.[85] Rotor noise that was a problem with older wind systems has been more or less attenuated through technological advances.[86] However, the impacts on wildlife of the incessant swishing sound from modern rotors and shadow flicker caused by the oscillating blades are not yet fully understood. In the absence of better understanding, wind development planners should avoid developments in areas that are inhabited by endangered species.

2.6.4 Offshore Wind Farms and Ecological Concerns

Threats to habitat viability apply to offshore wind power developments as well. However, the contention that ecologically sensitive site planning can avert many ecological problems is true for offshore wind developments as well.[87] Mitigation measures can be designed to avoid damaging the health of reefs, marine breeding grounds, and aquatic foraging areas. The marine habitat can be highly resilient. For example, research indicates that although the noise emitted and the turbidity caused during the process of tower construction can scare off marine mammals-postconstruction, the marine mammals tend to return to the area.[88] Research also indicates that the base of wind turbine towers can potentially act as artificial reefs for benthic fauna; and as such, can positively contribute to the marine habitat.[89]

Overall, extant research in regard to ecosystem management of offshore wind energy developments generally indicates that informed environmental planning can avert most threats to the marine habitat. However, as is the case with onshore developments and animal habitats, more research still needs to be done on the effect of operational noise and vibrations on aquatic creatures.

2.6.5 The Importance of Environmental Impact Assessments

Ecological threats and appropriate mitigation measures are site-specific, because flora and fauna profiles vary. Accordingly, to fully anticipate the impact of wind power projects on a given ecosystem, environmental impact assessments (EIAs) should be undertaken. EIAs are detailed assessments of ecological impacts associated with specific projects.[90]

The first step of an EIA is to establish the baseline. The baseline represents the state of the ecosystem prior to any development. The next step is to conflate ecological and engineering principles to predict and evaluate impacts that will occur at the site preparation, construction and operation stages. Finally, the EIA typically concludes by recommending mitigation measures that will minimize the impact of the project on the ecosystem.[91] In short, EIAs are site-specific blueprints for mitigating ecological damage associated with wind energy projects.

If the intention is to ensure that ecological damage is minimized, it is imperative for the EIA to be a part of the project approval process.[92] Furthermore, the development of standardized EIA templates helps to ensure every project site is evaluated according to the same criteria with the same depth of analysis.[93] Almost counterintuitively, as opposed to an absence of standards, research indicates that regulatory standardization of EIA criteria is greatly appreciated by environmental and corporate stakeholders alike.[94] This is because standardization allows environmental watchdog groups to influence what goes into an EIA through political lobbying and more effectively evaluate EIA submissions from project developers. Standardization also insulates project development firms from public criticism that the EIAs they carry out lack an acceptable standard of rigor.[95]

There are three caveats associated with the management of EIA policy. First, projects which have been planned in an ecologically sensitive matter should not be delayed by red tape associated with an inefficient EIA review process, because unpredictability deters investment.[96] This implies that authorities that are responsible for vetting the assessments and granting approval must have the resources, competencies, and operational obligation to expediently carry out effective, timely evaluation of submissions.[97]

Second, a degree of flexibility should be built into the EIA process in order to allow amendments to be made to EIAs as characteristics of projects change, new technology emerges, and project finances fluctuate.[98] Mechanisms should exist to allow project developers to make minor amendments to project designs and have these amendments approved in a fast-track manner without the entire EIA being resubmitted. An EIA should be an advisory tool that helps to make wind energy projects more environmentally sound; it should not be used to delay or derail projects that are beneficial to the community.[99] In order to achieve economic and environmental balance, many nations draw a distinction between small and large wind energy developments. Larger developments require more detailed EIAs. In Germany, projects involving 20 turbines or more require significantly more due diligence and preparation of a mandatory EIA. In Sweden, any installation over 25 MW requires a comprehensive EIA.[100]

Third, policy should ensure that EIAs are prepared and disseminated for stakeholder evaluation and input well before project approval is given.[101] For wind energy project developers and civic sponsors, one of the main purposes of preparing an EIA is to minimize the threat of public protest caused by poor planning. Without giving stakeholders a voice, EIAs created with even the best intentions may still fuel protest.[102] This caveat may seem like a trite observation; however, all too often EIAs are prepared in isolation from stakeholders and appended to projects as afterthoughts.[103]

2.6.6 Aesthetics

Any cost-benefit analysis related to the siting of a wind power project in a community should attempt to estimate the financial impact the development will have on the economic fortunes of the community. Research indicates that one source of opposition to wind power developments stems from concerns over the potential adverse economic impact the development might have on property values and in some cases, tourism revenue.[104] Financial impact assessments can be approximated by using hedonic pricing which is an environmental impact estimation technique that uses experience in one community to estimate impact in another community.[105] The scant research that does exist in regard to assessing the impact of wind power projects on property prices and tourism indicate that any adverse impact that does exist is likely short lived.[106] Property values have been shown to bounce back after wind farms become operational,[107] and research indicates that wind turbines can actually be a boon to tourism in some cases.[108] However, the limited amount of research in this area suggests that formal financial impact studies might be warranted in communities where this might be a concern.

If indirect costs associated with a wind energy project are to be estimated, indirect savings associated with the same project should also be estimated. In fossil fuel dominant societies, electricity generated by wind energy projects would otherwise likely come from fossil fuel power sources, which inflict both health and environmental damages on communities. High concentrations of sulphur dioxide associated with coal combustion have been linked to the degradation of buildings and monuments, as well as to the acidification of lakes and waterways.[109] CO_2 from fossil fuel combustion is the main anthropic contributor to climate change.[110] Furthermore, pollution from coal-fired power plants has been linked to respiratory diseases. To put the damage into perspective, the Ontario Medical Association estimated that health problems in the late 1990s stemming from pollution attributed primarily to fossil fuel-fired power generation annually cost Ontario over C$1 billion in health costs and contributed to over 1900 premature deaths.[111]

2.7 CONCLUSION

It is perhaps no exaggeration to say that the realizable potential of wind power depends on how well developers manage social and environmental issues.[112] As more wind farms are developed, threats to both social and ecological endowments will increase.[113] As available sites become scarcer, the impetus to build wind farms on socially and ecologically sensitive areas will also increase.[114] However, by developing and overseeing improved standards for managing threats to social and ecological endowments, policymakers can play a role in ensuring that the benefits derived from wind energy are not realized at the expense of a community's natural habitat.

NOTES

1. Ackermann, Thomas, and Lennart Söder. 2002. "An Overview of Wind Energy-Status 2002." *Renewable and Sustainable Energy Reviews* 6 (1–2): 67–127.
2. For a highly readable introduction to wind power technology, see Wizelius, Tore. 2007. *Developing Wind Power Projects: Theory and Practice*. Oxford: Earthscan.
3. Celik, A. N., T. Muneer, and P. Clarke. 2007. "An Investigation into Micro Wind Energy Systems for their Utilization in Urban Areas and their Life Cycle Assessment." Proceedings of the Institution of Mechanical Engineers, Part A: *Journal of Power and Energy* 221 (8): 1107–1117.
4. Wizelius, Tore. 2007. *Developing Wind Power Projects: Theory and Practice*. Oxford: Earthscan.
5. Ackerman, Thomas. 2005. *Wind Power in Power Systems*. West Sussex: John Wiley and Sons. This book provides comprehensive coverage of wind power technology.
6. For a very useful introduction to the world of renewable energy, see Boyle, Godfrey. 2004. *Renewable Energy: A Power for a Sustainable Future*. 2nd ed. Oxford: Oxford University Press.
7. Ackermann, Thomas, and Lennart Söder. 2002. "An Overview of Wind Energy-Status 2002." *Renewable and Sustainable Energy Reviews* 6 (1–2): 67–127.
8. Wizelius, Tore. 2007. *Developing Wind Power Projects: Theory and Practice*. Oxford: Earthscan.
9. Ibid.
10. Ackermann, Thomas, and Lennart Söder. 2002. "An Overview of Wind Energy-Status 2002." *Renewable and Sustainable Energy Reviews* 6 (1–2): 67–127.
11. DONG Energy. 2008. *Final Report on Offshore Wind Technology*. Fredericia, Denmark: Risø National Sustainable Energy Laboratory, New Energy Externalities Developments for Sustainability Consortium.
12. This theme is prominent in many influential studies including the works by Wizelius, DONG Energy, and Ackermann and Söder that have been referenced in this chapter. Also see Dismukes, John P., Lawrence K. Miller, Andrew Solocha, Sandeep Jagani, and John A. Bers. 2007. "Wind Energy Electrical Power Generation: Industrial Life Cycle of a Radical Innovation." Paper read at the Portland International Center for Management of Engineering and Technology 2007 Symposium, August 5–9, in Portland, Oregon.
13. For example, see: Brown, Brit T, and Benjamin A Escobar. 2007. "Wind Power: Generating Electricity and Lawsuits." *Energy Law Journal* 28 (2): 489–516; and DeCarolis, Joseph F., and David W. Keith. 2006. "The Economics of Large-Scale Wind Power in a Carbon Constrained World." *Energy Policy* 34 (4): 395–410.
14. DeCarolis, Joseph F., and David W. Keith. 2006. "The Economics of Large-Scale Wind Power in a Carbon Constrained World." *Energy Policy* 34 (4): 395–410.

15. Wizelius, Tore. 2007. *Developing Wind Power Projects: Theory and Practice.* Oxford: Earthscan.

16. Ibid.

17. Blakeway, Darrell, and Carol Brotman White. 2005. "Tapping the Power of the Wind: FERC Initiatives to Facilitate Transmission." *Energy Law Journal* 26 (2): 393–423.

18. Wizelius, Tore. 2007. *Developing Wind Power Projects: Theory and Practice.* Oxford: Earthscan.

19. Boyle, Godfrey. 2004. *Energy Systems and Sustainability: Power for a Sustainable Future.* Oxford: Oxford University Press.

20. Wizelius, Tore. 2007. *Developing Wind Power Projects: Theory and Practice.* Oxford: Earthscan.

21. Boyle, Godfrey. 2004. *Renewable Energy: A Power for a Sustainable Future. 2nd ed.* Oxford: Oxford University Press.

22. For more on this see Ackerman, Thomas. 2005. *Wind Power in Power Systems.* West Sussex: John Wiley and Sons; and Karki, Rajesh, and Roy Billinton. 2001. "Reliability/Cost Implications of PV and Wind Energy Utilization in Small Isolated Power Systems." *IEEE Transactions on Energy Conversion* 16 (4): 368–373.

23. This has been extensively researched. See for example: Holttinen, Hannele. 2008. "Estimating the Impacts of Wind Power on Power Systems." *Summary of IEA Wind Collaboration. Environmental Research Letters* 3: 1–6; and Denholm, Paul, Gerald L. Kulcinski, and Tracey Holloway. 2005. "Emissions and Energy Efficiency Assessment of Baseload Wind Energy Systems." *Environmental Science & Technology* 39 (6): 1903–1911.

24. World Wind Energy Association (WWEA). 2008. Press Release: "Wind Turbines Generate More Than 1% of the Global Electricity." Bonn: World Wind Energy Association.

25. Danish Energy Agency. 2012. *Accelerating Green Energy Towards 2020.* Copenhagen: Danish Energy Agency.

26. Ackermann, Thomas, and Lennart Söder. 2002. "An Overview of Wind Energy-Status 2002." *Renewable and Sustainable Energy Reviews* 6 (1–2): 67–127.

27. See for example: Denholm, Paul, Gerald L. Kulcinski, and Tracey Holloway. 2005. "Emissions and Energy Efficiency Assessment of Baseload Wind Energy Systems." *Environmental Science & Technology* 39 (6): 1903–1911; and DeCarolis, Joseph F., and David W. Keith. 2006. "The Economics of Large-Scale Wind Power in a Carbon Constrained World." *Energy Policy* 34 (4): 395–410.

28. Karki, Rajesh, and Roy Billinton. 2001. "Reliability/Cost Implications of PV and Wind Energy Utilization in Small Isolated Power Systems." *IEEE Transactions on Energy Conversion* 16 (4): 368–373.

29. Boyle, Godfrey, Bob Everett, and Janet Ramage, eds. 2004. *Energy Systems and Sustainability: Power for a Sustainable Future.* Oxford: Oxford University Press.

30. Denholm, Paul, Gerald L. Kulcinski, and Tracey Holloway. 2005. "Emissions and Energy Efficiency Assessment of Baseload Wind Energy Systems." *Environmental Science & Technology* 39 (6): 1903–1911.

31. Ibid.

32. Macintosh, Andrew, and Christian Downie. 2006. *Wind Farms: The Facts and the Fallacies.* Canberra, Australia: The Australia Institute.

33. Diesendorf, Mark. 2003. "Why Australia Needs Wind Power." *Dissent* 13 (Summer 2003): 43–48.

34. Wizelius, Tore. 2007. *Developing Wind Power Projects: Theory and Practice.* Oxford: Earthscan.

35. Firestone, Jeremy, and Willett Kempton. 2007. "Public Opinion About Large Offshore Wind Power: Underlying Factors." *Energy Policy* 35 (3): 1584–1598.
36. Zamot, Hector Rene, Efrain O'Neill-Carrillo, and Agustin Irizarry-Rivera. 2005. "Analysis of Wind Projects Considering Public Perception and Environmental Impact." Paper read at 37th Annual North American Power Symposium, 2005, 20–October 23–25, in Ames, Iowa.
37. Thompson, Robert. 2005. "Reporting Offshore Wind Power: Are Newspapers Facilitating Informed Debate?" *Coastal Management* 33 (3): 247–262.
38. The following sources provide support that popular perception does not reflect reality: Zamot, Hector Rene, Efrain O'Neill-Carrillo, and Agustin Irizarry-Rivera. 2005. "Analysis of Wind Projects Considering Public Perception and Environmental Impact." Paper read at 37th Annual North American Power Symposium, 2005, 20–October 21, in City Missouri; and NWCC. 2001. "Avian Collisions with Wind Turbines: A Summary of Existing Studies and Comparisons to Other Sources of Avian Collision Mortality in the United States." City: National Wind Coordinating Committee (NWCC); and a good summary in Wizelius, Tore. 2007. *Developing Wind Power Projects: Theory and Practice*. Oxford: Earthscan.
39. Wizelius, Tore. 2007. *Developing Wind Power Projects: Theory and Practice*. Oxford: Earthscan.
40. Thompson, Robert. 2005. "Reporting Offshore Wind Power: Are Newspapers Facilitating Informed Debate?" *Coastal Management* 33 (3): 247–262.
41. Ibid.
42. Wizelius, Tore. 2007. *Developing Wind Power Projects: Theory and Practice*. Oxford: Earthscan.
43. For a specific example see: Rodman, Laura C., and Ross K. Meentemeyer. 2006. "A Geographic Analysis of Wind Turbine Placement in Northern California." *Energy Policy* 34 (15): 2137–2149.
44. Wizelius, Tore. 2007. *Developing Wind Power Projects: Theory and Practice*. Oxford: Earthscan.
45. Wolsink, Maarten. 1988. "The Social Impact of a Large Wind Turbine." *Environmental Impact Assessment Review* 8 (4): 323–334.
46. Ladenburg, Jacob. 2008. "Attitudes Towards On-Land And Offshore Wind Power Development In Denmark; Choice Of Development Strategy." *Renewable Energy* 33 (1): 111–118.
47. Wizelius, Tore. 2007. *Developing Wind Power Projects: Theory and Practice*. Oxford: Earthscan.
48. Thompson, Robert. 2005. "Reporting Offshore Wind Power: Are Newspapers Facilitating Informed Debate?" *Coastal Management* 33 (3): 247–262.
49. Support for this contention can be found in: Komor, Paul. 2004. *Renewable Energy Policy*. Lincoln: iUniverse; and Thompson, Robert. 2005. "Reporting Offshore Wind Power: Are Newspapers Facilitating Informed Debate?" *Coastal Management* 33 (3): 247–262; and Wolsink, Maarten. 2000. "Wind Power and the NIMBY-Myth: Institutional Capacity and the Limited Significance of Public Support." *Renewable Energy* 21 (1): 49–64.
50. Ladenburg, Jacob. 2008. "Attitudes Towards On-Land And Offshore Wind Power Development In Denmark; Choice Of Development Strategy." *Renewable Energy* 33 (1): 111–118.
51. Wolsink, Maarten. 2000. "Wind Power and the NIMBY-Myth: Institutional Capacity and the Limited Significance of Public Support." *Renewable Energy* 21 (1): 49–64.

52. Wizelius, Tore. 2007. *Developing Wind Power Projects: Theory and Practice*. Oxford: Earthscan.

53. Wolsink, Maarten. 2000. "Wind Power and the NIMBY-Myth: Institutional Capacity and the Limited Significance of Public Support." *Renewable Energy* 21 (1): 49–64.

54. Wizelius, Tore. 2007. *Developing Wind Power Projects: Theory and Practice*. Oxford: Earthscan.

55. Wizelius, Tore. 2007. *Developing Wind Power Projects: Theory and Practice*. Oxford: Earthscan.

56. This stems from performance management theory. For background, see: Neely, Andy, Chris Adams, and Mike Kennerley. 2002. *The Performance Prism*. Essex: Prentice Hall Financial Times Publishing.

57. Wizelius, Tore. 2007. *Developing Wind Power Projects: Theory and Practice*. Oxford: Earthscan.

58. Komor, Paul. 2004. *Renewable Energy Policy*. Lincoln: iUniverse.

59. Wizelius, Tore. 2007. *Developing Wind Power Projects: Theory and Practice*. Oxford: Earthscan.

60. Ibid.

61. Ibid.

62. This is emphasized in: McKinsey, John Arnold. 2007. "Regulating Avian Impacts Under the Migratory Bird Treaty Act and Other Laws: The Wind Industry Collides With One of Its Own, the Environmental Protection Movement." *Energy Law Journal* 28 (1): 71–93; and Wolsink, Maarten. 2000. "Wind Power and the NIMBY-Myth: Institutional Capacity and the Limited Significance of Public Support." *Renewable Energy* 21 (1): 49–64.

63. Komor, Paul. 2004. *Renewable Energy Policy*. Lincoln: iUniverse.

64. Dismukes, John P., Lawrence K. Miller, Andrew Solocha, Sandeep Jagani, and John A. Bers. 2007. "Wind Energy Electrical Power Generation: Industrial Life Cycle of a Radical Innovation." Paper read at the Portland International Center for Management of Engineering and Technology 2007 Symposium, August 5–9, in Portland, Oregon.

65. Wolsink, Maarten. 2000. "Wind Power and the NIMBY-Myth: Institutional Capacity and the Limited Significance of Public Support." *Renewable Energy* 21 (1): 49–64.

66. Firestone, Jeremy, and Willett Kempton. 2007. "Public Opinion About Large Offshore Wind Power: Underlying Factors." *Energy Policy* 35 (2007) 35 (3): 1584–1598.

67. Zamot, Hector Rene, Efrain O'Neill-Carrillo, and Agustin Irizarry-Rivera. 2005. "Analysis of Wind Projects Considering Public Perception and Environmental Impact." Paper read at 37th Annual North American Power Symposium, 2005, 20– October 23–25, in Ames, Iowa.

68. A superb book that describes the notion of keystone species is Miller, G. Tyler. 2004. *Environmental Science: Working with the Earth. 5th ed.* Pacific Grove: Brooks Cole Publishers.

69. Zamot, Hector Rene, Efrain O'Neill-Carrillo, and Agustin Irizarry-Rivera. 2005. "Analysis of Wind Projects Considering Public Perception and Environmental Impact." Paper read at 37th Annual North American Power Symposium, 2005, 20– October 23–25, in Ames, Iowa.

70. Firestone, Jeremy, and Willett Kempton. 2007. "Public Opinion About Large Offshore Wind Power: Underlying Factors." *Energy Policy* 35 (2007) 35 (3): 1584–1598.

71. For more background on avian impacts, see McKinsey, John Arnold. 2007. "Regulating Avian Impacts Under the Migratory Bird Treaty Act and Other

Laws: The Wind Industry Collides With One of Its Own, the Environmental Protection Movement." *Energy Law Journal* 28 (1): 71–93.

72. Wizelius, Tore. 2007. *Developing Wind Power Projects: Theory and Practice.* Oxford: Earthscan.

73. Boyle, Godfrey. 2004. *Renewable Energy: A Power for a Sustainable Future. 2nd ed.* Oxford: Oxford University Press.

74. Wizelius, Tore. 2007. *Developing Wind Power Projects: Theory and Practice.* Oxford: Earthscan.

75. Zamot, Hector Rene, Efrain O'Neill-Carrillo, and Agustin Irizarry-Rivera. 2005. "Analysis of Wind Projects Considering Public Perception and Environmental Impact." Paper read at 37th Annual North American Power Symposium, 2005, 20– October 23–25, in Ames, Iowa.

76. de Lucas, Manuela, Guyonne F. E. Janss, and Miguel Ferrer, eds. 2007. *Birds and Wind Farms: Risk Assessment and Mitigation.* London: Quercus Publishing.

77. Magoha, Paul. 2002. "Footprints in the Wind? Environmental Impacts of Wind Power Development." *Re-Focus* 3 (5): 30–33.

78. Ackermann, Thomas, and Lennart Söder. 2002. "An Overview of Wind Energy Status 2002." *Renewable and Sustainable Energy Reviews* 6 (1): 67–128.

79. For more on factors influencing habitat viability, see Begon, Michael, Colin A. Townsend, and John L. Harper. 2006. *Ecology: From Individuals to Ecosystems. 4th ed.* Oxford: Wiley-Blackwell Publishing.

80. Harrop, D. Owen, and J. Ashley Nixon. 1999. *Environmental Assessment in Practice.* New York: Routledge.

81. Wizelius, Tore. 2007. *Developing Wind Power Projects: Theory and Practice.* Oxford: Earthscan.

82. Harrop, D. Owen, and J. Ashley Nixon. 1999. *Environmental Assessment in Practice.* New York: Routledge.

83. Begon, Michael, Colin A. Townsend, and John L. Harper. 2006. *Ecology: From Individuals to Ecosystems. 4th ed.* Oxford: Wiley-Blackwell Publishing.

84. Harrop, D. Owen, and J. Ashley Nixon. 1999. *Environmental Assessment in Practice.* New York: Routledge Publishing.

85. More on this topic is included in: Ackermann, Thomas, and Lennart Söder. 2002. "An Overview of Wind Energy Status 2002." *Renewable and Sustainable Energy Reviews* 6 (1): 67–128; and Ardente, Fulvio, Marco Beccali, Maurizio Cellura, and Valerio Lo Brano. 2008. "Energy Performances and Life Cycle Assessment of an Italian Wind Farm." *Renewable and Sustainable Energy Reviews* 12 (1): 200–217.

86. See both: Magoha, Paul. 2002. "Footprints in the Wind? Environmental Impacts of Wind Power Development." *Re-Focus*, 3 (5): 30–33; and Zamot, Hector Rene, Efrain O'Neill-Carrillo, and Agustin Irizarry-Rivera. 2005. "Analysis of Wind Projects Considering Public Perception and Environmental Impact." Paper read at 37th Annual North American Power Symposium, 2005, October 23–25, in Ames, Iowa.

87. This is illustrated in: Magoha, Paul. 2002. "Footprints in the Wind? Environmental Impacts of Wind Power Development." *Re-Focus*, 3 (5): 30–33; and Ardente, Fulvio, Marco Beccali, Maurizio Cellura, and Valerio Lo Brano. 2008. "Energy Performances and Life Cycle Assessment of an Italian Wind Farm." *Renewable and Sustainable Energy Reviews* 12 (1): 200–217; and McKinsey, John Arnold. 2007. "Regulating Avian Impacts Under the Migratory Bird Treaty Act and Other Laws: The Wind Industry Collides With One of Its Own, the Environmental Protection Movement." *Energy Law Journal* 28 (1): 71–93.

88. DONG Energy. 2008. *Final Report on Offshore Wind Technology*. Fredericia, Denmark: Risoe National Sustainable Energy Laboratory, New Energy Externalities Developments for Sustainability Consortium.

89. Ibid.

90. For more complete overviews of EIAs, see: Harrop, D. Owen, and J. Ashley Nixon. 1999. *Environmental Assessment in Practice*. New York: Routledge Publishing; and Lawrence, David P. 2003. *Environmental Impact Assessment: Practical Solutions to Recurrent Problems. 1st ed*. Hoboken, NJ: John Wiley and Sons.

91. Harrop, D. Owen, and J. Ashley Nixon. 1999. *Environmental Assessment in Practice*. New York: Routledge Publishing.

92. Brown, Brit T., and Benjamin A Escobar. 2007. "Wind Power: Generating Electricity and Lawsuits." *Energy Law Journal* 28 (2): 489–516.

93. Magoha, Paul. 2002. "Footprints in the Wind? Environmental Impacts of Wind Power Development." *Re-Focus*, 3 (5): 30–33.

94. McKinsey, John Arnold. 2007. "Regulating Avian Impacts Under the Migratory Bird Treaty Act and Other Laws: The Wind Industry Collides With One of Its Own, the Environmental Protection Movement." *Energy Law Journal* 28 (1): 71–93.

95. McKinsey, John Arnold. 2007. "Regulating Avian Impacts Under the Migratory Bird Treaty Act and Other Laws: The Wind Industry Collides With One of Its Own, the Environmental Protection Movement." *Energy Law Journal* 28 (1): 71–93.

96. Ibid.

97. Lawrence, David P. 2003. *Environmental Impact Assessment: Practical Solutions to Recurrent Problems. 1st ed*. Hoboken, NJ: John Wiley and Sons.

98. This process is described in both: McKinsey, John Arnold. 2007. "Regulating Avian Impacts Under the Migratory Bird Treaty Act and Other Laws: The Wind Industry Collides With One of Its Own, the Environmental Protection Movement." *Energy Law Journal* 28 (1): 71–93; and Wizelius, Tore. 2007. *Developing Wind Power Projects: Theory and Practice*. Oxford: Earthscan.

99. Lawrence, David P. 2003. *Environmental Impact Assessment: Practical Solutions to Recurrent Problems. 1st ed*. Hoboken, NJ: John Wiley and Sons.

100. Wizelius, Tore. 2007. *Developing Wind Power Projects: Theory and Practice*. Oxford: Earthscan.

101. Ibid.

102. de Lucas, Manuela, Guyonne F.E. Janss, and Miguel Ferrer, eds. 2007. *Birds and Wind Farms: Risk Assessment and Mitigation*. London: Quercus Publishing.

103. Lawrence, David P. 2003. *Environmental Impact Assessment: Practical Solutions to Recurrent Problems. 1st ed*. Hoboken, NJ, USA: John Wiley and Sons.

104. Firestone, Jeremy, and Willett Kempton. 2007. "Public Opinion About Large Offshore Wind Power: Underlying Factors." *Energy Policy* 35 (3): 1584–1598.

105. Turner, R. Kerry, David. W. Pearce, and Ian Bateman. 1994. *Valuing Concern for Nature, Environmental Economics: An Elementary Introduction*. Baltimore: Johns Hopkins University Press.

106. Firestone, Jeremy, and Willett Kempton. 2007. "Public Opinion About Large Offshore Wind Power: Underlying Factors." *Energy Policy* 35 (2007) 35 (3): 1584–1598.

107. Wolsink, Maarten. 1988. "The Social Impact of a Large Wind Turbine." *Environmental Impact Assessment Review* 8 (4): 323–334.

108. Maruyama, Y., M. Nishikido, and T. Iida. 2007. "The Rise of Community Wind Power in Japan: Enhanced Acceptance through Social Innovation." *Energy Policy* 35 (5): 2761–2769.

109. Miller, G. Tyler. 2004. *Environmental Science: Working with the Earth. 5 ed.* Canada: Brooks Cole Publishers.

110. IPCC. 2007. *Climate Change 2007: Synthesis Report.* Geneva: Intergovernmental Panel on Climate Change (IPCC).

111. Rowlands, Ian H. 2007. "The Development of Renewable Electricity Policy in the Province of Ontario: The Influence of Ideas and Timing." *Review of Policy Research* 24 (3): 185–207.

112. McKinsey, John Arnold. 2007. "Regulating Avian Impacts Under the Migratory Bird Treaty Act and Other Laws: The Wind Industry Collides With One of Its Own, the Environmental Protection Movement." *Energy Law Journal* 28 (1): 71–93.

113. DONG Energy. 2008. Final Report on Offshore Wind Technology. Fredericia, Denmark: Risoe National Sustainable Energy Laboratory, New Energy Externalities Developments for Sustainability Consortium.

114. For a specific example, see Markevicius, Antanas, Vladislovas Katinas, and Mantas Marciukaitis. 2007. "Wind Energy Development Policy and Prospects in Lithuania." *Energy Policy* 35 (10): 4893–4901.

CHAPTER 3

The Policy SET Model

3.1 THE VALUE OF A COMMON ANALYTICAL FRAMEWORK

As outlined in Chapter 1, the main intention of this book is to identify and explain forces that catalyze or prevent a more robust diffusion of wind power in the electricity sector. This chapter lays the groundwork for such an analysis by introducing and describing the main features of a common framework that can be used for guiding analysis and development of wind power development policy.

The main merit of applying a common framework to case study analysis is that it makes it possible to compare wind power policies in different nations and highlight similarities and differences. In the best case scenario, comparative analysis will unearth sufficient commonalities to construct theory to help us better understand what causes wind power to flourish in one nation and flounder in another. Even if sufficient commonalities are not uncovered, a comprehensive analysis using a common framework will at least provide insight into which issues are of greatest importance in a given national context and the scale and scope of how influential variables interconnect to shape wind power development prospects.

3.2 THE SEAMLESS WEB

In 1983, Thomas Hughes published a book titled *Networks of Power* in which he described the evolution of electrification in Western society from 1880 to 1930. In undertaking his analysis, Hughes observed that the diffusion of electrification occurred amidst a "seamless web" of social, technical, economic, and political causal factors that engender the development of a specific technological regime. In his own words:

> Electric power systems embody the physical, intellectual and symbolic resources of the society that constructs them. Therefore, in explaining changes in the configuration of power systems, the historian must examine the changing resources and aspirations of organizations, groups and individuals. Electric power systems made in different societies—as well as in different times—involve certain basic technical components and connections, but variations in the basic essentials often reveal variations in resources, traditions, political arrangements, and economic practices from one society to another and from one time to another. In a sense, electric power systems, like so much other technology, are both causes and effects of social change. . . . Power systems reflect and influence the context, but they also develop an internal dynamic. Therefore, the history of the evolving power systems requires attention not only to the forces at work within a given context but to the internal dynamics of a developing technological system as well.[1]

The complexity that Hughes attributed to electric power system evolution has clearly escalated since the turn of the twentieth century—the era during which Hughes undertook his analysis. In the late 1800s and early 1900s, electric power systems were dominated by rudimentary coal-fired and hydroelectric generation plants. Today, power system developers are confronted with a much broader array of technological options that include electricity generation from biomass, waste, wind power, solar energy, geothermal energy, nuclear fission, and tidal energy. Even conventional technologies—hydropower, coal-fired power, oil combustion, and natural gas combustion—have evolved, now exhibiting a myriad of technological features designed to facilitate greater levels of generation efficiency, minimize environmental impact or maximize economic efficiency. Underpinning all these options are networks of stakeholders that have vested interests in terms of supporting one technology over another. Moreover, the market for electric power has grown to a multibillion dollar industry which amplifies the competitiveness and financial stakes of electric power markets. In a phrase, the electric power industry can now be described as a "complex adaptive market."[2]

A complex adaptive market is a market in which there are a number of influential variables that shape the scale, scope, and direction of evolution.[3] The influential variables are so numerous and the interdependencies between the variables (the seamless webs) are so capricious that that the prediction of technological diffusion and evolution is fraught with incalculable risk.[4] Most major technology industries are distinguished by complex adaptive markets.

In complex adaptive markets, the traditional linear approach to market analysis is unworkable. This is because linear models for analyzing the evolution of markets require a degree of certainty in regard to causal relationships. In the presence of causal certainty, affecting a change in one influential variable will lead to a predictable change in another variable. For example, a gasoline retailer that lowers its price to undercut a neighborhood competitor will, in the short run, generate more revenue that may or may not make up for the lower profit margin induced by the lower price. The two variables that govern success in such a market are customer response and competitive response times. If enough customers are coaxed away from the neighborhood competitor and the neighborhood competitor's response is slow enough, this strategy might be profitable. If an insufficient number of customers switch or the competitor responds by immediately matching prices, this strategy likely will not be profitable. In any case, the strategic challenge is linear, involving only a couple of key variables that can be modeled and predicted to a fairly high degree of accuracy.

In complex adaptive markets, such as modern electric power markets, the seamless web of social, technological, economic, and political forces that influence the effectiveness of a given energy policy is too interdependent and too complex to accurately model. Therefore, predicting how a market will react to a given policy is an exercise in fallibility. Moreover, the contextual differences between national (and even regional) markets further convolute the design of universal policy frameworks that can predict industry evolution in any sociocultural setting. So if the traditional linear approach to market analysis is not viable, what is the solution?

Strategic management theory provides some guidance for addressing this challenge. Historically, strategic management theory also embraced linear models for guiding corporate strategy development. Broadly speaking, there were two traditional camps of thought. An *industrial organization* camp embraced a perspective that advocated the exploitation of favorable market conditions as a central tenet.[5] A *resources-based view* camp embraced the perspective that a firm's internal resources should be manipulated to create competitive advantage in a given market.[6] Both are linear perspectives in that they exhibit a belief that if a firm implements initiatives X and Y, a particular result, Z, will ensue. In the 1990s this linear perspective came under challenge from complexity theorists who pointed out that the dynamic nature of modern, global markets was such that predicting the emergence of trends (which is instrumental to the success of a linear strategy) was unrealizable. Eric Beinhocker, a complexity theorist, argued that since precise trends could not be predicted in complex adaptive markets, the best strategy for a given firm would be to attempt to predict more general trends—which he called *emergent land-scapes*—in order to position the firm in the general vicinity of an emergent trend and then structure the organization in a way to permit rapid strategic adjustment as trends became clearer.[7] Although the application of this notion can be fraught with difficulties in competitive, technological markets, it is conceptually not much different than the practice of sending out armed scouting parties to identify favorable settlement sites during the Western conquest of the Americas. When a favorable site was discovered, some armed members of the scouting party would remain and defend the site while others were dispatched to beckon the main body of settlers.

3.3 COMPLEX ADAPTIVE MARKET THEORY IN WIND POWER MARKET ANALYSIS

Intuitively, the notion of predicting and responding to emergent landscapes is logical when developing strategy in complex adaptive markets. After all, if one cannot accurately predict actual developments due to complexity, the next best thing is to try to position one's firm in the vicinity of emergent trends and structure the firm in a way to ensure that the organization can respond to emergent trends in a competitively expedient manner. But does the notion of emergent landscapes have relevance for policy analysis? As the case studies included in this book will demonstrate, the answer is assuredly yes.

In nations with successful track records in facilitating wind power diffusion, policies and initiatives have differed. However, there have been two common threads connecting successful wind power development programs—a political will to signal progressive commitment to support capacity development, and comparative political-bureaucratic prowess in monitoring market development and adjusting policy to respond to emergent market challenges. Using the language of complexity theory, the political will to progressively support capacity development is akin to the strategic decision a firm makes to compete in specific market niches—this is the act of defining the boundaries of the landscape within which activities will take place. Political-bureaucratic prowess in monitoring market development and adjusting policy to meet emergent needs is akin to the structural and systematic changes that a firm makes in order for it to respond expediently to emergent market developments. In short, conceptualizing complex markets from a complexity theory lens has as much relevance to policymaking as it does to strategic management.

Relevance aside, adopting complexity theory as, what Joseph Schumpter has termed, a "pre-analytic vision"[8] gives rise to a key challenge associated with the notion of emergent landscapes—how does one identify the forces which actuate emergent landscapes? If one is to have any hope of predicting emerging landscapes, one needs to begin to map out the variables and interdependencies which most influence market evolution.[9] This is as true for developing wind power policy as it is for strategic management.

As the case studies in this text will demonstrate, there are numerous variables which influence the evolution of wind power in a given nation. Considering a few of the objections to wind power raised in academic journals supports the verity of this observation. There have been claims made that wind power is simply not commercially competitive.[10] If this is true, there are economic factors that influence the evolution of wind power. There have been claims that one of the reasons why wind power is not commercially competitive stems from the stochastic nature of wind power flows—sometimes the wind blows and sometimes it does not.[11] If this is true, there are technological factors that influence the evolution of wind power. Recently, in many industrialized nations, there have been reports that wind turbines defile the aesthetics of a community; and this plays a role in blocking wind power diffusion.[12] If this is true, there are social factors that influence the evolution of wind power. Finally, there have been contentions that the structure of wind power development policy in a given nation plays a major role in facilitating wind power diffusion.[13] If this is true, there are political and policy-specific factors that influence the fortunes of wind power. In order to effectively analyze wind power development policy in any given nation, a host of diverse yet influential variables must be documented, interdependencies must be highlighted and general trends must be extrapolated from this analysis. Addressing these challenges represents the key steps for applying complexity theory to wind power politics and policy analysis. The question is, how does one develop an analytical framework that is comprehensive enough to ensure key variables are not omitted from the analysis? The next section addresses this challenge.

3.4 STEP ANALYSIS

Once again, the field of strategic management has a contribution to make in terms of guiding the development of an analytical framework for comprehensively rooting out influential variables—the STEP analysis. STEP is an acronym that stands for *social, technological, economic,* and *political.* In strategic management, when a firm wishes to analyze the strengths, weaknesses, opportunities, and threats associated with a given market, it also conducts a STEP analysis.

The social category incorporates all issues related to public perception and behavior. This is a broad category that naturally exhibits a degree of overlap with the other STEP categories. For example, the affluence of a nation (an economic variable) tends to shape public perspectives on environmental governance and this in turn influences voting patterns and civic behavior. Simply put, richer nations are less willing to accept environmental degradation as a necessary cost associated with economic development. Moreover, affluent societies tend to exhibit a higher willingness to pay for clean energy technologies. So although affluence is an economic variable, affluence also shapes social behavior and when it does, the behavioral manifestation becomes a social variable. For the purposes of this study, common social influences include elements such as perspectives on wind power technology, cultural fit, environmental philosophies, strength of community activism, education levels, worker profiles, and mass media modus operandi, to name but a few.

The technology category incorporates all issues pertaining to the evolution of energy technology. Obviously, this category too exhibits a degree of overlap with other STEP categories. For example, the presence of a pool of experienced nuclear engineers (a social variable—education) tends to increase the chances that nuclear power will receive government support. However, the presence of nuclear engineers also enables evolution of the technology, and therefore influences the technological progression of nuclear power systems. In this sense, it becomes a technological variable. For the purposes of this study, common technological influences include elements such as the availability of skilled labor, geographic conditions that favor one technology over another, domestic resource availability, electricity sector structuring, and the presence of domestic manufacturing capability in given technologies.

The economic category incorporates issues of general economic concern (i.e., a nation's industrial profile) and forces that influence the economics of the energy sector. For example, some nations (i.e., Japan) host energy-intensive industries that depend on cheap energy flows to support international competitiveness. This gives rise to an industrial lobby that has extensive influence on energy policy. The economics of the energy sector are also influenced by a number of forces from other STEP categories such as the presence of government subsidies (a political variable) or the importance of a given energy technology in terms of employment (a social variable). For the purposes of this study, common economic factors include the general economic climate in the nation, the national industry profile, energy industry subsidies, and dominant economic development patterns.

The political (and policy) category includes an analysis of four main subcategories. The first investigates national political structure. How a nation is politically structured influences the initiatives that reach the policy agenda and how emergent policies are designed and implemented. The second subcategory involves analysis of the characteristics of the governing regime because the ideological make-up of the governing regime influences what types of technologies are prioritized and how energy policy is conceptualized and implemented. The third subcategory examines fiscal health, because this influences the ability of the ruling regime to finance change. The final subcategory investigates the make-up of the policy regime, itself. How energy policy is designed and which technologies are prioritized exhibit links to past policy decisions.

As suggested above, the four STEP categories exhibit considerable overlap. This is to be expected within a framework for understanding a complex adaptive system—a system with numerous influential variables and countless interconnections. Therefore, in some cases, during the STEP analysis presented in this book there will be some choices that the reader may disagree with in regard to which forces have been assigned to which categories. However, these types of judgments are far less important than ensuring comprehensive analysis—ensuring that influential forces have not been overlooked. In the final analysis, the goal of a STEP analysis is to provide a clear methodology for identifying agents of change and predicting emergent trends.

An important feature of a STEP analysis is that it is contextually adaptable. For firms in different industries the social, technological, economic, and political variables that most influence market dynamics vary depending on the context. For example, religious factors (part of the social sphere) may influence the market for some products (i.e., contraceptives) and have very little influence on the market for other products (i.e., skateboards). Moreover, even within the same industrial sector, national contextual differences will influence market dynamics. For example, in Canada the provision of electricity falls within provincial jurisdiction, suggesting that provincial policy holds significant sway over market developments;[14] on the other hand, in Taiwan the provision of electricity is coordinated at the national level, suggesting that counties (the Taiwanese version of a province) have far less sway over market developments.[15]

This all suggests that each country will exhibit a unique STEP profile in terms of what factors most influence wind power diffusion. Nevertheless, the STEP methodology will be constant throughout, enabling comparative analysis. During the literature review phase of the research, four scoping questions were used to guide the analysis. They include:

1. What social forces appear to have influenced wind power development policy in this nation?
2. What technological forces appear to have influenced wind power development policy in this nation?
3. What economic forces appear to have influenced wind power development policy in this nation?
4. What political forces appear to have influenced wind power development policy in this nation?

Once the influences were identified in each of the four STEP categories, further analysis was undertaken in order to try to prioritize which of these forces tended to most influence wind power development. This task was accomplished by aggregating primary and secondary sources of information and then identifying any factors that tended to stand out in terms of centrality of analysis (i.e., how central the factor was in regard to the theme of the case study). Variables which displayed higher frequency of occurrence in the research literature were also highlighted for further analysis to determine the extent to which these variables contributed to influencing wind power development.

Admittedly, this type of meta-analysis is subject to experimenter selection bias in that the source of primary research data is dependent on the author's contact with expert sources—in this case, engendering a bias toward experts from the policy field. Furthermore, although the secondary research data was extracted from leading academic databases (Science Direct, Scopus, and ABI Inform), it is acknowledged that the bulk of the literature selected for analysis tended to feature political or policy-related discourse. It is possible that a sociologist undertaking the same type of research study might place more emphasis on social influences affecting wind power development. Similarly, an engineer undertaking the same type of research study might place higher weighting on technological influences and an economist undertaking the same type of research might focus more on exploring economic influences to a greater depth. With that said, although this study is decidedly biased toward politics and policy, this bias is deemed justifiable given the policy-centered theme of this book. It should be further noted that much of the research presented in the case studies has been published in other forms in peer-reviewed academic journals. Therefore, there has been a degree of academic scrutiny applied to the insights presented in each of the case study chapters.

3.5 THE *POLITICAL SET* MODEL

The emphasis on politics and policy as the core theme of this wind power diffusion study prompted the development of a modified STEP framework that serves as the structural shell for each of the case study chapters. The analytical framework is called the *Political SET* model, with the acronym SET standing for social, economic, and technological. In short, it is a framework based on STEP analysis theory that places politics and policy at the center of the analysis. The model is depicted in Figure 3.1, and each of the components will be described in further detail.

In Figure 3.1, the tips of the three SET categories are tapered into triangles to graphically illustrate the notion that there is a hierarchy of importance within each category. For example, for each nation examined in this book, some social factors are more important than other social factors in regard to influencing wind power development. For example, in Japan, concerns over aesthetic impairment caused by wind power projects arguably represent the most significant social barrier to wind power development. It is important to note that the hierarchy of importance in one nation may not be the same as the hierarchy of importance in another nation. To illustrate, in many rural regions of China, concerns over access to electricity trump aesthetic concerns.[16] Moreover, the importance of forces within a given STEP category changes over time in response to market dynamics. In China, for instance,

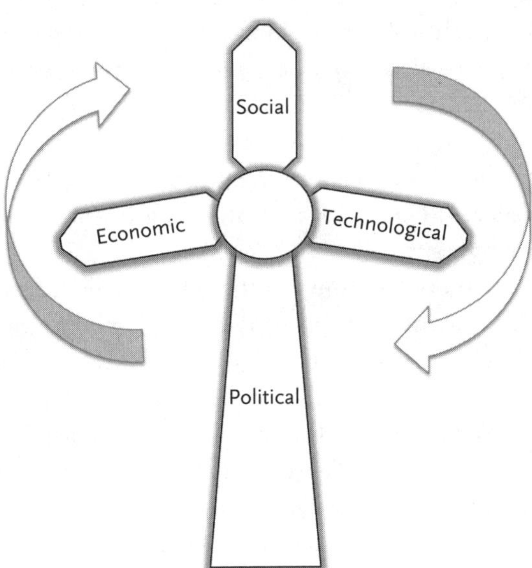

Figure 3.1. The *Political SET* Model

as more communities are connected to the electricity grid, concerns over access to electricity will give way to other concerns such as the perceived health threats attributed to wind farms.

A central core connects the dynamically evolving STEP elements (depicted by circular arrows) presented in Figure 3.1. This is intended to graphically highlight the interrelationships between these four categories. For example, influences within the social sphere both influence and are influenced by factors in the technological sphere. Again to illustrate: in China, the world's fastest growing wind power market, a progressively expanding pool of electrical engineers not only enable the diffusion of wind power projects, they provide the human capital necessary to drive wind power research and development programs and have supported the emergence of internationally competitive Chinese wind turbine manufacturers. In return, China's wind turbine manufacturers not only provide a source of employment to China's growing corps of electrical engineers; these firms have also supported the gestation of wind system component manufacturers, providing welcome employment opportunities in rural regions.

The category at the base of Figure 3.1 is the political sphere. The premise of this book is that both politics and policy (the components of the political sphere) influence and are influenced by forces from the social, economic, and technological spheres. In other words, politics and policy do not unilaterally dictate wind power market development, even in a nation such as China with a one-party political system. The *Political SET* model suggests that policymakers design policy under the influence of social, economic, technological and political forces. Moreover, the foundational placement of the political sphere also graphically illustrates the premise that politics and policy represent the most influential STEP category in terms of fostering wind power development. As all six case studies will demonstrate, how politicians and policymakers choose to interpret signals from the SET environment often serves as the critical catalyst in influencing the trajectory of wind power development.

The *Political SET* model epitomizes the interconnections found within a complex adaptive system. Within these four spheres, there are numerous forces that have been identified in each of the case studies and that impact wind power diffusion. Each variable of importance influences and is influenced by other variables within the same STEP category. For example, the existence of farmers' co-ops in Denmark (a social variable) facilitates a high degree of community consensus over what constitutes acceptable site planning (another social variable). Each variable of importance also influences and is influenced by forces arising in other STEP spheres, catalyzing changes to a number of influential variables. In other words, a significant change to an influential factor can initiate a series of cascading reverberations that result in widespread systematic change. For example, a robust feed-in tariff (a policy variable) attracts investment in wind power systems (an economic variable), which amplifies the scale and scope of development, engendering community concerns over the aesthetic degradation of landscapes (a social variable), prompting wind system manufacturers to develop techniques to minimize the aesthetic encumbrance of wind systems (a technological barrier). These interdependencies are not only nation-specific; the interrelationships are also *temporally* dynamic—they change over time.

The complexities within the *Political SET* model highlights the limitations of this framework for historical analysis (past), formative or summative policy evaluation (present), or predictive application (future). For historical analysis, the descriptive accuracy of the *Political SET* model depends on the accuracy of the analyst in identifying and weighing the impact of influential variables. This also holds true for formative or summative policy evaluation; however, in formative or summative policy evaluation, using the *Political SET* model is further complicated by the need to identify comparatively more effective policies. This of course depends on the accuracy with which an analyst can predict the market response, assuming alternative policies were applied. All of these limitations also serve as caveats when using the *Political SET* model for predictive application; however, yet another level of complexity emerges when trying to apply this framework to predictive analysis because one must try to anticipate the evolution of influential variables and how these changes will influence the rest of the variables within the confines of the model.

The *Political SET* model cannot be considered to be a diagnostic tool like the diagnostic equipment for analyzing the performance of an automobile engine. Rather, the value of the *Political SET* model is that it promotes a formalized, guided analysis of complex adaptive markets, which if undertaken in a comprehensive manner will produce enhanced understanding of the landscape that influences development patterns. It forces analysts to provide empirically verifiable support for conclusions that are being made. Employing the *Political SET* model in this way is akin to an art critic standing in front of a Picasso painting and employing agreed-upon evaluative criteria for abstract art to describe and critique the subject under study. Although others may disagree with the conclusions of such a critique or even on the criteria used for analysis, one should not be able to find fault with quality of the analysis—it should be both comprehensive and well-supported. It is with this spirit in mind that the *Political SET* model has been applied for analyzing wind power developments policy in the six case study nations introduced in this book.

Each of the case studies presented in this book begins with a general overview to the energy situation in the country under investigation and a chronological overview of the history of wind power development in that country. Once the national context has been explicated, the conditions which support or hinder wind power development will be analyzed by applying the *Political SET* model.

For each chapter, the *Political SET* analysis is conducted in the same four-step sequence. It begins by identifying key social, economic, and technological trends in the case study nation, under the thematic title "Understanding the General Forces for Change." This initial section is intended to provide the reader with a general perspective of the national landscape that affects wind power development. It describes the general socioeconomic conditions into which wind power is being introduced.

The second step of the *Political SET* analysis examines how the influential social, economic, and technological (SET) elements have influenced political ideology and policy. This section bears the thematic title "Influences on Government Policy." In essence, it examines the governance conduits—the channels through which stakeholder influence is conveyed to policymakers. This second stage of the analysis is an extension of the national landscape section in that it specifically describes how influential SET variables have influenced government policy behavior.

The third step of the *Political SET* analysis is undertaken under the theme "Political Influences on Policy," and examines the regime characteristics that influence policymaking in the nation under investigation. As alluded to earlier, a review of policy research has highlighted four regime characteristics that have been chosen for inclusion in this framework. The first characteristic is the national political structure, because how a nation is structured politically influences the autonomy that a governing party has over policymaking, which in turn influences choice of policy instruments.[17] The second characteristic is the governing party's political ideology, because ideology also influences the nature of policymaking and the choice of policy instruments that are implemented.[18] The third characteristic is national fiscal health. Fiscal health frames the types of responses a ruling party has at its discretion for solving policy challenges. The fourth characteristic is the policy regime itself. Policymaking in any given country tends to exhibit a high degree of path dependency; past approaches to resolving policy challenges in certain areas tend to be replicated in other areas, if successful.[19]

Taken together, the second and third steps of the *Political SET* analysis reflect a widely accepted view that politics causes policy.[20] Key stakeholders,

Table 3.1 ORGANIZATION OF CASE STUDY CHAPTERS		

Section	General Theme
Section 1	Case Study Introduction
Section 2	An Overview of Energy in the Case Study Nation
Section 3	History of Wind Power Development in the Case Study Nation
Section 4	Understanding the General Forces for Change (National Landscape)
Section 5	Influences on Government Policy (the Governance Conduits)
Section 6	Political Influences on Policy (Regime Characteristics)
Section 7	The Culmination of Influences (the Policy Conduits)
Section 8	What to Expect Going Forward

whether they come from within civil society, business, politics, academia, or some other definable group, have the ability to influence policy through a number of mechanisms. Therefore, analyzing wind power development policy cannot be comprehensively carried out without considering how these stakeholders affect policy.

The fourth and final step of the *Political SET* analysis, titled the "Culmination of Influences," examines how energy politics and policy influence the social, economic and technological spheres. This part reflects Theodore Lowi's opposing notion that policy causes politics.[21] In this last stage of the *Political SET* analysis, the influence that government energy policy has on shaping the SET landscape is considered. This step also considers potential social, economic and technological responses to political influence and emergent policy. It also lays the contextual groundwork for a concluding section in every case study chapter, "What to Expect Going Forward," that draws from the *Political SET* analysis to speculate on emergent trends in the given nation. Table 3.1 summarizes the manner in which each of the case study chapters has been organized.

3.7 TYING THEORY TO THE *POLITICAL SET* MODEL

The preanalytic vision that went into the development of the *Political SET* model is underpinned by an array of well-traveled policy theories. This merits discussion because the links between the *Political SET* model and prevalent policy theory suggests that the model is more than just a cross-disciplinary application of a strategic management model, it also represents an extension of well-entrenched policy theory to guide policymaking in complex adaptive markets.

The main premise of the *Political SET* model reflects a key tenet of discursive politics—numerous stakeholders influence the design and success of policy.[22] As the case studies in this book will demonstrate, the needs and aspirations of all relevant stakeholders must be understood if one is to design effective policy for supporting wind power development. *Political SET* analysis promotes proactive efforts to better understand key players in industry, influential civic groups, influential political groups, and dominant social trends that influence civic behavior. This focus on significant influencers is rooted in Heclo's concept of issue networks, groups of interested stakeholders that coalesce around certain issues and exert significant influence on the policymaking process.[23] It also affirms the potential influence of advocacy coalitions, which are like issue networks but can also form from outside the policy regime to challenge undesirable policies.[24]

Although *Political SET* analysis encourages stakeholder analysis as part of the methodology, it should not be misconstrued as neglecting prevalent theories of power. Analysis of political influence is placed at the base of the *Political SET* model to acknowledge that stakeholders possess varying degrees of influence on wind power policy and of all stakeholders, politicians, and policymakers are in the unique position of possessing the means to manipulate the policy environment, in order to catalyze change. This reflects Harold Lasswell's contention that the study of politics is the study of influence and the influential.[25] Moreover, isolation of the economic sphere (the "E" in the SET analysis) in the *Political SET* model incapsulates a tenet from iron triangle theory that certain special interest groups can capture the policy process through economic might.[26] As Bardach has observed, policy is an extension of power.[27]

Deeper analysis of how the variables within the *Political SET* model interact to influence policy also sheds light on the linkage between the concept of bounded rationality and stakeholder theory. Bounded rationality is a notion put forth by Herbert Simon that argues that decisions are never fully rational; people make decisions that are bounded by cognitive and environmental constraints.[28] Consequently, the rational decision-making capacity of any individual is bounded by that individual's breadth and depth of knowledge. It follows that the process of including a greater diversity of stakeholders in the policymaking process elicits outcomes that are not only more socially acceptable, but are potentially more rational in that they are based on a broader pool of knowledge.

This line of reasoning also conflates the concept of path dependency with stakeholder theory. As the case studies produced in this book demonstrate, there is empirical evidence to support the intuitive observation that wind power development policy in any given nation tends to mirror policy strategies that the government has successfully applied in the past. However, when one digs deeper to understand the nature of past successes, it becomes apparent that policies stand greater chance of succeeding if they simultaneously garner acceptance from the most powerful stakeholders and avoid alienating other stakeholders that can obstruct policy implementation. This explains why policies tend to be path dependent and why revolutionary change only occurs when policies are significantly disconnected from stakeholder expectations.[29]

The path dependent nature of the policy process also ties in with the notion of incrementalism. Lindblom has described policymaking as a process of "continually building out from the current situation, step-by-step and by small degrees."[30] In other words, although the policymaking process is path-dependent in that it tends to draw from the success of previous policies, the process is also dynamic in that the formative evaluation of policies to enhance effectiveness promotes incremental change. In short, policymaking is a dynamic process, but it is bounded by historical and stakeholder influences.

In summary, the *Political SET* model incorporates many prominent tenants stemming from policy theory. It recognizes the importance of understanding stakeholder influences, yet it also recognizes that power within policy networks vary. It recognizes that policy regimes tend to embrace certain approaches to policy design, implementation, and evaluation, yet these approaches are subject to subtle evolutionary change. Lastly, it recognizes that technology policy takes place within complex adaptive markets; consequently, as the needs and aspirations of stakeholders evolve, the nature of public policy will undoubtedly evolve in response. Failure to anticipate change in the *Political SET* environment results in stakeholder dissonance and ineffective public policy.

NOTES

1. Hughes, Thomas P. 1983. *Networks of Power: Electrification in Western Society 1880–1930*. Baltimore: Johns Hopkins University Press.
2. Valentine, Scott Victor. 2013. "Wind Power Policy in Complex Adaptive Markets." *Renewable and Sustainable Energy Reviews* 19 (1): 1–10.
3. This is the essence of chaos theory as described in Brown, Shona L., and Kathleen M. Eisenhardt. 1998. *Competing on the Edge: Strategy as Structured Chaos.* Cambridge, MA: Harvard Business School Press.
4. Axelrod, Robert, and Michael D. Cohen. 1990. *Harnessing Complexity: Organizational Implications of a Scientific Frontier*. New York: The Free Press.
5. Tsai, Stephen D., Hong-quei Chiang, and Scott Valentine. 2003. "An Integrated Model for Strategic Management in Dynamic Industries: Qualitative Research from Taiwan's Passive-Component Industry." *Emergence* 5 (4): 34–56.

6. Barney, Jay B. 1995. "Looking Inside for Competitive Advantage." *The Academy of Management Executive* 9 (4): 49–61.

7. Beinhocker, Eric. 1999. "Robust Adaptive Strategies." *Sloan Management Review* 4 (3): 95–106.

8. Schumpter, Joseph A. 2010/1943. *Capitalism, Socialism and Democracy*. London: Routledge.

9. Beinhocker, Eric. 1999. "Robust Adaptive Strategies." *Sloan Management Review* 4 (3): 95–106.

10. DeCarolis, Joseph F., and David W. Keith. 2006. "The Economics of Large-Scale Wind Power in a Carbon Constrained World." *Energy Policy* 34 (4): 395–410.

11. Díaz-González, Francisco, Andreas Sumper, Oriol Gomis-Bellmunt, and Roberto Villafáfila-Robles. 2012. "A Review of Energy Storage Technologies for Wind Power Applications." *Renewable and Sustainable Energy Reviews* 16 (4): 2154–2171.

12. Firestone, Jeremy, and Willett Kempton. 2007. "Public Opinion About Large Offshore Wind Power: Underlying Factors." *Energy Policy* 35 (3): 1584–1598.

13. For example, see the following for a review on the importance of properly structured feed-in tariffs: Mendonca, Miguel, David Jacobs, and Benjamin Sovacool. 2009. *Powering the Green Economy: The Feed-in Tariff Handbook*. Oxford: Earthscan.

14. Valentine, Scott Victor. 2010. "Canada's Constitutional Separation of (Wind) Power." *Energy Policy* 38 (4): 1918–1930.

15. Valentine, Scott Victor. 2010. "Disputed Wind Directions: Reinvigorating Wind Power Development in Taiwan." *Energy for Sustainable Development* 14 (1): 22–34.

16. Valentine, Scott Victor. 2013. "Enhancing Climate Change Mitigation Efforts through Sino-American Collaboration." *Chinese Journal of International Politics* 6 (2): 159-182.

17. Howlett, Michael, and M. Ramesh. 1995. *Studying Public Polic: Policy Cycles and Policy Subsystems*. Oxford: Oxford University Press.

18. Lasswell, Harold Dwight. 1935. *Politics: Who Gets What, When, How*. York, PA: McGraw-Hill.

19. Peters, B. Guy, Jon Pierre, and Desmond S. King. 2005. "The Politics of Path Dependency: Political Conflict in Historical Institutionalism." *Journal of Politics* 67 (4): 1275–1300.

20. Bardach, Eugene. 2006. "Policy Dynamics." In *The Oxford Handbook of Public Policy*, edited by M. Moran, M. Rein and R. E. Goodin, pp. 336–366. Oxford: Oxford University Press.

21. Salamon, Lester M. (ed.). 2002. *The Tools of Government: A Guide to the New Governance*. Oxford: Oxford University Press.

22. For more of discursive politics, see: Fischer, Frank. 2003. *Reframing Public Policy: Discursive Politics and Deliberative Practices*. Oxford: Oxford University Press; and Rhodes, R.A.W. 2006. "Policy Network Analysis." In *The Oxford Handbook of Public Policy*, edited by M. Moran, M. Rein and R. E. Goodin, pp. 425–447. Oxford: Oxford University Press.

23. Heclo, Hugh. 1978. "Issue Networks and the Executive Establishment." In *The New American Political System*, edited by A. King, pp. 444–472. Washington, DC: American Enterprise Institute.

24. Sabatier, Paul A., and Hank C. Jenkins-Smith (eds.). 1993. *Policy Change and Learning: An Advocacy Coalition Approach, Theoretical Lenses on Public Policy*. Boulder, CO: Westview Press.

25. Lasswell, Harold Dwight. 1935. *Politics: Who Gets What, When, How*. York, PA: McGraw-Hill.

26. Cater, Douglas. 1964. *Power in Washington: A Critical Look at Today's Struggle in the Nation's Capital*. New York: Random House.

27. Bardach, Eugene. 2006. "Policy Dynamics." In *The Oxford Handbook of Public Policy*, edited by M. Moran, M. Rein and R. E. Goodin, pp. 336–366. Oxford: Oxford University Press.

28. Simon, Herbert. 1982. *Models of Bounded Rationality. Vol. 1, Economic Analysis and Public Policy*. Cambridge, MA: MIT Press.

29. Peters, B. Guy, Jon Pierre, and Desmond S. King. 2005. "The Politics of Path Dependency: Political Conflict in Historical Institutionalism." *Journal of Politics* 67 (4): 1275–1300.

30. Lindblom, Charles E. 1959. "The Science of 'Muddling Through.'" *Public Administration Review* 19 (2): 79–88.

CHAPTER 4

Wind Power in Denmark

4.1 INTRODUCTION

People are trapped in history, and history is trapped in them.
—James Baldwin, from *Notes of a Native Son*

In technological policy literature, the term "path dependency" frequently emerges in attempts to explain why a given technological track develops. The premise behind the notion of technological path dependency is that historical social, technological, economic, and political forces foster conditions for a particular technology to thrive. Once a technology becomes dominant, vested interests—which profit from the technology—hinder radical change, because change carries an implicit threat that those benefitting from the status quo might suffer an erosion of economic benefits.

To illustrate path dependency, consider the history of the QWERTY keyboard (referring to the sequencing of letters from left to right on the top row of a standard computer keyboard).[1] Keyboards on typewriters were designed in this way to reduce mechanical type hammers from clashing with each other. Over time, type hammers were made obsolete by type-balls. Nevertheless, the QWERTY keyboard remained unchanged (even in this day of computerized word processing)—despite the fact that research has shown the QWERTY layout to be inferior in terms of optimizing typing speed. This layout has perpetuated because legions of typists have learned on the QWERTY keyboard; therefore, technological familiarity has insulated this design feature from change.

The notion of path dependency is relevant to the story of wind power development in Denmark because, as will be described in this chapter, a number of social, economic, technological, and political forces shepherded

Denmark's ascent to the top position as the nation with the world's highest percentage of wind power contributing to national electricity generation.

In addition to illustrating the influence of technological momentum, there are two other contemplative policy insights to be gleaned from studying wind power diffusion in Denmark. First, Denmark's wind power development experience demonstrates that grassroots support mechanisms which engage communities and individuals in the development process bolster the effectiveness of economic incentives.[2] Second, Denmark's wind power story demonstrates that establishing a technological foothold is never a guarantee of uncontested market entrenchment. As any technology matures, its impact on society, business and political fortunes evolves. In Denmark, wind power began its ascent as a technology for the rural masses. By 2001, 80% of all wind power turbines installed in Denmark were owned by community cooperatives, comprising as many as 150,000 individuals.[3] Yet as wind power in Denmark enters the second decade of the twenty-first century, mega projects are increasingly spearheaded by utilities and major wind power development firms. As this pattern of ownership has evolved, public support for wind power has tailed off. Concerns over the cost of wind power and the impact that this technology is having on the Danish landscape has cast a degree of uncertainty over whether or not the government can achieve its laudable goal of generating 50% of Denmark's electricity through wind power by 2020.[4]

As this chapter will document, a unique confluence of STEP conditions set Denmark on a path toward wind power diffusion. Therefore, replication of Denmark's policy model will not guarantee similar levels of success in other nations that seek to emulate Denmark's wind power achievements by replicating Danish policies. Moreover, there is no guarantee that the Danish approach to wind power development will continue to be successful as the government struggles to adjust policy to address the social, economic, and technical challenges associated with higher levels of wind power diffusion.

4.2 AN OVERVIEW OF ELECTRICITY GENERATION IN DENMARK

Table 4.1 highlights an interesting trend in Denmark. Between 2000 and 2010, global primary energy consumption increased by 28%, yet primary energy consumption in Denmark actually *decreased* by 3%, despite economic growth of 6.7%. Moreover, this is not a recent development. Compared to 1980, by 2010, GDP in Denmark nearly doubled, yet primary energy consumption grew by less than 7%.[5] Simply put, Denmark has achieved a degree of success in decoupling economic growth from energy consumption. This is a result of efforts on the part of the government, industry, utilities,

Table 4.1 PRIMARY ENERGY CONSUMPTION TRENDS IN DENMARK 2000–2010

Mt Oil Equivalent	2000	2001	2002	2003	2004	2005	2006	2007	2008	2009	2010
Denmark	20.1	20	19.9	21.4	20.5	19.8	21.9	20.9	20.1	18.8	19.5
World	9382.4	9465.6	9651.8	9997.8	10,482	10,800.9	11,087.8	11,398.4	11,535.8	11,363.2	12,002.4
% of Global Total	0.21%	0.21%	0.21%	0.21%	0.20%	0.18%	0.20%	0.18%	0.17%	0.17%	0.16%

Source of data: BP. 2011. _Statistical Review of World Energy 2011_. London: British Petroleum (BP).

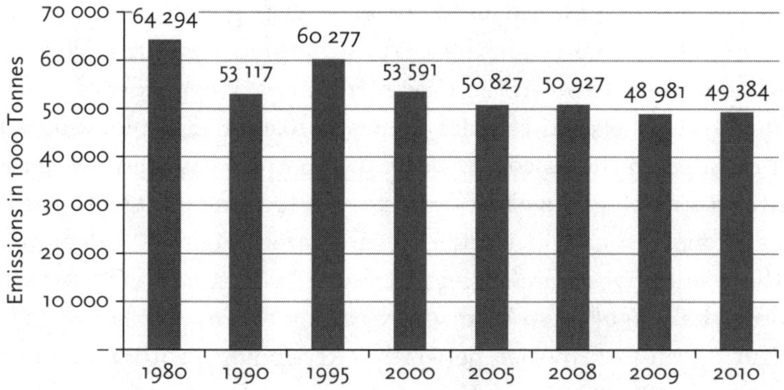

Figure 4.1. CO_2 Emissions Trends in Denmark

Source: Danish Energy Agency. 2011. *Annual Energy Statistics 2010*. Copenhagen: Danish Energy Agency.

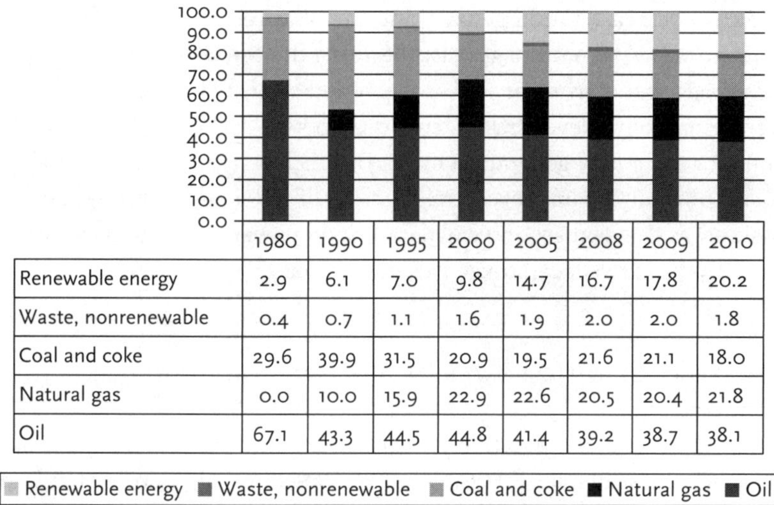

	1980	1990	1995	2000	2005	2008	2009	2010
Renewable energy	2.9	6.1	7.0	9.8	14.7	16.7	17.8	20.2
Waste, nonrenewable	0.4	0.7	1.1	1.6	1.9	2.0	2.0	1.8
Coal and coke	29.6	39.9	31.5	20.9	19.5	21.6	21.1	18.0
Natural gas	0.0	10.0	15.9	22.9	22.6	20.5	20.4	21.8
Oil	67.1	43.3	44.5	44.8	41.4	39.2	38.7	38.1

Figure 4.2. Primary Energy Consumption by Fuel Source (in %)

Source: Danish Energy Agency. 2011. *Annual Energy Statistics 2010*. Copenhagen: Danish Energy Agency.

communities, and consumers to reduce Denmark's sizable carbon footprint through energy efficiency improvements.

Energy efficiency improvement tells only part of the story in Denmark's energy sector. In 2010, CO_2 emissions were 23% lower than 1980 levels and 7% lower than 1990 levels. As Figure 4.1 illustrates, Denmark's CO_2 emissions reached a peak in 1995 before beginning a progressive decline.

As Figure 4.2 suggests, the diminishing role of oil in Denmark's primary energy mix explains part of the nation's progress in reducing CO_2 emissions.

In the early 1970s, oil accounted for as much as 90% of Denmark's primary energy because it was also used for electricity generation.[6] The two oil price shocks that occurred in the 1970s prompted the Danish government to initiate policies to wean the nation from heavy dependence on foreign oil. One strategy was to escalate domestic fossil fuel exploration efforts, which resulted in the discovery of oil and natural gas fields in the North Sea. Another strategy involved facilitating a technological shift away from oil as a primary source of electricity generation. Figure 4.2 demonstrates that these initiatives were successful. Since 1980, Denmark has managed to reduce the role of oil in its primary energy mix by nearly half (43%) by substituting oil-fired power generation technology with renewable and gas-fired power technologies. The vast majority of oil that is still consumed in Denmark now comes from domestic supply sources.

Another contributing factor to Denmark's improving CO_2 profile is a reduced portfolio in coal-fired electricity generation. As Figure 4.3 illustrates, the role of coal has decreased in large part due to enhanced reliance on natural gas (thanks to domestic natural gas discoveries in the North Sea) and wind power for electricity generation, as well as expanded capacity in combined heat and power technology. Nevertheless, coal-fired power still contributed 43.8% to Denmark's electricity generation in 2010, which suggests that Denmark still has a lot of room for improvement in terms of reducing energy-related GHG emissions; as the next section will detail, wind power stands to play a key role in further efforts to decarbonize the energy mix.

4.3 HISTORY OF WIND POWER DEVELOPMENT IN DENMARK

Wind power has a comparatively long history in Denmark. As early as the 1900s, purportedly over 30,000 wind turbines provided power for Danish farms and homes. During World War 1, Denmark continued to encourage expansion of wind power in order to circumvent wartime fossil fuel supply disruptions. It has been estimated that by 1918, approximately 120 local utilities in Denmark integrated wind power into the electricity grid, totaling approximately 3 MW or about 3% of the Danish electricity supply.[7] Between world wars the security impetus for further wind power development diminished; however, the outbreak of World War II and the reemergence of fossil fuel supply disruptions prompted the Danes to once again revisit wind power development. During this period wind power turbines with rated capacities of 45 kW were developed.

There is clear logic behind the allure of wind power in Denmark—the nation is ideally positioned to exploit the power of wind. Its 5,000 miles of

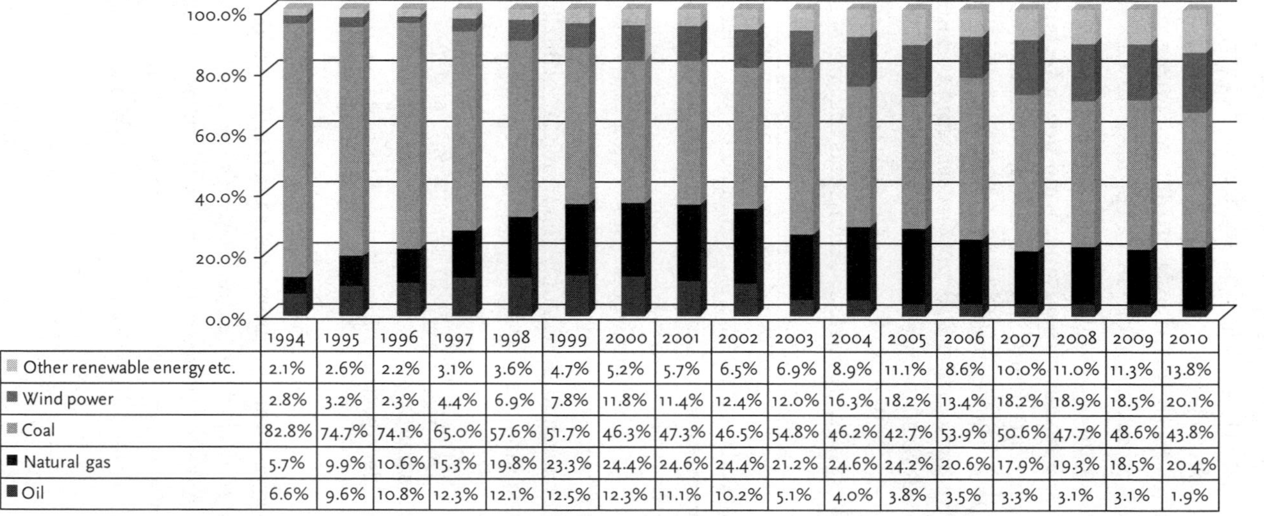

	1994	1995	1996	1997	1998	1999	2000	2001	2002	2003	2004	2005	2006	2007	2008	2009	2010
Other renewable energy etc.	2.1%	2.6%	2.2%	3.1%	3.6%	4.7%	5.2%	5.7%	6.5%	6.9%	8.9%	11.1%	8.6%	10.0%	11.0%	11.3%	13.8%
Wind power	2.8%	3.2%	2.3%	4.4%	6.9%	7.8%	11.8%	11.4%	12.4%	12.0%	16.3%	18.2%	13.4%	18.2%	18.9%	18.5%	20.1%
Coal	82.8%	74.7%	74.1%	65.0%	57.6%	51.7%	46.3%	47.3%	46.5%	54.8%	46.2%	42.7%	53.9%	50.6%	47.7%	48.6%	43.8%
Natural gas	5.7%	9.9%	10.6%	15.3%	19.8%	23.3%	24.4%	24.6%	24.4%	21.2%	24.6%	24.2%	20.6%	17.9%	19.3%	18.5%	20.4%
Oil	6.6%	9.6%	10.8%	12.3%	12.1%	12.5%	12.3%	11.1%	10.2%	5.1%	4.0%	3.8%	3.5%	3.3%	3.1%	3.1%	1.9%

Figure 4.3. Electricity Production by Fuel Source

Source: Danish Energy Agency. 2011. *Annual Energy Statistics 2010*. Copenhagen: Danish Energy Agency.

coastline, flat expanses of agricultural land, and blustery North Sea location has prompted wind power expert Paul Gipe to liken Denmark's wind power potential to that of the American Great Plains.[8]

After suffering two major fossil fuel supply disruptions in the span of three decades, the Danish government deemed it prudent to ensure the nation was better insulated from external supply disruptions and commissioned feasibility studies and research to advance wind power technology. Research efforts would continue from 1947 to 1968,[9] ultimately laying the technological foundation for Denmark's emergence as a world leader in wind power. During this period, the 200 kW Gedser wind turbine was developed (in 1956), characterized by three blades on a horizontal axis in an upwind position, a design that still dominates commercial wind power systems to this day.[10] For many years, the Gedser turbine was the largest in the world.[11] However, between 1945 and the early 1970s, economic justification for the commercialization of emergent wind power technology was weak. Western Europe had enjoyed three decades of peace and oil was cheap and readily available. Consequently, by 1972, oil constituted a whopping 93% of Denmark's primary energy supply.[12]

In 1973, the decision by the Organization of Arab Petroleum Exporting Countries (OAPEC) to embargo shipments of oil to the United States and allies in response to the US decision to resupply the Israeli army during the Yom Kippur War led to the inflation of oil prices, and economic perturbation in Danish energy circles. This catalyzed a surge of smaller wind power system research and development (R&D) in Denmark,[13] and prompted the government to commission a 1975 study by the Dutch Academy of Technical Sciences to consider the feasibility of a utility-scale wind power development program.[14]

In 1976, with oil still lurking around the $60 US per barrel level (in 2010 adjusted prices), the Danish government announced a new energy policy that was geared toward reducing oil dependence. This plan included support for the addition of nuclear power and alternative energy generation capacity.[15] On the heels of this announcement a national wind energy program was unveiled, featuring plans to build a test center for small wind turbines at the Risø National Laboratory for Sustainable Energy (Risø Laboratory). The objective of this program was to centralize R&D to support aspiring manufacturers and to provide a standardized system for certifying the quality of Danish-made wind turbines, thereby providing investors with a higher degree of quality assurance.[16] Between 1976 and 1995, 10% of the government funds dedicated to energy research were channeled into wind energy R&D.[17]

The initial research strategy championed by the national government was to create technologically advanced wind turbines designed by consortiums of large Danish firms for placement in large wind farms that would be owned and operated by utilities.[18] In fact, by 1977 plans were already underway to build two 630 kW prototype turbines. This project alone took up 40% of the wind energy program's budget and culminated in completion of the two test units in 1978 and 1980, respectively.[19] The trend toward targeting R&D for the development of large wind turbines would continue up to 1989.[20]

Providentially, the commitment of large Danish firms to this new industry never materialized. Instead, by 1978 there were only a score of wind turbine manufacturers, all specializing in small wind turbines. Most of these firms were manufacturers of agricultural equipment looking for ways to diversify product lines for agricultural consumers. With technical assistance provided by Risø Laboratories and other government-sponsored wind power R&D programs, companies such as Vestas, which was established in 1898 as a blacksmith foundry, began to produce uncomplicated yet reliable turbines based more on accumulated manufacturing knowledge than advanced aerodynamic principles.[21] So while many other nations were pursuing advanced wind turbine R&D, Denmark's manufacturers were establishing a reputation for quality that would prove to be the decisive factor in global market competition. The sequence of starting with simpler technology, learning by interacting with government-sponsored research organizations, and designing to rigid government-enforced standards and honing quality through experience is seen as the critical evolutionary path underpinning Denmark's highly successful wind power turbine industry.[22]

Domestically, however, the structure of Denmark's utility sector inhibited the government's vision of encouraging the diffusion of wind power through mega projects. In the 1970s, Denmark's utilities were predominantly small local or regional utilities that were generally community-owned and operated as nonprofit organizations. The utilities were typically vertically integrated, controlling electricity generation from production through to transmission and distribution (T&D). Furthermore, Denmark's electricity grid was divided into western (Jutland, Fynen, and smaller islands) and eastern (Zealand, Lolland-Falster, and smaller islands) systems, and in the 1970s, was not interconnected. The process of coordinating the activities of 87 utilities in the western sector and 28 utilities in the eastern sector in order to dampen the stochastic flows of wind power represented a daunting challenge to the state regulatory body, the Danish Energy Authority.[23] Consequently, up to 1979, although there was a significant amount of research being done in wind energy, installations were limited to a few pioneering farm communities. Denmark's fledgling wind energy system manufacturers were having a hard time finding a market for their wares. This was about to change due to a development thousands of kilometers away.

The 1979 energy crisis that emerged as a consequence of the Iran Revolution consolidated political will to proactively drive wind power development.[24] That year, a Danish Ministry of Energy was established in order to proactively direct national energy strategy,[25] and two subsidies were introduced that would alter the course of wind power development in Denmark. The first subsidy allowed investors in wind turbines to claim up to 30% of total investment costs (including installation and connection costs);[26] however, only turbines approved by the Risø Laboratories were eligible. Once wind turbines were operational, a second subsidy permitted a tax-deduction on the sale of surplus wind power to the grid. Moreover, utilities were required to purchase all such surplus.[27]

There was a special stipulation attached to the 1979 policy initiatives that would prove to be instrumental in facilitating public support for wind power development. Only individuals (or cooperative groups of individuals) living within 3 km of a given project were eligible for the investment subsidy.[28] This condition ensured that project investors and host communities were often one and the same, significantly mitigating community opposition to these new projects. However, uncertainty over revenue flows prevented broader scale investment and, as Figure 4.4 illustrates, it wasn't until the mid-1980s that wind power really took off.

In hindsight, the wind power subsidy programs launched in 1979 can be considered to be policy reconnaissance that provided the feedback necessary for formulating a bolder set of policy initiatives in 1985. While the government was monitoring the impact of its investment subsidy program on wind power adoption, it was also conducting a national study to assess the feasibility and potential of large-scale wind power diffusion. Based on an assumption that the average wind turbine in the near future would possess a rated capacity of 2.5 MW, the study concluded that the potential for wind

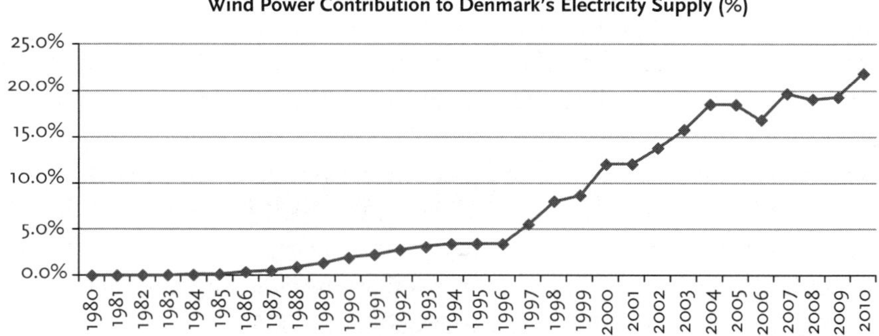

Figure 4.4. The Rise of Wind Power

Source: Danish Energy Agency. 2011. *Annual Energy Statistics 2010*. Copenhagen: Danish Energy Agency.

power generation in Denmark was in the neighborhood of 30 TW hours per year, roughly equal to total Danish electricity consumption at the time.[29] In order to conduct this study it was necessary to develop a wind atlas for Denmark which, when published in 1981 by Risø Laboratories, would go on to serve as a valuable resource for wind power project developers.[30]

In aggregate, the government-led initiatives undertaken between 1979 and 1984 laid the foundation for supporting larger scale wind power development. The national wind power potential study confirmed the contribution that wind power could make to domestic energy security. The creation of the wind atlas enabled developers to target the most profitable sites. The establishment of a wind energy department at the Risø Laboratories provided a one-stop shop of technical support for aspiring wind turbine manufacturers. Market response to the development subsidies gave the government insight into what it would take to catalyze investment and gave host communities a chance to evaluate the pros and cons of small scale wind power projects. In fact, the only glaring omission from this array of support initiatives was the creation of a domestic wind power market that was large enough to nurture globally competitive wind turbine manufacturers. In the absence of other opportunities, the slow market build-up exhibited in Denmark between 1979 and 1985 would not have been sufficient to support the emergence of global wind turbine manufacturing giants such as Vestas and Bonus. Fortunately, fate had a role to play in temporarily rectifying this omission.

In 1978, the United States passed the National Energy Act (USNEA) and the US Public Utilities Regulatory Policies Act (PURPA), which paved the way for private electricity generators to enter the US electricity market by guaranteeing grid access at avoided costs (the cost the utility would face if it were to build a new generation plant itself). The USNEA also established an investment tax credit of 40% for residential wind power systems and 10% for business investment in wind. These were modest incentives that had limited efficacy, except in California—the reason being, on the heels of this act, California mandated its electric utilities to offer contracts to wind power providers at standard terms and conditions, which meant that wind power developers could receive 30 year contracts with guaranteed fixed prices for the first 10 years.[31] These policies, along with inflated oil prices, produced a boom in California for wind power and Danish wind power turbines emerged as the preferred choice for fulfilling this demand thanks to quality assurances provided by Risø's certification system. Although policies changed in the United States (and California) and the bottom fell out of the US wind power market in 1985, Danish wind turbine firms had already enjoyed a five-year period of prosperity and learning by doing.

Meanwhile, back in Denmark, with oil prices beginning to soften, it was apparent that new policies would be required if wind power development was to reach a level of commercialization that would provide Denmark with domestic energy security. Therefore, in 1984, legislation was passed which significantly liberalized the wind power generation market. Under the terms of the legislation, private wind power producers were permitted a tax refund of €0.037 per kWh.[32] Wind power generators that wished to sell power into the grid were to receive guaranteed access, with utilities committing to wind power purchase contracts of at least ten years in duration to purchase wind power at a price that was equal to 85% of the retail price. Utilities were also compelled to pay 35% of any grid connection costs.[33] These policies amounted to payments to wind power providers of approximately €0.08 per kWh, ensuring a substantial return on investment.[34] As Table 4.2 illustrates, by 1985 these policies were fueling progressive market expansion.

It bears mention that wind power projects up to this time could be characterized as small projects owned by individuals or collective groups of individuals, because of regulations which encouraged local investment. With hundreds of wind power turbines springing up around the country, concerns over the adverse aesthetic impacts of these turbines were beginning to emerge.[35] Consequently, the government decided to take steps to encourage concentrated development through wind farms. To support private wind farm development, an enhanced subsidy was announced that provided up to 50% of the capital costs for approved wind power projects.[36]

In 1985, an agreement between Denmark's utilities and the Danish Ministry of Energy that had been under negotiation since the late 1970s was concluded. It compelled utilities to develop 100 MW of wind power between 1986 and 1990, effectively tripling installed capacity in Denmark.[37] This voluntary agreement, which received no financial support from the government,[38] underpinned an expectation that the government envisaged utilities playing a greater role in supporting wind power development.[39] There was incentive for utilities to cooperate in this voluntary manner in order to stave-off formal government mandates—and given that the majority of Denmark's utilities were not-for-profit entities owned by municipalities and communities, there was a degree of reluctant acquiescence, because the adverse financial impact of premium-priced wind power would eventually be passed along to the end-consumer.

In 1986, with the pace of wind power development heating up, the government introduced two curious initiatives that seemed directed at reining in the pace of development. First, it reduced the tax credit subsidy that was initiated in 1979 from 20% to 15% for both individual projects and wind farms.[40] Second, it introduced revised legislation that narrowed the

Table 4.2 WIND POWER DEVELOPMENT IN DENMARK IN THE 1980S

	1980	1981	1982	1983	1984	1985	1986	1987	1988	1989
Wind Power Onshore Capacity, MW	3	6	11	14	20	47	72	112	190	247
Wind Power Offshore Capacity, MW	0	0	0	0	0	0	0	0	0	0
Wind Power's Share of Domestic Electricity	0.0%	0.0%	0.1%	0.1%	0.1%	0.2%	0.4%	0.6%	1.1%	1.4%

Source: Danish Energy Agency. 2011. *Annual Energy Statistics 2010*. Copenhagen: Danish Energy Agency.

eligibility criteria for these reduced subsidies. A wind turbine owner was required to live close to the turbine site and could only receive tax credit for electricity generated from the person's investment that was equal to the lesser of 150% of the person's annual consumption or 9000 kWh.[41] These new restrictions can be considered to be a manifestation of policy learning. The government was responding to the realization that it was more effective to provide incentives for electricity generation (the 1984 production subsidies) than to simply construct a wind turbine (the 1979 investment subsidies).

The repercussions of this seemingly minor policy adjustment illustrate the precarious nature of policy setting in multistakeholder networks. The reduced investment subsidy turned out to be a severe blow for Denmark's wind turbine industry. In conjunction with the expiration of California's renewable energy support scheme and the declining cost of oil, which reduced the feed-in tariff (FIT) paid to wind power providers,[42] the market for wind power systems collapsed. By the end of 1986, many Danish wind turbine manufacturers had either declared bankruptcy or merged with other firms. Even Vestas filed for bankruptcy in October 1986 and was only saved through a restructuring program.[43] The government's response to this unintended consequence of the new policy was to establish the Wind Turbine Guarantee Company, which guaranteed long-term financing to large export projects provided that the turbines met rigid government standards for quality.[44] This type of policy riposte—responding to emergent problems with policy adjustments—would become a regular feature of Danish wind power policy in subsequent years. As an interesting aside, some analysts have contended that the diminished investment subsidy was actually beneficial in the long run, as it provided the impetus for Denmark's surviving wind turbine firms to diversify into international markets, establishing competitive footholds in international markets that diversified risk and enhanced competitive advantage.[45]

The following year, the government continued to send seemingly mixed messages in regard to wind power development intentions. On the one hand, the central government requested regional authorities to develop regional plans for the siting of prospective wind farms and most municipalities began to actively undertake these efforts.[46] On the other hand, the government further reduced the 1979 capital investment subsidy from 15% to 10%.[47] This further reduction was met with a degree of understandable angst in wind power development circles. In hindsight, these actions reflected a shift in developmental focus wherein a policy preference was emerging which favored larger wind farms over small community developments. It is understandable that a reduction in front-end subsidizes would be viewed as a valid strategy for reining in excessive subsidization of larger wind power projects, given that larger wind farms were capable of more cost effective electricity generation. It also bears emphasizing that despite criticism that the tightening of subsides to wind power producers undermined market confidence,[48] installed capacity grew fivefold between 1986 and 1989.

In 1989, government policy continued to transition toward support for wind farms over single turbines. The investment subsidy for wind turbines that began in 1979 at 20%—and was reduced to 15% and 10% in 1986 and 1987 respectively—was eliminated altogether in 1989. This made the pursuit of economies of scale a critical factor for wind power profitability. However, there was also clear evidence that the government was aware that concentrated development of wind turbines might undermine community support for wind power. Consequently, regional plans began to emerge that mandated standards clearly designed to minimize the social impact associated with larger wind power developments. Standards promoted the construction of tubular wind turbine towers to replace lattice towers, which were shown to adversely impact birdlife. Standards were also set to minimize the adverse aesthetic impact of rotor blades by requiring them to be coated with nonreflective paints and to rotate in a uniform clockwise pattern. For wind farms, prescribed standards were published to govern size, appearance and placement of turbines.[49]

In 1990, with installed wind power capacity at 326 MW, the Ministry of Energy published *Energy 2000*, which announced intentions to meet a goal of 1500 MW of installed wind power capacity by 2005. This would equate to 10% of Denmark's projected electricity consumption.[50] In pursuit of this goal, the government negotiated a second agreement with Denmark's utilities compellingly utilities to build another 100 MW of wind power capacity over the subsequent five-year period.[51] This was despite the fact that utilities were still striving to fulfill the first 100 MW target (which eventually would be met two years later than planned, in 1992). As Table 4.3 illustrates, the 1500 MW target would eventually be surpassed by 1999; however, meeting this goal would not be without new challenges, necessitating new riposte strategies.

For starters, the government strategy of regulating cooperative investment in wind turbines while encouraging utility-led wind farm projects engendered an unanticipated degree of social dissonance. By 1990, community resistance resulted in more than 10% of wind power project approvals being rescinded after public appeals to the Ministry of Environment.[52] In fact, social opposition was a key factor behind the two-year delay in utilities meeting the first 100 MW wind power installation obligation.[53] In response, rather than reviving policy to encourage further community-led wind power development, the government began to shift strategic sights to offshore wind farm development, culminating in the development of the world's first offshore wind farm, consisting of eleven 450 kW turbines located near Vindeby.[54]

Table 4.3 WIND POWER DEVELOPMENT IN DENMARK IN THE 1990S

	1990	1991	1992	1993	1994	1995	1996	1997	1998	1999
Wind Power Onshore Capacity, MW	326	388	431	463	516	590	804	1113	1428	1743
Wind Power Offshore Capacity, MW	0	5	5	5	5	10	10	10	10	10
1990s Wind Power's Share of Domestic Electricity	1.9%	2.3%	2.8%	3.1%	3.4%	3.5%	3.4%	5.6%	8.1%	8.7%

Source: Danish Energy Agency. 2011. *Annual Energy Statistics 2010*. Copenhagen: Danish Energy Agency.

An additional challenge was that the overall pace of wind power development in Denmark was decidedly declining. As Table 4.3 illustrates, in 1990, 79 MW of installed capacity was added. Between 1991, 1992, and 1993, the annual pace of expansion tailed off to 62 MW, 43 MW, and 32 MW, respectively. Irrespective of whether this decline can be attributed to elimination of the investment subsidy,[55] increased public opposition to wind power,[56] or cheaper oil, this trend did not go unnoticed in government circles. Consequently, near the end of 1992, the government adopted a new policy riposte.[57]

Under a new subsidy program, in addition to receiving 85% of the retail electricity price from utilities, wind power generators were entitled to receive €0.013 per kWh as a carbon tax reimbursement and €0.023 per kWh as a production incentive. Generation facilities owned by electricity utilities were ineligible for the production incentive.[58] These additional subsidies represented an attempt to replace the termination of the investment subsidy, which rewarded turbine construction with a policy tool (a feed-in tariff) that would best encourage the desired goal of enhanced wind power production. The comparative effectiveness of this tool would prove to be highly influential in regard to the formulation of wind power policy in other nations.

In 1993, a number of further initiatives were announced which indicated that government planners were beginning to take a strategically methodological approach to wind power development. The Ministry of Environment and Energy ordered all Danish municipalities to undertake wind power potential studies, assigning a deadline of June 1995.[59] It also commissioned an updated economic survey of privately owned wind turbines, a study into the external costs of wind power from a social perspective, an exercise charting conditions for offshore wind turbine installation, an evaluation of R&D effectiveness, and a study investigating more effective ways to promote rural wind power development.

It merits repeating that 1993 was a mediocre year for wind power development in Denmark with only 32 MW added. Some have attributed the sluggish market response to the policy reforms of 1992 to tax reforms that disrupted market dynamics.[60] Although there may have been a degree of truth to this, it is equally possible that 1993's flaccid performance can be explained by a lag in market response to the policy change in late 1992. Simply put, policy change takes time to affect behavior.[61]

Guided by findings of a specially commissioned study released in February 1994, which favored bolstering rural wind power development, the government also announced a new turbine replacement investment subsidy program in May 1994. The scheme offered a tax credit of up to 15% of the original investment cost for upgrading existing turbines to larger capacity models. In conjunction with the subsidies announced in 1992, this program would catalyze a widespread upgrade of existing turbines, although like the 1992 subsidies, the impact of this new turbine replacement scheme exhibited a time lag and would not influence the pace of development until 1996. Yet there were signs that the 1992 subsidies were having some effect: 53 MW of wind power capacity was added in 1994, expanding national capacity by 11%.

Over the subsequent two-year period—1995 and 1996—the catalytic effect of the new government wind power policies became clear. In 1995, a record 79 MW of wind power capacity was added. In 1996, an astonishing 214 MW of installed capacity was added, surpassing the cumulative total for the four previous years. As an illustration of the success of the turbine replacement scheme, the 416 wind turbines that were constructed in 1996 generated the same amount of electricity as the 3000 wind turbines constructed prior to 1990.[62] By this stage the pace of wind power development was brisk enough that the government was emboldened to reiterate its commitment to achieving 1500 MW of installed capacity by 2005 and announce a longer-term goal of meeting 50% of Denmark's electricity needs through wind power by 2030, targeting 4000 MW of offshore wind power and 1500 MW of onshore power—a threefold expansion from 1996 levels.[63]

In subsequent years, market momentum continued to build; from 1997 to 1999, installed wind power capacity expanded by 309 MW, 315 MW, and 315 MW, respectively. By the end of 1999, national wind power generation capacity stood at 1753 MW (including 10 MW in offshore wind power capacity), surpassing the 2000-target of 1500 MW that was originally set in 1990. Moreover, thanks to a new agreement between the government and Denmark's utilities to install 750 MW offshore wind turbines before 2008, the dawn of a new era for offshore wind development appeared imminent.[64]

In 1999, two new pro-wind policies were unveiled. First, given the success of the 1994 turbine replacement scheme, the government was prompted to announce an updated subsidy that mirrored the general trend toward replacing development subsidies with production subsidies. The replacement scheme guaranteed a payment of €0.081 for the first 12,000 full-load hours of production, which equated to approximately five years of operation.[65] This scheme allowed existing turbines of less than 100 kW to be replaced by turbines of up to three times the discarded capacity, while turbines between 100 kW and 1500 kW could be replaced by twice the capacity until the end of 2003.[66] The offer was so attractive that most small wind power systems that were not upgraded between 1994 and 1999 were upgraded in the first two years of the new millennium. Second, the government signaled intentions to bolster wind power diffusion by announcing a new short-term target of achieving 20% contribution to electricity generation from renewable sources by the end of 2003—a lofty goal given that the contribution of wind power to Denmark's electricity grid had only reached 8.7% by 1999 and there were no other substantial renewable technologies contributing to electricity generation.[67]

Lamentably, in the same year, the government made what would turn out to be a forced, ill-fated policy decision that was largely precipitated by a European Union (EU) led initiative to establish a common green certificate market throughout the European Union. In 1999, the Danish government announced an intention to replace the wildly successful FIT system with a carbon trading system. According to the proposed scheme, a national CO_2 emissions quota for 2000 would be set at 23 million tons and electricity utilities would then be assigned individual CO_2 emission quotas to contribute to achieving the national target. The national CO_2 quota would be gradually lowered to 20 million tons in 2003. Utilities that exceeded assigned emission quotas would be fined €5.4 per ton of CO_2. Conversely, utilities that did not fully utilize their quotas could bank the credits for subsequent years or resell them on the open carbon market.[68]

There were two seemingly sound justifications for initiating such a change. First, a system was needed to encourage further reduction of CO_2 emissions in order to meet Denmark's Kyoto Protocol commitment to reduce emissions of greenhouse gases by 21% of 1990 levels over the 2008–2012 commitment period. This necessitated a policy to encourage utilities to reduce coal-fired electricity generation. Ironically, the wind power FIT was adversely affecting this quest because utilities were continuing to utilize coal-fired plants in part to offset the premium cost of wind power. Second, the EU-wide green certificate initiative presented financial opportunities for nations with high capacities in renewable energy, and therefore integrating with EU policy in an expedient fashion was desirable. However, as will be detailed in the next paragraph, it was also obvious from the financial structuring of the new policy that the government felt a need to gradually pare back subsidization of wind power. In 1998 alone, over €77 million was purportedly paid to wind power generators in the form of production incentives.[69]

The proposed shift to a carbon trading scheme was initially supported by the Danish wind power lobby thanks to some favorable front-end benefits and prospects of a much larger market for wind power throughout the European Union. However, the attractiveness of the program came with risks. Initially, the subsidization of wind power in Denmark would not be drastically affected; over time the subsidies would be decreased, to be replaced by green certificates that would vary in value depending on progress within the European Union in achieving aggregate emission reduction targets. According to this new policy, for projects initiated before 1999, wind power producers would receive a FIT of €0.044/kWh and a production incentive of €0.036/kWh for up to 25,000 full-load hours. Projects initiated between January 2000 and December 2002 would receive a FIT of €0.044/kWh plus a reduced production incentive of €0.013/kWh (for up to 25,000 full-load hours). Projects initiated after January 2003 would not receive any subsidies and instead receive market price for electricity generated plus €0.013—€0.036 per kWh under the green certificate scheme.[70]

In order to ensure optimal market conditions for supporting carbon trading, the government also introduced an energy act amendment which deregulated electricity generation and forced the breakup of vertically integrated utility activities. The amended act further stipulated that management of the high-voltage section of the power grid was to be transferred to a state owned company.[71]

The announcement of these intended changes had predictable consequences in regard to market activity. Wind power developers responded by fast-tracking projects to take advantage of the higher front-end incentives, and in conjunction with the new incentives for turbine replacement, this catalyzed the installation of a record 637 MW of wind power (including 40 MW of offshore wind power) in 2000, bringing total installed capacity up to 2390 MW (see Table 4.4).[72] Ironically, amidst this flurry of activity it became apparent near the end of 1999 that stakeholders were ill-prepared to launch the green certificate scheme as planned in January 2000. Therefore, a decision was made to postpone the launch until January 2002.

As often happens in markets where expiring subsidies are greater than incoming subsidies, the development rush of 2000 produced relatively flaccid market conditions in 2001, only 107 MW of capacity was added. However, in 2001, the Danish government also announced a commitment under the 2001 European RES-electricity directive to achieve a 29% share of renewable energy contribution to gross electricity consumption by 2010.[73] This announcement infused renewable energy developers with a degree of confidence that if the new green certificate program fell short of catalyzing the capacity development necessary to meet the 2010 target, there would be further government intervention. When the launch of the green certificate program was postponed once again to 2003 (in September 2001),[74] the wind power development market once again began to heat up as developers fast-tracked more projects in order to take advantage of existing subsidies.

Unfortunately, in November 2001, optimism that the government would eventually sort out its policy quandary was attenuated by the election of a right-wing Liberal Party—the Conservative People's Party coalition, led by Prime Minister Anders Forgh Rasmussen. The Rasmussen administration supported a political platform that decried excessive government wind subsidies.[75] In 2002, market fears were confirmed as the Rasmussen

Table 4.4 WIND POWER DEVELOPMENT IN DENMARK IN THE 2000S

	2000	2001	2002	2003	2004	2005	2006	2007	2008	2009	2010
Wind power onshore capacity, MW	2340	2447	2676	2692	2700	2704	2712	2700	2739	2821	2934
Wind power offshore capacity, MW	50	50	214	423	423	423	423	423	423	661	868
Wind power's share of domestic electricity	12.1%	12.1%	13.8%	15.8%	18.5%	18.5%	16.9%	19.7%	19.1%	19.3%	21.9%

Source: Danish Energy Agency. 2011. *Annual Energy Statistics 2010.* Copenhagen: Danish Energy Agency.

government moved to downgrade support for renewable energy development. The annual budget for the Energy Technology Program was reduced by nearly 70%, and the Development Program for Renewable Energy Sources was eliminated altogether.[76] Nevertheless, since the prevailing FIT and turbine replacement subsidy schemes were still in place in early 2002, developers continued to fast-track projects.[77] As Table 4.4 illustrates, 393 MW of capacity was added in 2002 (including 164 MW in offshore capacity, thanks to the 160 MW Horns Rev project).[78]

In June 2002, the Rasmussen administration announced that it had shelved plans for introducing the green certificate program and instead announced a new but reduced subsidy program which provided wind power generators with an added premium of €0.013 per kilowatt hour (guaranteed for 20 years) on top of the Nordpool pool spot market price for electricity.[79] A separate subsidy added a scrap premium of €0.023 per kWh for upgrading existing turbines.[80] Keeping true to its political promise to minimize the impact of subsidies on fiscal health, these new wind power subsidies were to be passed on to electricity consumers as a Public Service Obligation (PSO) tariff.[81]

The scale of reductions in subsidies virtually stopped wind power development overnight. In 2003, only 16 MW of onshore capacity was added. However, thanks to offshore wind farm installations—catalyzed by the agreement made with the utilities in 1998 to add 750 MW by 2008—209 MW of offshore capacity was added (including the huge 166 MW Nysted offshore wind farm). Despite elevated offshore activity, that same year, the Rasmussen administration announced that utilities would not be bound to invest in further wind power development as per the 1998 agreement.[82] Consequently, by 2004, the wind power development market in Denmark was in freefall. That year, only 8 MW of wind power capacity was added.

The collapse of Denmark's wind power market suddenly became a political hot potato for the ruling coalition due to the fact that wind power manufacturing was Denmark's third-largest export industry, providing direct employment to 6,600 people, indirect employment to a further 15,000 people, and investment income to an estimated 125,000 Danish households that owned shares in wind turbines.[83] In response, the Rasmussen administration was forced to backtrack on some of its cutbacks. It partially reinstated plans for offshore wind power development by agreeing with utilities that two new offshore wind farms of 200 MW capacity each would be installed by 2007.[84] To facilitate this, a tendering system would be developed with the winning bid guaranteed a fixed tariff for the equivalent of up to 50,000 full load hours (approximately 12 years).[85] The government also partially restored funding for the Energy Technology Program, with a compromise being that the direction of research would focus on sustainable energy systems rather than the development of specific technologies.[86] These minor concessions would prove to be insufficient for reinvigorating the flagging wind power market. As Table 4.4 illustrates, between 2005 and 2007, there was actually net zero growth in wind power capacity. However, thanks to other government initiatives to improve energy efficiency and reduce demand, the contribution of wind power to Denmark's electricity grid actually rose to 19.7% in 2007 (from 18.5% in 2004).

Given the fact that virtually no wind power development occurred between 2004 and 2007, it is easy to understand why critics felt justified in intimating that the Rasmussen administration's wind power policy was damaging to the industry.[87] However, in hindsight, this developmental lull was perhaps necessary in order to rein in excessive subsidies and reengineer Denmark's grid for bolder renewable energy initiatives in years to come. By 2004, wind power's share of domestic electricity production had actually risen to 18.5%, compared to 12.1% in 2001 (see Figure 4.8). Not only had this resulted in Denmark's electricity consumers paying substantially higher prices for electricity (in comparison to neighboring EU countries), but there were amplified concerns that integration of such high levels of wind power was producing oversupply during certain periods which had to off-loaded to the Nord Pool market at a heavy loss. Moreover, claims that the new policies were threatening the commercial viability of the wind power manufacturing sector may have been more emotive than logical because 90% of the revenues earned by Danish wind power firms were from export sales.[88]

There is little evidence to support the contention that the Rasmussen administration was opposed to renewable energy. Indeed, the overall contribution of renewable energy to Denmark's electricity supply continued to increase under the Rasmussen administration's policies. This was partly due to enhanced support for fledgling renewable energy technologies that the government truly felt needed government support. For example, a subsidy was announced for biogas that would provide €0.08 per kWh as a FIT for 10 years and €0.05 per kWh for the subsequent 10 years. Other subsidies were announced to support special plants using energy sources or technologies of major importance to the future exploitation of renewable energy, such as wave power, fuel cells, solar energy and biomass.[89] In fact, evidence suggests that the government was proactively engaged in initiatives designed to avert some of the technical barriers to further wind power development. For example, in 2005 the Danish government, in adhering to terms of an EU agreement to liberalize electricity markets in order to support further renewable energy diffusion, initiated a major restructuring. East and West transmission networks were merged under the management of a new state-owned grid operator, Energinet Danmark. Energinet Danmark was also appointed to oversee operation of the nation's gas network, an initiative that provided the technological foundation for rectifying supply fluctuations from wind power through gas-fired peak-load power plants.[90]

In fact, some analysts have pointed out that the Rasmussen administration's wind power subsidies announced in 2002 to replace the abandoned green certificate program amounted to precisely the same amount that would have been provided under later stages of the green certificate program that was proposed by the preceding Social Democrat–Social Liberal coalition that held power between 1993 and 2001.[91] Furthermore, new policies that would eventually emerge suggests that the new government was not as opposed to wind power as critics suggested, rather this period of policy entrenchment represented a period of learning by doing where the government was trying to introduce a degree of financial restraint into a sector (renewable energy) that it nevertheless supported.

After five years of stagnant market activity, the government announced a new national energy strategy on January 19, 2007 that signaled the start of a new developmental push. Under the proposal, which was agreed to by all the parliamentary parties except the far-left Red-Green Alliance, Denmark would aim to expand renewable energy capacity to satisfy at least 20% of total energy consumption by 2011 and 30% of total energy consumption by 2025.[92]

In the same year, the government began to mobilize new policies to achieve these goals. It was recognized that many of the nation's wind power systems were aging and in need of replacement. Consequently, the Rasmussen administration announced a new wind turbine substitution scheme. The goal was to replace approximately 900 turbines (450 kW or less) with 150 to 200 turbines in the 2 MW range.[93] In 2008, the Rasmussen administration further announced intentions to encourage a 1300 MW increase in wind power capacity by the end of 2012.[94] According to the plan, 800 MW of this total would come from three new offshore wind parks. To facilitate this increase, an additional balancing cost subsidy of €0.03 per kWh was offered on top of the 2002 subsidy of €0.013 per kWh that was tacked on to the spot price.[95] Moreover, a compensation package was announced to financially reward communities for hosting onshore wind farms. On top of all this, the government announced that its CO_2 tax would be increased to a level that would equate with the expected price of carbon in 2008–2012—estimated at approximately €20 per ton—and a new nitric oxide tax of approximately €670 per metric ton would be introduced from January 1, 2010.[96]

These new policies reinvigorated wind power development. In 2009, 320 MW of new installed wind power capacity was added (238 MW in offshore developments) and in 2010, another 320 MW was added (207 MW in offshore developments). After an additional 178 MW was installed in 2011, increasing total installed capacity to 3871 MW, wind power was supplying 28% of Danish electricity consumption.

In the remainder of this chapter, we will employ the *Political SET* framework to examine the main elements that have catalyzed support for wind power in Denmark. Section 4 will examine the sociocultural, economic, and technological conditions that have provided a positive national landscape for supporting Denmark's wind power boom. In section 5, the governance conduits—the influence of sociocultural, technological, and economic forces on political behavior—will be highlighted in order to demonstrate how political behavior was influenced by SET forces. In section 6, key regime characteristics of the Danish system of governance will be examined in order to understand the political forces that influenced wind power diffusion. Section 7 reviews how the STEP forces conflated to enable the success Denmark has enjoyed in wind power diffusion. The chapter will conclude with a discussion of what might be in store for wind power developers in Denmark in the near future.

4.4 UNDERSTANDING THE GENERAL FORCES FOR CHANGE

4.4.1 Sociocultural Landscape

As outlined earlier in the chapter, wind power development in Denmark was prominent during both world wars. Consequently, for many older members of Denmark's farming communities, the presence of wind turbines was by no means unnatural. As wind power began to expand in the 1980s, historical affinity helped to reduce resistance to a technology that could be aesthetically invasive.

Another key sociocultural element that engendered public support for wind power stems from the fact that Denmark is a sparsely populated, agriculturally vibrant country. One reason for Denmark's thriving agricultural sector is that Danish farmers have a history of cooperating to innovate and improve agricultural returns. In the early 1980s, the prospect of enhancing profitability by investing in wind power projects hosted on agricultural land enticed a number of cooperatives to get involved. Thanks to high levels of community investor participation in wind power projects, public opinion toward wind power remained positive throughout the first two decades of growth—surveys conducted in the 1990s indicated that approximately 80% of the Danish population supported wind power. By 2000, approximately 150,000 Danish households were registered investors in wind power projects. As an expert on wind power development in Denmark summarized, "it is much easier to accept a little noise and the view of a turbine if it reminds you of the fact that the turbine gives you money when the wind blows."[97]

Economic affluence in Denmark has fostered progressively lofty environmental expectations. In energy circles, this has manifested itself through the emergence of pro-wind lobby groups such as the Wind Mill Owners Association, which has substantially influenced national energy policy. Moreover, in the 1970s, two highly influential anti-nuclear power groups emerged—the Organization Against Nuclear Power and the Organization for Renewable Energy—and they proved to be successful in dissuading Denmark's policymakers from acting on nuclear power development aspirations.[98] At its zenith, the anti-nuclear power movement had attracted over 30,000 members and contributed to wind power assuming a mantle as the preferred alternative energy technology.[99]

Public support for wind power in Denmark extended beyond rhetoric. A majority of citizens were willing to pay extra to support environmentally benign energy systems.[100] This allowed the government to pass through the cost of wind power subsidization to the end-consumer in the form of environmental premiums, which made electricity in Denmark approximately 50% more expensive than neighboring countries were paying.[101] Public acquiescence in regard to accepting higher energy costs proved to be instrumental to wind power development in the 1980s, when the decline of oil prices would have made it hard for wind power to compete on a level market basis. Similarly, without public economic concessions that enabled the incorporation of carbon pricing into the cost of electricity, the resurgence of wind power in the late 2000s might not have been possible.

4.4.2 Economic Landscape

The two oil shocks of the 1970s hit Denmark particularly hard, because over 90% of its electricity generation came from oil-fired power plants. Not only did the shocks catalyze a search for alternative forms of energy, it also helped entrench public acquiescence to accept higher energy costs in order to transition away from dependence on an energy source (oil) over which there was no domestic control. Ideologically, the choice came down to enduring the capricious nature of fossil fuel costs or accepting the more stable yet comparatively higher costs of wind power.

In contrast to oil, which prior to the discoveries in the North Sea was mostly imported, there were also direct economic side benefits associated with wind power. Not only would Danish individuals and groups have an opportunity to invest in (and financially benefit from) wind energy projects, but domestic support for wind power would eventually sire a domestic wind power manufacturing industry that would become Denmark's third-largest export sector. These cumulative benefits placed wind power in a decidedly attractive light.

By the end of 1990s (the second decade of wind power development in Denmark), electricity production costs for the most efficient wind turbines (600 kW) at average wind quality sites were estimated to be approximately €0.043 per kWh, including €0.006 per kWh for backup. In comparison, generation costs at conventional power plants were running in the range of €0.032-€0.034 per kWh.[102] With the price of carbon credits under the European Emission Trading Scheme ranging between €8–12 per ton of carbon, a gradual convergence between the cost of wind power and the cost of coal took place in the 1990s—wind power was becoming an economically rational alternative.

4.4.3 Technological Landscape

Favorable geographic conditions in Denmark enable rich exploitation of wind power technology. As mentioned earlier, Denmark is a sparsely populated nation with large expanses of treeless agricultural land that is ideal for wind turbine siting. Furthermore, Denmark's windswept North Sea location and its undulating coastline make it ideal for offshore wind power. One study has estimated that enough economically exploitable offshore wind power potential exists to provide at least 20 TWh per year[103]—approximately 60% of 2009 electricity consumption levels.

It is conceivable that wind power in Denmark would be underexploited if it wasn't for the fact that prior to the mid-1980s, the nation also lacked conventional energy resources. Consternation over the precarious dependence on foreign fossil fuel supplies catalyzed efforts aimed at improving self-sufficiency in energy. This included providing funding for alternative energy R&D and financing fossil fuel exploration efforts that would eventually discover rich deposits in the North Sea.[104]

Aside from favorable geographic conditions, a major contributing factor to wind power development in Denmark stemmed from the technological development strategies exhibited by domestic wind turbine manufacturers. As mentioned earlier, Denmark's wind turbines were predominantly designed by firms that were accustomed to manufacturing reliable, easy to maintain machinery for farms.[105] These firms took a safe technical path in regard to turbine development—producing machines that were both reliable and technologically unsophisticated.[106] The result was that Denmark's wind turbines delivered what they promised in terms of performance.

Another technological characteristic that benefitted wind power development was a culture of collaboration between industry, academia, and government.[107] Not only did this help centralize R&D and improve research efficiency, it also enabled learning by interacting, which resulted in a virtuous circle that provided expedient technological responses to emergent challenges.[108]

A final significant technological element in support of enhanced wind power diffusion was grid resilience. Electricity grid interconnections to Germany and the European Continental Grid in Western Denmark and to the NORDEL synchronous area for Eastern Denmark provided the grid resiliency necessary for integrating wind power. When the two Danish grids were interconnected in 2009 by a high-voltage current known as the Great Belt Link,[109] resilience was further enhanced.[110]

4.5 INFLUENCES ON GOVERNMENT POLICY

As will be demonstrated in the Canadian and Japanese case study chapters which describe phlegmatic wind power markets, government temerity in responding to landscape influences such as those described in the previous section can engender stakeholder opposition or result in squandered opportunity. Conversely, governments that respond effectively to landscape trends smooth the developmental path of wind power. Denmark is a case in point. This section will briefly summarize how social, economic and technological forces in Denmark directly influenced government policy. This is not to say that the influence was coercive in all cases. Indeed, as this section will demonstrate, in many instances the government demonstrated acute awareness of market dynamics and adjusted policy to respond to emergent stakeholder needs.

4.5.1 Sociocultural → Political

As outlined in the historical review of Danish wind power policy, the national energy plan of 1976 aimed to wean the nation from a reliance on oil by considering alternative energy forms, including nuclear power, that the government was bullish about. However, well-organized opposition from two anti-nuclear NGOs resulted in a government decision in 1985 to scrap plans to use nuclear power.[111] Any latent support for nuclear power that still existed in policy circles evaporated with the Chernobyl disaster in 1986. The government responded to social disapproval of nuclear power by refocusing efforts on wind power, a technology that was deftly promoted by two proactive trade organizations—the Danish Wind Turbine Owners Association and the Danish Wind Turbine Manufacturers Association.[112]

Denmark is also characterized by a unique pattern of utility ownership that the government exploited in its wind power development policy. From the 1970s through to the turn of the century, Denmark's electricity system was predominantly consumer-owned and operated through either cooperatives or municipal ownership. The utilities were nonprofit organizations, with operational surpluses channeled back to the consumers in the form of lower electricity prices.[113] Consequently, when the government sought to enhance wind power diffusion, it adopted policies that supported cooperative ownership. At the same time, it attempted to coerce utilities into voluntarily investing in wind power by appealing to the fiduciary duty that such organizations have to the communities in which they operate. Intriguingly, this strategy produced a pool of wind power investors that was estimated in 2000 to range between 120,000 and 150,000 individuals.[114] In short, Denmark's wind turbine investment policies attracted a base of entrenched supporters that served to attenuate any consumer dissonance associated with higher than average electricity prices.

4.5.2 Economic → Political

As will be demonstrated in all the case studies included in this book, energy economics exerts perhaps the greatest influence over a given nation's propensity to support wind power. In almost every nation, the decision to support wind power is predicated upon assessments of the impact that comparatively high wind power costs will have on industrial competitiveness and consumer willingness to pay higher energy prices to support greater contributions from wind power.

In Denmark, industrial concerns typically opposed wind power, arguing that it would inflate the cost of energy and undermine industrial competitiveness. However, the lobbying efforts of industry associations were eclipsed by the lobbying efforts of Denmark's wind power associations and the government's desire to wean the nation from an overdependence on imported fossil fuels. One of the main reasons the lobbying efforts of industrial associations were ineffectual stems from the observation that preserving the status quo did not necessarily represent a preferred economic alternative. In the 1970s, the capricious nature of fossil fuel prices proved to be damaging to Danish industry; consequently, arguments in favor of preserving the status quo for economic reasons were weakened. As an illustration of this, electricity costs in Denmark in the early 1980s were about three times greater than the prices found in neighboring Sweden.[115] This was largely because of the dominant role of high-priced oil in Denmark's electricity mix. Succinctly put, the economic case that is typically used to oppose wind power was severely undermined in Denmark by an overdependence on oil during a period of capricious oil prices.

As Denmark's wind power capacity expanded, the number of investors in wind power projects and the employment attributed to wind power manufacturing increased substantially, and this also strengthened the economic appeal of wind power. As opposed to oil imports, where industry profits went to foreign suppliers, Danish businesses and the general public were beginning to reap investment returns from the growing wind power sector. This indirect economic payback associated with wind power enhanced its economic appeal and offset the comparably higher electricity costs.

4.5.3 Technological → Political

The economic promise of wind power was also underpinned by a technological attribute. The government recognized that the nation possessed some core technological competencies to justify earmarking wind turbine manufacturing as a desirable activity within the nation's economic development strategy.[116] At the outset, most wind turbine developers were machine-shop type firms with experience in producing agricultural equipment;[117] however, policy was necessary to provide these small entrepreneurs with the technological support needed to evolve. The governmental response was the creation of a tight network for wind power innovation that involved industry, academia, and government sponsored think tanks.[118] This centralization of research helped fledgling wind turbine manufacturers to learn by interacting, which over time yielded cost improvements.[119]

In addition to providing centralized support for research, the government also centralized quality control by establishing a turbine quality certification scheme at Risø National Laboratories. Only turbines that were certified by this scheme were eligible for government subsidies. This created a virtuous circle wherein manufacturers were increasingly pushed to elevate standards, while at the same time, were also exposed to best practice through the government's research support programs. These two factors—access to technological research and centralized quality control—were recognized as fundamental drivers of success for Denmark's wind turbine manufacturers in California during the heyday of wind power expansion. America's wind turbine manufacturers, which were focusing on technological shock and awe, could not compete with the reliability of Denmark's wind power systems.

It is notable that the government learned from its mistakes in the mid-1980s, when a domestic downturn in installed wind power capacity resulted in the bankruptcy of many domestic wind power manufacturers. In the 1990s, it created an agency that was responsible for supporting wind turbine exports, and in doing so helped smaller firms ride out domestic market downturns by diversifying into foreign markets. By 2002, Danish wind turbine producers reported revenues of over €3 billion, with 90% of the revenue coming from overseas markets.[120]

Finally, it should be noted that the government also recognized the finite nature of its North Sea oil and natural gas fields. Indications that production declines in these fields would likely occur by 2030 encouraged the government to continue support for renewable energy.[121]

4.6 POLITICAL INFLUENCES ON POLICY

4.6.1 National Political Structure

Denmark is a democratic, constitutional monarchy that embraces a parliamentary system of governance wherein the prime minister is elected by any political coalition that can claim a majority in parliament. This means that the party with the most votes is not guaranteed a position in the ruling coalition. For example, after the national elections in 2011, the Liberal party emerged with the greatest percentage of votes (26.7%); however, the Social Democrats who received the second most votes (24.8%) were able to unite to form a parliamentary majority called the Red Alliance and its leader, Helle Thorning-Schmidt, became prime minister. Denmark's parliament consists of 175 seats: 135 seats are determined based on majority vote within 135 constituencies, and the remaining 40 seats are allocated to each party in proportion to total votes garnered in the national election.

Since the beginning of the twentieth century, no single party has been able to claim an absolute majority in Denmark's parliament; with any party receiving more than 2% of the popular vote guaranteed parliamentary representation, there are a number of diverse political perspectives evident. For example, in the 2011 election, eight parties were represented in the Danish Parliament. This has engendered a political system wherein governance is typically undertaken through negotiation and compromise. Therefore, even when one coalition government is replaced by another, there is a high degree of policy continuity.

Finally, it is worth noting that as a member of the European Union (EU), Denmark is also obligated to adhere to EU laws and to comply with EU directives; consequently, when agreements are made within the European Union in regard to climate change mitigation strategies or EU energy policy, Denmark is obligated to harmonize domestic policy to support EU initiatives. For example, in 2001 the European Union announced a renewable energy directive that targeted 20% contribution from renewable energy to the EU electricity grid by 2010.[122] Consequently, Denmark was obliged to draft national policy for supporting this directive. Therefore, even if a degree of antipathy were to exist in regard to renewable energy in Denmark, the force of an EU directive would generally be sufficient to elevate the probability of compliance.

4.6.2 Governing Party Ideology

In the late 1970s, when wind power was being investigated as a possible solution to overreliance on oil-fired power, the Danish government was led by a Social Democrat, Anker Jorgensen. Jorgensen's coalition supported active involvement of the government in influencing social welfare, and indeed it was subsequently criticized for policies that produced large state budget deficits. Accordingly, it should come as no surprise that in response to the oil crises of the 1970s, the government was prepared to play an active role in facilitating a transition away from oil-fired electricity generation.

Interestingly, when policy was being formulated to respond to the oil crises, there was a great deal of support for nuclear power development both within Denmark's parliament and within the ruling coalition.[123] In fact many analysts believe that it was only due to well-coordinated social opposition to nuclear power that Denmark did not adopt a more rigorous nuclear power development agenda. The capacity of social dissent to sway political behavior in this manner underscores the high level of responsiveness to popular opinion that characterizes the nation's unique parliamentary system.

In 1982, the Jorgensen administration was succeeded by a coalition led by Conservative party leader, Poul Schlüter, who would lead the government until 1993. Even though the political ideology of Schlüter's administration was further right on the political spectrum, government financial support for wind power development actually increased substantially during this period. This highlights an important characteristic of Danish politics—although there may be a number of parties vying for political power in Denmark, most support slightly left-of-center policies in order to appeal to the Danish voter's high environmental ethic. Even parties that purport to embrace right-wing policies—such as the Liberal-led coalition of Anders Fogh Rasmussen that came to power in 2001—have eventually been forced to backtrack on radical changes to energy policy.

4.6.3 Fiscal Health

Denmark is one of the most socially progressive nations in the world. As the Ministry of Social Affairs and Integration sums up, "the Danish welfare model is based on the principle that all citizens shall be guaranteed certain fundamental rights in case they encounter social problems such as unemployment, sickness or dependency."[124] This commitment to universal welfare has produced one of the most economically equitable nations in the world, as measured by the Gini coefficient.

Since the 1970s there has been an ever-increasing demand for enhanced social services to support the evolving needs of Denmark's baby boomers. With the exception of a four year period between 1986 and 1989, Denmark's government posted fiscal deficits every year between 1970 and 1997. Progressive fiscal deficits of this type engender political conditions wherein all new policies become thoroughly vetted and compared to competing funding requests. Therefore, given the breadth of demand for social welfare services in Denmark, direct subsidization of wind power through fiscal budgeting has not been the norm.

4.6.4 Policy Regime

Given government fiscal constraints between 1970 and 2000, government support for wind power diffusion has largely centered on strategic planning and regulatory support. The only substantial subsidy that the government provided directly from the fiscal budget was support for R&D. Over the first three decades of the program, public support for R&D averaged about 4% of total global public funding spent on wind energy R&D.[125] The rationale behind this bold R&D commitment was that government saw wind system manufacturing as a technological sector that meshed with Denmark's aspirations to nurture knowledge-based technological industries. Government support was as much of an investment designed to nurture internationally competitive wind power manufacturers as an investment in wind power diffusion in Denmark.

In terms of domestic wind power development, the role of the government was more facilitative than catalytic. It coordinated research, mandated regional and municipal development plans, commissioned the development of wind power potential studies, and established the framework necessary to encourage community investment in wind power projects.[126]

The finances necessary for facilitating development rarely threatened fiscal health. The investment subsidies that were offered to individuals and cooperatives during the fledgling years of the wind power development program were structured as tax incentives; therefore, the fiscal impact was deferred, to be theoretically offset by profits associated with the investment. The feed-in tariffs (or price premiums) were financed mostly by passing the cost through to end-consumers, which explains in part why electricity consumers in Denmark have historically paid such high prices for electricity. Even grid reinforcement was a cost that the government forced onto the utilities.

4.7 THE CULMINATION OF INFLUENCES

All of the STEP forces discussed in this chapter have interacted in a manner that has produced a policy soup that exhibits a unique flavor which no other nation shares. It is for this reason that investigating Denmark's wind power support policies in isolation of the aforementioned social, technological, economic, and political influences would likely yield suboptimal results when trying to determine which of Denmark's policies are transferrable to other national contexts. A brief review of how wind power development policy has evolved to Denmark illustrates the ineluctable influence of STEP elements.

If it weren't for social opposition to nuclear power back in the early 1980s, it may very well be that today's scholars might be critically evaluating the rise of Denmark's thriving nuclear power program. It was only through well-organized social opposition that political support for nuclear power was derailed, opening a window of opportunity for wind power technology development.

If Denmark's agricultural manufacturing sector was not the first to exhibit commercial interest in wind power development, the domestic evolution of wind turbine manufacturing may have centered around the challenge of developing large-scale, high-tech turbines, as has been the model in many other European nations. In such a case Danish firms may not have been able to establish the quality reputation that allowed them to profit from California's wind power boom. Moreover, if a national culture did not exist in Denmark wherein academia, government, and industry were comfortable with collaborative R&D, wind power manufacturing initiatives may have become too fragmented to compete on an international scale.

To the contrary, Denmark's agricultural firms *were* interested in producing wind turbines and they *were* able to leverage centralized support for R&D, thanks to a government that recognized that an opportunity existed for nurturing a wind power manufacturing industry.[127] Arguably, without centralized research (incorporating government bodies, academia, and industry) and the government-sponsored turbine certification scheme,[128] Denmark's manufacturers may have been eradicated during the wind power market downturn of the mid-1980s.[129]

In Denmark, as in other nations, the threat of NIMBY (not in my backyard) opposition to wind power existed despite the nation's rich history in wind power development. However, the government was effective in recognizing this threat and in proactively establishing policies that would mitigate public opposition.[130] The central government wisely delegated the planning of wind power siting to municipalities, which had a better grasp of community sensitivities.[131] Furthermore, to ensure that projects were developed in a manner that would minimize community dissent, rigid siting standards were developed to mitigate complaints of noise and shadow flicker and to improve aesthetic design of wind power parks.[132] Wind power development subsidies were initially directed at encouraging private or cooperative community investment in wind power projects.[133] The government went as far as to establish concentric geographic boundaries around projects, beyond which investor tax incentives would not be available. In aggregate, all of these government initiatives established high levels of community support for wind power development and arguably attenuated any dissonance that might have otherwise existed.[134] If policymakers of other nations wish to learn anything from Denmark in terms of effective wind power policy design, these initiatives for alleviating public opposition most certainly represent areas of best practice.

Another notable characteristic of wind power development policy in Denmark is that it was rarely as consistent as some may think. Rather, the Danish government exhibited a willingness to experiment with different wind power development policies as new needs and concerns arose. The results were often inconsistent, eliciting both growth spurts and periods of stagnation and retrenchment.[135] As one scholar aptly summarized, "Danish wind power support policy has been diverse. It has included long-term government support for research, development and demonstration, national tests and certification of wind turbines, government-sponsored wind energy resource surveys, feed-in tariffs and regulations, investment subsidies, government energy planning and targets, local ownership of wind turbines and careful selection of sites."[136] As another analyst adds, "wind power support policy in Denmark has been eclectic. It has included a mixture of technology push and demand pull policies. Examples of technology push policies include R&D funding, establishment of and financial support for a test station at Risø and other regulations like approval schemes and standardization. Examples of demand pull instruments include financial incentives and different forms of regulations particularly over ownership."[137]

The malleability of wind power development policy was most evident as the program matured and it became obvious that higher concentrations of installed wind power capacity were desirable in order to minimize connection costs and try and mitigate growing public concern over adverse aesthetic impacts associated with Denmark's numerous small wind system installations. On two separate occasions, the government announced supplemental policies to upgrade smaller capacity turbines with larger models. In a similar vein, the notion that wind farms might be an effective way to mitigate the aesthetic impact of wind turbine siting catalyzed enhanced government pressure for utilities to drive development.[138] Yet as an indication of the fluid nature of government policy, when it became apparent that onshore wind farms were inflaming community dissonance, the government shifted its focus to encourage offshore developments. In fact, the government created a Committee for Offshore Wind Turbines specifically for the purpose of optimizing offshore developments.[139] The seemingly erratic nature of Danish wind power policy can be considered to be a manifestation of *gradualism*—policy design that evolves in response to dynamic market conditions and changing expectations of stakeholders. Given the conciliatory nature of politics in Denmark, a gradualist policy has been effective because evolving policy better supports emerging market needs and stakeholder expectations.

Finally, is worth emphasizing that the main government strategy of subsidizing wind power generation through tax credits and FITs (that were financed by passing through costs to end-consumers) was a product of high fiscal debt loads and social willingness to pay higher energy prices in return for environmentally benign technologies. If a government surplus had existed, the subsidization approach might have been different because Danish politics tends to favor direct government support. Similarly, if electricity consumers would have been less willing to accept higher electricity prices, the government might have had to resort to a different policy instrument to finance wind power development.

This all points to one conclusion—a unique conflation of social, technological, economic, and political factors all conspired to sire the types of wind power development policies exhibited in Denmark. If any of these influences would have been dramatically different, the story of wind power development in Denmark might have also been substantially different.

4.8 WHAT TO EXPECT GOING FORWARD

In March 2012, Denmark's new ruling coalition under the leadership of Helle Thorning-Schmidt dropped a bombshell on the Danish energy community.

On the heels of an eight-year period during which installed wind power capacity in Denmark grew by an annual average of only 114 MW (an annual growth rate of slightly more than 3%), the government announced a green energy acceleration program which established a target of 100% renewable energy in the energy and transport sectors by 2050 and an interim target of achieving 50% contribution to electricity consumption by wind power by 2020.[140] Over the next eight years, policies will be needed to double existing capacity—an expansion rate of almost 8% per year.

Efforts will apparently focus on exploiting offshore wind potential. Although 500 MW of the targeted 3000 MW in additional capacity is slated to come from onshore wind facilities, the vast majority of new wind power capacity will come from large offshore wind farms such as the ones planned at Kriegers Flak (600MW) and Horns Rev (400MW).[141] This represents an intrepid initiative given the emergence of concerns over the cost implications of adding more wind power and the emergence of pockets of public opposition. Accordingly, it is worth drawing on our understanding of the STEP environment in Denmark to speculate on why such a bold policy has come about.

There is clearly a national security theme underlying Danish support for renewable energies. According to one study, based on proven reserves and existing technology, self-sufficiency in natural gas and petroleum will only exist for another decade.[142] There is also international political pressure for Denmark to reduce its per capita GHG emissions, which at about 9.6 tons per year rank among the highest in the world and double the EU average.[143]

There is also an emerging economic justification for supporting ambitious wind power development targets. Given the capricious nature of fossil fuel prices over the past five years and the declining cost profile for wind power, it has become obvious to Denmark's government that any substantial carbon pricing regime will tip the balance, making wind power an economically superior alternative to coal-fired power. Technically, research indicates that exploiting existing wind power potential could satisfy 100% of Denmark's electricity needs by 2050.[144] Consequently, it seems clear that the Danish government is confident that carbon pricing will be a permanent feature of EU energy policy and in making an early and substantial commitment to wind power, Denmark will position itself to amass carbon credits and give Denmark's wind power firms a domestic financial boost prior to an antici-pated market boom.

Given the lofty targets for offshore wind power capacity, there is always the possibility that further development will increase community opposition to wind projects in general. However, in a 2007 study by a researcher at the University of Copenhagen, it was found that although there is a social preference to offshore wind farms, respondents "living close to either on-land or offshore wind turbines did not display a more negative attitude toward wind power generation when compared to respondents who are not living close to wind turbines."[145] In short, although there are indications that some dissonance exists toward specific wind power projects, claims that opposition to wind power in Denmark have reached prohibitive levels seem to be exaggerated.

These ambitious offshore wind power development targets indicate that the government is aware of the economic efficiencies associated with larger wind farms and that it sees the development of offshore wind farms as representing the least contentious way to catalyze the development of larger wind power projects. The government's large-scale offshore development strategy also makes perfect business sense because larger projects require greater amounts of capital and incur higher levels of risk. This implies that investors need to have higher levels of business savvy than the individuals, farmers, cooperatives, and independent project developers that dominated onshore wind power development in the early years of wind diffusion in Denmark.[146]

This is not to say that achieving these ambitious targets will be easy. By 2020, well over 50% of the wind turbines that are currently in existence will be in need of replacement.[147] This suggests that a new round of onshore wind system upgrades will be forthcoming and result in more visually invasive wind systems being erected in communities. Combining new onshore developments with these mega-offshore developments will result in a greater visual presence of wind turbines in Denmark. Back in 1988, a wind power researcher remarked that "wind farms as big as those found in California would never be acceptable in the Danish landscape when we look at it from an aesthetic point of view."[148] It appears that this premise is about to be tested.

Technically and economically, there are already signs that wind power production is occasionally exceeding power demands. This places increased importance on the interconnections that Denmark has with neighboring countries, because at present sufficient electricity storage does not exist in Denmark. Some critics charge that the generation of surplus wind power is economically undesirable because spot purchase prices are never certain, and historically, low spot prices have resulted in wind power being sold at a loss. Amplifying contributions from wind power will inevitably increase the likelihood of more instances of surplus power being generated—suggesting that political support for wind power could face challenges by fossil fuel special interest groups on economic grounds.[149] Therefore, the evolution of a large-scale EU market for green energy is a critical factor for supporting Danish wind power targets.

It has been said that four factors have been instrumental to Denmark's success in wind power: i) successful design and implementation of siting rules, ii) robust financial support mechanisms, iii) the existence of organizations concerned with environmental protection, and iv) the establishment of socially sensitive ownership patterns of wind farms.[150] In targeting a 50% contribution by wind power to electricity supply by 2020, the challenge of managing all four of these factors will be amplified. Danish philosopher Søren Kierkegaard is purported to have said, "Life has its own hidden forces which you can only discover by living." It remains to be seen whether or not the forces that have supported stellar wind power development in Denmark will continue to propel wind power diffusion to new heights or whether some of these other hidden forces alluded to in this chapter will rise up and alter Denmark's energy landscape.

NOTES

1. David, Paul A. 1985. "Clio and the Economics of QWERTY." *American Economic Reviews* 75 (2): 332–337.
2. Pettersson, Maria, Kristina Ek, Kristina Söderholm, and Patrik Söderholm. 2010. "Wind Power Planning and Permitting: Comparative Perspectives from the Nordic Countries." *Renewable and Sustainable Energy Reviews* 14 (9): 3116–3123.
3. Meyer, Niels I. 2004. "Renewable Energy Policy in Denmark." *Energy for Sustainable Development* 8 (1): 25–35; and Ibenholt, Karin. 2002. "Explaining Learning Curves for Wind Power." *Energy Policy* 30 (13): 1181–1189.
4. Danish Energy Agency. 2012. *Accelerating Green Energy Towards 2020*. Copenhagen, Denmark: Danish Energy Agency.
5. European Renewable Energy Council. 2009. "Renewable Energy Policy Review: Denmark." In *RES 2020: Monitoring and Evaluation of the RES Directives Implementation in EU27 and Policy Recommendations to 2020*, pp. 1–10. Brussels: European Renewable Energy Council. www.erec.org/fileadmin/erec_docs/ Projcet_Documents/RES2020/DENMARK_RES_Policy_Review__09_Final.pdf.
6. Carlman, I. 1988. "Wind Power in Denmark! Wind Power in Sweden?" *Journal of Wind Engineering and Industrial Aerodynamics* 27 (1–3): 337–345.
7. Meyer, Niels I. 2004. "Renewable Energy Policy in Denmark." *Energy for Sustainable Development* 8 (1): 25–35; and Nelson, Vaughn. 2009. *Wind Energy: Renewable Energy and the Environment*. Boca Raton, Florida: CRC Press.
8. Gipe, Paul. 1991. "Wind Energy Comes of Age: California and Denmark." *Energy Policy* 19 (8): 756–767.
9. Nelson, Vaughn. 2009. *Wind Energy: Renewable Energy and the Environment*. Boca Raton, Florida: CRC Press.
10. Meyer, Niels I. 2004. "Renewable Energy Policy in Denmark." *Energy for Sustainable Development* 8 (1): 25–35.
11. Christensen, P., and H. Lund. 1998. "Conflicting Views of Sustainability: The Case of Wind Power and Nature Conservation in Denmark." *European Environment* 8 (1): 1–6.
12. Carlman, I. 1988. "Wind Power in Denmark! Wind Power in Sweden?" *Journal of Wind Engineering and Industrial Aerodynamics* 27 (1–3): 337–345.
13. Danielsen, O. 1995. "Large-Scale Wind Power in Denmark." *Land Use Policy* 12 (1): 60–62.

14. Meyer, Niels I. 2004. Renewable energy policy in Denmark. *Energy for Sustainable Development* 8 (1): 25–35.

15. Meyer, Niels I. 2004. "Renewable Energy Policy in Denmark." *Energy for Sustainable Development* 8 (1): 25–35.

16. Carlman, I. 1988. "Wind Power in Denmark! Wind Power in Sweden?" *Journal of Wind Engineering and Industrial Aerodynamics* 27 (1–3): 337–345.

17. Klaassen, G., A. Miketa, K. Larsen, and T. Sundqvist. 2005. "The Impact of R&D on Innovation for Wind Energy in Denmark, Germany and the United Kingdom." *Ecological Economics* 54 (2–3): 227–240.

18. Kamp, L. M. 2004. "Wind Turbine Development 1973–2000: A Critique of the Differences in Policies between the Netherlands and Denmark." *Wind Engineering* 28 (4): 341–354.

19. Carlman, I. 1988. "Wind Power in Denmark! Wind Power in Sweden?" *Journal of Wind Engineering and Industrial Aerodynamics* 27 (1–3): 337–345.

20. Klaassen, G., A. Miketa, K. Larsen, and T. Sundqvist. 2005. "The Impact of R&D on Innovation for Wind Energy in Denmark, Germany and the United Kingdom." *Ecological Economics* 54 (2–3): 227–240.

21. Kamp, L. M. 2004. "Wind Turbine Development 1973–2000: A Critique of the Differences in Policies between the Netherlands and Denmark." *Wind Engineering* 28 (4): 341–354; and Wüstenhagen, R. 2003. "Sustainability and Competitiveness in the Renewable Energy Sector: The Case of Vestas Wind Systems." *Greener Management International* 44: 105–115.

22. Kamp, L. M. 2004. "Wind Turbine Development 1973–2000: A Critique of the Differences in Policies between the Netherlands and Denmark." *Wind Engineering* 28 (4): 341–354.

23. Meyer, Niels I. 2004. "Renewable Energy Policy in Denmark." *Energy for Sustainable Development* 8 (1): 25–35.

24. Carlman, I. 1988. "Wind Power in Denmark! Wind Power in Sweden?" *Journal of Wind Engineering and Industrial Aerodynamics* 27 (1–3): 337–345.

25. Meyer, Niels I. 2004. "Renewable Energy Policy in Denmark." *Energy for Sustainable Development* 8 (1): 25–35.

26. Kristinsson, K., and R. Rao. 2008. "Interactive Learning or Technology Transfer as a Way to Catch-Up? Analysing the Wind Energy Industry in Denmark and India." *Industry and Innovation* 15 (3): 297–320.

27. Meyer, Niels I. 2004. "Renewable Energy Policy in Denmark." *Energy for Sustainable Development* 8 (1): 25–35.

28. Buen, J. 2006. "Danish and Norwegian Wind Industry: The Relationship Between Policy Instruments, Innovation and Diffusion." *Energy Policy* 34 (18): 3887–3897.

29. Meyer, Niels I. 1995. "Danish Wind Power Development." *Energy for Sustainable Development* 2 (1): 18–25.

30. Meyer, Niels I. 2004. "Renewable Energy Policy in Denmark." *Energy for Sustainable Development* 8 (1): 25–35.

31. Mallon, Karl (ed.). 2006. *Renewable Energy Policy and Politics: A Handbook for Decision Making*. Oxford: Earthscan.

32. Meyer, Niels I. 2004. "Renewable Energy Policy in Denmark." *Energy for Sustainable Development* 8 (1): 25–35.

33. Kristinsson, K., and R. Rao. 2008. "Interactive Learning or Technology Transfer as a Way to Catch-Up? Analysing the Wind Energy Industry in Denmark and India." *Industry and Innovation* 15 (3): 297–320; and Buen, J. 2006. "Danish and Norwegian Wind Industry: The Relationship Between Policy Instruments, Innovation and Diffusion." *Energy Policy* 34 (18): 3887–3897.

34. Morthorst, P. E. 1999. "Capacity Development and Profitability of Wind Turbines." *Energy Policy* 27 (13): 779–787.

35. Madsen, Birger T. 1988. "Windfarming in Denmark." *Journal of Wind Engineering and Industrial Aerodynamics* 27 (1–3): 347–358.

36. Ibid.

37. Meyer, N. I. 2004. "Development of Danish Wind Power Market." *Energy and Environment* 15 (4): 657–673.

38. Madsen, Birger T. 1988. "Windfarming in Denmark." *Journal of Wind Engineering and Industrial Aerodynamics* 27 (1–3): 352.

39. Mitchell, C. 1995. "Comparison of the Means and Cost of Subsidizing Wind Energy." *Proceedings of the Institution of Mechanical Engineers, Part A: Journal of Power and Energy* 209 (3): 185–188.

40. Carlman, I. 1988. "Wind Power in Denmark! Wind Power in Sweden?" *Journal of Wind Engineering and Industrial Aerodynamics* 27 (1–3): 337–345; and Madsen, Birger T. 1988. "Windfarming in Denmark." *Journal of Wind Engineering and Industrial Aerodynamics* 27 (1–3): 347–358.

41. Frandsen, Sten, and Per D. Andensen. 1996. "Wind Farm Progress in Denmark." *Renewable Energy* 9 (1–4): 848–852.

42. This is because the FIT was tied to the retail price of electricity (85% of retail prices), so when the declining cost of oil reduced the overall cost of electricity, the FIT also declined.

43. Wüstenhagen, R. 2003. "Sustainability and Competitiveness in the Renewable Energy Sector: The Case of Vestas Wind Systems." *Greener Management International* 44: 105–115.

44. Kamp, L. M. 2004. "Wind Turbine Development 1973–2000: A Critique of the Differences in Policies between the Netherlands and Denmark." *Wind Engineering* 28 (4): 341–354.

45. Kristinsson, K., and R. Rao. 2008. "Interactive Learning or Technology Transfer as a Way to Catch-Up? Analysing the Wind Energy Industry in Denmark and India." *Industry and Innovation* 15 (3): 297–320.

46. Christensen, P., and H. Lund. 1998. "Conflicting Views of Sustainability: The Case of Wind Power and Nature Conservation in Denmark." *European Environment* 8 (1): 1–6.

47. Buen, J. 2006. "Danish and Norwegian Wind Industry: The Relationship Between Policy Instruments, Innovation and Diffusion." *Energy Policy* 34 (18): 3887–3897.

48. Ibid.

49. Christensen, P., and H. Lund. 1998. "Conflicting Views of Sustainability: The Case of Wind Power and Nature Conservation in Denmark." *European Environment* 8 (1): 1–6.

50. Danielsen, O. 1995. "Large-Scale Wind Power in Denmark." *Land Use Policy* 12 (1): 60–62.

51. Ibid.

52. Christensen, P., and H. Lund. 1998. "Conflicting Views of Sustainability: The Case of Wind Power and Nature Conservation in Denmark." *European Environment* 8 (1): 1–6.

53. Buen, J. 2006. "Danish and Norwegian Wind Industry: The Relationship between Policy Instruments, Innovation and Diffusion." *Energy Policy* 34 (18): 3887–3897.

54. Meyer, Niels I. 2004. "Renewable Energy Policy in Denmark." *Energy for Sustainable Development* 8 (1): 25–35; and Breton, Simon-Philippe, and Geir Moe. 2009. "Status, Plans and Technologies for Offshore Wind Turbines in Europe and North America." *Renewable Energy* 34 (3): 646–654.

55. Buen, J. 2006. "Danish and Norwegian Wind Industry: The Relationship Between Policy Instruments, Innovation and Diffusion." *Energy Policy* 34 (18): 3887–3897.

56. Danielsen, O. 1995. "Large-Scale Wind Power in Denmark." *Land Use Policy* 12 (1): 60–62.

57. Lemming, J. 1994. "New and Important Initiatives Promoting the Wind Energy Development in Denmark." *Renewable Energy* 5 (1–4): 551–555.

58. Agnolucci, Paolo. 2007. "Wind Electricity in Denmark: A Survey of Policies, Their Effectiveness and Factors Motivating Their Introduction." *Renewable and Sustainable Energy Reviews* 11 (5): 951–963.

59. Meyer, Niels I. 2004. "Renewable Energy Policy in Denmark." *Energy for Sustainable Development* 8 (1): 25–35.

60. Buen, J. 2006. "Danish and Norwegian Wind Industry: The Relationship Between Policy Instruments, Innovation and Diffusion." *Energy Policy* 34 (18): 3887–3897.

61. Moran, Michael, Martin Rein, and Robert E. Goodin (eds.). 2006. *The Oxford Handbook of Public Policy*. Edited by R. E. Goodin, The Oxford Handbooks of Political Science. New York: Oxford University Press.

62. Christensen, P., and H. Lund. 1998. "Conflicting Views of Sustainability: The Case of Wind Power and Nature Conservation in Denmark." *European Environment* 8 (1): 1–6.

63. Buen, J. 2006. "Danish and Norwegian Wind Industry: The Relationship Between Policy Instruments, Innovation and Diffusion." *Energy Policy* 34 (18): 3887–3897.

64. Agnolucci, Paolo. 2007. "Wind Electricity in Denmark: A Survey of Policies, Their Effectiveness and Factors Motivating Their Introduction." *Renewable and Sustainable Energy Reviews* 11 (5): 951–963.

65. Meyer, Niels I., and Anne Louise Koefoed. 2003. "Danish Energy Reform: Policy Implications for Renewables." *Energy Policy* 31 (7): 597–607.

66. Meyer, Niels I. 2004. "Renewable Energy Policy in Denmark." *Energy for Sustainable Development* 8 (1): 25–35.

67. Buen, J. 2006. "Danish and Norwegian Wind Industry: The Relationship Between Policy Instruments, Innovation and Diffusion." *Energy Policy* 34 (18): 3887–3897.

68. Morthorst, P. E. 2000. "The Development of a Green Certificate Market." *Energy Policy* 28 (15): 1085–1094.

69. Agnolucci, Paolo. 2007. "Wind Electricity in Denmark: A Survey of Policies, Their Effectiveness and Factors Motivating Their Introduction." *Renewable and Sustainable Energy Reviews* 11 (5): 951–963.

70. Agnolucci, Paolo. 2007. "Wind Electricity in Denmark: A Survey of Policies, Their Effectiveness and Factors Motivating Their Introduction." *Renewable and Sustainable Energy Reviews* 11 (5): 951–963.

71. Meyer, N. I. 2004. "Development of Danish Wind Power Market." *Energy and Environment* 15 (4): 657–673.

72. Agnolucci, Paolo. 2007. "Wind Electricity in Denmark: A Survey of Policies, Their Effectiveness and Factors Motivating Their Introduction." *Renewable and Sustainable Energy Reviews* 11 (5): 951–963.

73. European Renewable Energy Council. 2009. "Renewable Energy Policy Review: Denmark." In *RES 2020: Monitoring and Evaluation of the RES Directives Implementation in EU27 and Policy Recommendations to 2020*, pp. 1–10. Brussels: European Renewable Energy Council. www.erec.org/fileadmin/erec_docs/Projcet_Documents/RES2020/DENMARK_RES_Policy_Review__09_Final.pdf.

74. Agnolucci, Paolo. 2007. "Wind Electricity in Denmark: A Survey of Policies, Their Effectiveness and Factors Motivating Their Introduction." *Renewable and Sustainable Energy Reviews* 11 (5): 951–963.

75. Hvelplund, Frede. 2006. "Renewable Energy and the Need for Local Energy Markets." *Energy* 31 (13): 2293–2302.

76. Rasmussen, F., and P. H. Madsen. 2004. "Current Direction of Danish Wind Energy Research—The Researchers Point of View." *Journal of Solar Energy Engineering, Transactions of the ASME* 126 (4): 1105–1109.

77. Meyer, Niels I. 2004. "Renewable Energy Policy in Denmark." *Energy for Sustainable Development* 8 (1): 25–35.

78. Breton, Simon-Philippe, and Geir Moe. 2009. "Status, Plans and Technologies for Offshore Wind Turbines in Europe and North America." *Renewable Energy* 34 (3): 646–654.

79. Buen, J. 2006. "Danish and Norwegian Wind Industry: The Relationship Between Policy Instruments, Innovation and Diffusion." *Energy Policy* 34 (18): 3887–3897.

80. Agnolucci, Paolo. 2007. "Wind Electricity in Denmark: A Survey of Policies, Their Effectiveness and Factors Motivating Their Introduction." *Renewable and Sustainable Energy Reviews* 11 (5): 951–963.

81. European Renewable Energy Council. 2009. "Renewable Energy Policy Review: Denmark." In RES 2020: *Monitoring and Evaluation of the RES Directives Implementation in EU27 and Policy Recommendations to 2020*, pp. 1–10. Brussels: European Renewable Energy Council. www.erec.org/fileadmin/erec_docs/Projcet_Documents/RES2020/DENMARK_RES_Policy_Review__09_Final.pdf.

82. Smit, T., M. Junginger, and R. Smits. 2007. "Technological Learning in Offshore Wind Energy: Different Roles of the Government." *Energy Policy* 35 (12): 6431–6444.

83. Meyer, N. I. 2004. "Development of Danish Wind Power Market." *Energy and Environment* 15 (4): 657–673.

84. Ibid.

85. Pettersson, Maria, Kristina Ek, Kristina Söderholm, and Patrik Söderholm. 2010. "Wind Power Planning and Permitting: Comparative Perspectives from the Nordic Countries." *Renewable and Sustainable Energy Reviews* 14 (9): 3116–3123.

86. Rasmussen, F., and P. H. Madsen. 2004. "Current Direction of Danish Wind Energy Research—The Researchers Point of View." *Journal of Solar Energy Engineering, Transactions of the ASME* 126 (4): 1105–1109.

87. Buen, J. 2006. "Danish and Norwegian Wind Industry: The Relationship Between Policy Instruments, Innovation and Diffusion." *Energy Policy* 34 (18): 3887–3897.

88. Meyer, Niels I. 2004. "Renewable Energy Policy in Denmark." *Energy for Sustainable Development* 8 (1): 25–35.

89. European Renewable Energy Council. 2009. "Renewable Energy Policy Review: Denmark." In *RES 2020: Monitoring and Evaluation of the RES Directives Implementation in EU27 and Policy Recommendations to 2020*, pp. 1–10. Brussels: European Renewable Energy Council. www.erec.org/fileadmin/erec_docs/Projcet_Documents/RES2020/DENMARK_RES_Policy_Review__09_Final.pdf.

90. Mignard, D., G. P. Harrison, and C. L. Pritchard. 2007. "Contribution of Wind Power and CHP to Exports from Western Denmark During 2000–2004." *Renewable Energy* 32 (15): 2516–2528.

91. Agnolucci, Paolo. 2007. "Wind Electricity in Denmark: A Survey of Policies, Their Effectiveness and Factors Motivating Their Introduction." *Renewable and Sustainable Energy Reviews* 11 (5): 951–963.

92. European Renewable Energy Council. 2009. "Renewable Energy Policy Review: Denmark." In *RES 2020: Monitoring and Evaluation of the RES Directives Implementation in EU27 and Policy Recommendations to 2020*, pp. 1–10. Brussels: European Renewable Energy Council. www.erec.org/fileadmin/erec_docs/Projcet_Documents/RES2020/DENMARK_RES_Policy_Review__09_Final.pdf.

93. Ladenburg, Jacob. 2008. "Attitudes Towards On-Land and Offshore Wind Power Development in Denmark; Choice of Development Strategy." *Renewable Energy* 33 (1): 111–118.

94. Ekman, Claus Krog, and Søren Højgaard Jensen. 2010. "Prospects for Large-Scale Electricity Storage in Denmark." *Energy Conversion and Management* 51 (6): 1140–1147.

95. European Renewable Energy Council. 2009. "Renewable Energy Policy Review: Denmark." In *RES 2020: Monitoring and Evaluation of the RES Directives Implementation in EU27 and Policy Recommendations to 2020*, pp. 1–10. Brussels: European Renewable Energy Council. www.erec.org/fileadmin/erec_docs/Projcet_Documents/RES2020/DENMARK_RES_Policy_Review__09_Final.pdf.

96. European Renewable Energy Council. 2009. "Renewable Energy Policy Review: Denmark." In *RES 2020: Monitoring and Evaluation of the RES Directives Implementation in EU27 and Policy Recommendations to 2020*, pp. 1–10. Brussels: European Renewable Energy Council. www.erec.org/fileadmin/erec_docs/Projcet_Documents/RES2020/DENMARK_RES_Policy_Review__09_Final.pdf.

97. Meyer, N. I. 2004. "Development of Danish Wind Power Market." *Energy and Environment* 15 (4): 663.

98. Meyer, Niels I. 2004. "Renewable Energy Policy in Denmark." *Energy for Sustainable Development* 8 (1): 25–35.

99. Toke, David, Sylvia Breukers, and Maarten Wolsink. 2008. "Wind Power Deployment Outcomes: How Can We Account for the Differences?" *Renewable and Sustainable Energy Reviews* 12 (4): 1129–1147.

100. Meyer, Niels I. 1995. "Danish Wind Power Development." *Energy for Sustainable Development* 2 (1): 18–25.

101. Meyer, Niels I., and Anne Louise Koefoed. 2003. "Danish Energy Reform: Policy Implications for Renewables." *Energy Policy* 31 (7): 597–607.

102. Munksgaard, Jesper, and Anders Larsen. 1998. "Socio-Economic Assessment of Wind Power—Lessons from Denmark." *Energy Policy* 26 (2): 85–93.

103. Meyer, Niels I. 2004. "Renewable Energy Policy in Denmark." *Energy for Sustainable Development* 8 (1): 25–35.

104. Pettersson, Maria, Kristina Ek, Kristina Söderholm, and Patrik Söderholm. 2010. "Wind Power Planning and Permitting: Comparative Perspectives from the Nordic Countries." *Renewable and Sustainable Energy Reviews* 14 (9): 3116–3123.

105. Gipe, Paul. 1991. "Wind Energy Comes of Age: California and Denmark." *Energy Policy* 19 (8): 756–767.

106. Meyer, N. I. 2004. "Development of Danish Wind Power Market." *Energy and Environment* 15 (4): 657–673.

107. Kamp, L. M. 2004. "Wind Turbine Development 1973–2000: A Critique of the Differences in Policies between the Netherlands and Denmark." *Wind Engineering* 28 (4): 341–354.

108. Kamp, L. M. 2004. "Wind Turbine Development 1973–2000: A Critique of the Differences in Policies between the Netherlands and Denmark." *Wind Engineering* 28 (4): 341–354.

109. Breton, Simon-Philippe, and Geir Moe. 2009. "Status, Plans and Technologies for Offshore Wind Turbines in Europe and North America." *Renewable Energy* 34 (3): 646–654.

110. Hammons, T. J. 2008. "Integrating Renewable Energy Sources into European Grids." *International Journal of Electrical Power & Energy Systems* 30 (8): 462–475.

111. Carlman, I. 1988. "Wind Power in Denmark! Wind Power in Sweden?" *Journal of Wind Engineering and Industrial Aerodynamics* 27 (1–3): 337–345.

112. Buen, J. 2006. "Danish and Norwegian Wind Industry: The Relationship Between Policy Instruments, Innovation and Diffusion." *Energy Policy* 34 (18): 3887–3897.

113. Kristinsson, K., and R. Rao. 2008. "Interactive Learning or Technology Transfer as a Way to Catch-Up? Analysing the Wind Energy Industry in Denmark and India." *Industry and Innovation* 15 (3): 297–320.

114. Meyer, Niels I. 2004. "Renewable Energy Policy in Denmark." *Energy for Sustainable Development* 8 (1): 25–35.

115. Carlman, I. 1988. "Wind Power in Denmark! Wind Power in Sweden?" *Journal of Wind Engineering and Industrial Aerodynamics* 27 (1–3): 337–345.

116. Rasmussen, F., and P. H. Madsen. 2004. "Current Direction of Danish Wind Energy Research—The Researchers Point of View." *Journal of Solar Energy Engineering, Transactions of the ASME* 126 (4): 1105–1109.

117. Karnoe, Peter. 1995. "Competence as Process and the Social Embeddedness of Competence Building. *Academy of Management Journal* 38 (1): 427–431.

118. Kamp, L. M. 2004. "Wind Turbine Development 1973–2000: A Critique of the Differences in Policies Between the Netherlands and Denmark." *Wind Engineering* 28 (4): 341–354.

119. Smit, T., M. Junginger, and R. Smits. 2007. "Technological Learning in Offshore Wind Energy: Different Roles of the Government." *Energy Policy* 35 (12): 6431–6444.

120. Meyer, Niels I. 2004. "Renewable Energy Policy in Denmark." *Energy for Sustainable Development* 8 (1): 25–35.

121. Lund, H., and B. V. Mathiesen. 2009. "Energy System Analysis of 100% Renewable Energy Systems—The Case of Denmark in Years 2030 and 2050." *Energy* 34 (5): 524–531.

122. McLaren Loring, J. 2007. "Wind Energy Planning in England, Wales and Denmark: Factors Influencing Project Success." *Energy Policy* 35 (4): 2648–2660.

123. Buen, J. 2006. "Danish and Norwegian Wind Industry: The Relationship Between Policy Instruments, Innovation and Diffusion." *Energy Policy* 34 (18): 3887–3897.

124. Danish Ministry of Social Affairs and Integration. 2011. *Social Policy in Denmark.* Copenhagen: The Danish Ministry of Social Affairs and Integration.

125. Rasmussen, F., and P. H. Madsen. 2004. "Current Direction of Danish Wind Energy Research—The Researchers Point of View." *Journal of Solar Energy Engineering, Transactions of the ASME* 126 (4): 1105–1109.

126. Pettersson, Maria, Kristina Ek, Kristina Söderholm, and Patrik Söderholm. 2010. "Wind Power Planning and Permitting: Comparative Perspectives from the Nordic Countries." *Renewable and Sustainable Energy Reviews* 14 (9): 3116–3123.

127. Lemming, J. 1994. "New and Important Initiatives Promoting the Wind Energy Development in Denmark." *Renewable Energy* 5 (1–4): 551–555.

128. Rasmussen, F., and P. H. Madsen. 2004. "Current Direction of Danish Wind Energy Research—The Researchers Point of View." *Journal of Solar Energy Engineering, Transactions of the ASME* 126 (4): 1105–1109.

129. Meyer, N. I. 2004. "Development of Danish Wind Power Market." *Energy and Environment* 15 (4): 657–673.

130. Buen, J. 2006. "Danish and Norwegian Wind Industry: The Relationship Between Policy Instruments, Innovation and Diffusion." *Energy Policy* 34 (18): 3887–3897.

131. Pettersson, Maria, Kristina Ek, Kristina Söderholm, and Patrik Söderholm. 2010. "Wind Power Planning and Permitting: Comparative Perspectives from the Nordic Countries." *Renewable and Sustainable Energy Reviews* 14 (9): 3116–3123.

132. Meyer, N. I. 2004. "Development of Danish Wind Power Market." *Energy and Environment* 15 (4): 657–673.

133. Nielsen, Skjold R. 1994. "Wind Turbines—Localisation Strategy in Denmark." *Renewable Energy* 5 (1–4): 712–717.

134. Möller, Bernd. 2010. "Spatial Analyses of Emerging and Fading Wind Energy Landscapes in Denmark." *Land Use Policy* 27 (2): 233–241.

135. Buen, J. 2006. "Danish and Norwegian Wind Industry: The Relationship Between Policy Instruments, Innovation and Diffusion." *Energy Policy* 34 (18): 3887–3897.

136. Meyer, Niels I. 2004. "Renewable Energy Policy in Denmark." *Energy for Sustainable Development* 8 (1): 25–35.

137. Ibenholt, Karin. 2002. "Explaining Learning Curves for Wind Power." *Energy Policy* 30 (13): 1181–1189.

138. Nielsen, Skjold R. 1994. "Wind Turbines—Localisation Strategy in Denmark." *Renewable Energy* 5 (1–4): 712–717.

139. Ibid.

140. Danish Energy Agency. 2012. *Accelerating Green Energy Towards 2020.* Copenhagen: Danish Energy Agency.

141. Ibid.

142. Ekman, Claus Krog, and Søren Højgaard Jensen. 2010. "Prospects for Large-Scale Electricity Storage in Denmark." *Energy Conversion and Management* 51 (6): 1140–1147.

143. Ibid.

144. Lund, H., and B. V. Mathiesen. 2009. "Energy System Analysis of 100% Renewable Energy Systems—The Case of Denmark in Years 2030 and 2050." *Energy* 34 (5): 524–531.

145. Ladenburg, Jacob. 2008. "Attitudes Towards On-Land and Offshore Wind Power Development in Denmark; Choice of Development Strategy." *Renewable Energy* 33 (1): 111–118.

146. Markard, Jochen, and Regula Petersen. 2009. "The Offshore Trend: Structural Changes in the Wind Power Sector." *Energy Policy* 37 (9): 3545–3556.

147. Möller, B. 2006. "Changing Wind-Power Landscapes: Regional Assessment of Visual Impact on Land Use and Population in Northern Jutland, Denmark." *Applied Energy* 83 (5): 477–494.

148. Madsen, Birger T. 1988. "Windfarming in Denmark." *Journal of Wind Engineering and Industrial Aerodynamics* 27 (1–3): 348.

149. Ekman, Claus Krog, and Søren Højgaard Jensen. 2010. "Prospects for Large-Scale Electricity Storage in Denmark." *Energy Conversion and Management* 51 (6): 1140–1147.

150. Toke, David, Sylvia Breukers, and Maarten Wolsink. 2008. "Wind Power Deployment Outcomes: How Can We Account for the Differences?" *Renewable and Sustainable Energy Reviews* 12 (4): 1129–1147.

CHAPTER 5

Wind Power in Germany

5.1 INTRODUCTION

The question is not whether we are able to change, but whether we are changing fast enough.
—German Chancellor Angela Merkel[1]

In the previous chapter, the malleability of Danish energy policy was highlighted as a key factor behind the successful diffusion of wind power in Denmark. This chapter examines wind power diffusion in Germany, and in the process highlights a different, though equally successful policy ideology. Compared to policy of its Nordic neighbor, wind power development policy in Germany has been far more structured and invariable. In fact, the success of Germany's wind power development strategy often serves as an exemplar for proponents of consistent feed-in tariff regimes, which is considered by some to be the most effective strategy for driving wind power development.[2]

As this chapter will demonstrate, fostering wind power development in Germany is, like in other nations, a complex challenge involving dynamic interactions between government and nongovernment actors. As German wind power capacity expanded, there has been social dissent and utility opposition. Nevertheless, the German government has remained committed to aggressive wind power diffusion policies and has responded to emergent challenges in a remarkably unified manner wherein state, regional, and local government actors have formed integrated problem-solving networks.[3]

This chapter also highlights the seamless web of nation-specific STEP factors influencing wind power development that is apparent in Germany. As one pair of researchers observed, wind power development in Germany has been marked by "close interplay between the actors within the political system, technical and economic development, as well as social factors."[4]

As has been the case in most industrialized nations, forces in support of wind power development began to amass during the two energy crises of the 1970s. As the government began to evaluate its alternative energy technology options, nuclear power and wind power emerged as the two most viable utility-scale options.

In the 1970s, nuclear power in Germany enjoyed a modicum of developmental success. The nation's first commercial nuclear power plant commenced operation in 1969. By 2010, nuclear power contributed approximately 22% to Germany's electricity supply. However, nuclear power development has been contentious. Although there has been industrial support, there has also been strident public opposition, especially since Chernobyl. In the face of growing opposition, a plan to facilitate the phase-out of nuclear power has existed since the days of Chancellor Gerhard Schroeder (1998–2005). However, the Fukushima nuclear disaster in March 2011 catalyzed a drive to expedite plant closures. After the Fukushima disaster, German Chancellor Angela Merkel announced an intention to immediately close eight older reactors and phase out the remaining nine reactors by 2022. As one group of researchers concluded, "the need to get out of nuclear power seems to be overriding all other concerns."[5]

In 2012, the German government announced a bold plan to produce 80% of electricity through renewable sources by 2050. Central to this strategy is significant expansion of wind power capacity, particularly offshore. Yet as this chapter will document, wind power capacity expansion faces challenges.

5.2 AN OVERVIEW OF ELECTRICITY GENERATION IN GERMANY

Amidst the global imperative to reduce GHG emissions, Germany has established itself as a nation with a comparatively laudable track record of progressive decarbonization of its energy systems. For starters, energy efficiency and energy conservation programs in Germany have been relatively successful in reining in energy demand. As Table 5.1 illustrates (see next page), between 2001 and 2011, total primary energy consumption in Germany actually declined by 10%. By comparison, global total primary energy consumption *increased* by 24% during the same period.

Data in Figure 5.1 shows that Germany's energy related CO_2 emissions have also declined substantially since 1990 (the base year for Kyoto Protocol commitment targets). Since 1990, energy-related CO_2 emissions in Germany have declined by 21%, whereas globally, CO_2 emissions from energy actually *increased* by 38% over the same time period.

Table 5.1 TOTAL PRIMARY ENERGY CONSUMPTION IN GERMANY 2001–2011

Mt oil equivalent	2001	2002	2003	2004	2005	2006	2007	2008	2009	2010	2011	2001-11	2001-11
Germany	338.8	334.0	337.1	337.3	333.2	339.5	324.4	326.7	307.5	322.4	306.4	–32.4	–10%
World	9434.0	9613.9	9950.2	10449.6	10754.5	11048.4	11347.6	11492.8	11391.3	11977.8	12274.6	2840.6	24%

Source: BP. 2011. *Statistical Review of World Energy 2011*. London: British Petroleum (BP).

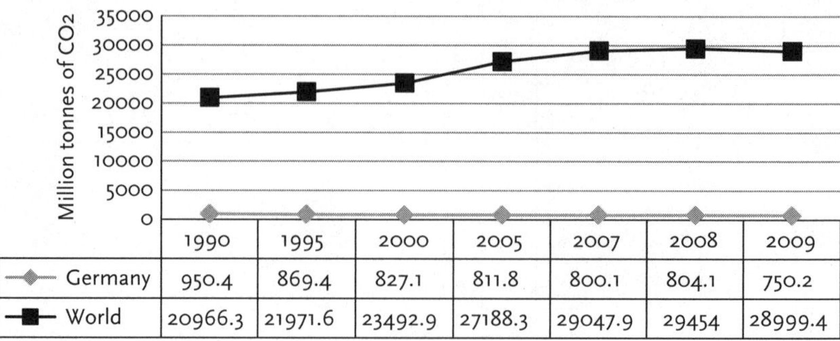

	1990	1995	2000	2005	2007	2008	2009
Germany	950.4	869.4	827.1	811.8	800.1	804.1	750.2
World	20966.3	21971.6	23492.9	27188.3	29047.9	29454	28999.4

Figure 5.1. German Progress in Reducing Energy-Related CO_2 Emissions
Source: IEA. 2011. *CO_2 emissions from fuel combustion: Highlights.* Paris: International Energy Agency.

Unfortunately, Germany's seemingly stellar track record is tempered somewhat when placed into context. The comparison in Figure 5.1 is slightly disingenuous in that the unification of East and West Germany catalyzed a technological transition in East Germany, which resulted in major energy efficiency improvements. The comparison in Figure 5.1 is also misleading in that the global CO_2 emission growth trend is heavily skewed by amplified emissions in developing nations such as China and India, which were not subject to the same CO_2 emission reduction obligations that Germany has been subject to since ratifying the Kyoto Protocol. Nevertheless, there has been progress. The data enable one to conclude that over the past decade, demand for energy has remained static in Germany while the energy infrastructure has improved technologically, reducing CO_2 emission intensity. This is not an insignificant achievement given Germany's historical reliance on coal as a major source of energy.

For centuries, coal has been an important resource in Germany. Evidence of coal mining dates back as early as the fourteenth century in the Ruhr area. By the eighteenth century, the use of coal as a preferred source of heating was advocated by the German government as a way to avoid deforestation. Thanks to abundant coal reserves, with the advent of modern electricity transmission networks, coal-fired power came to dominate the German electricity supply. As Figure 5.2 illustrates, despite efforts on the part of the German government to wean itself from thermal electricity technologies, as late as 2000, coal-fired (black and brown) power still accounted for 52% of the national electricity mix. Thanks in large part to expansion of renewable energy capacity, by 2010 contribution from coal-fired power had fallen to 41% of the national electricity.

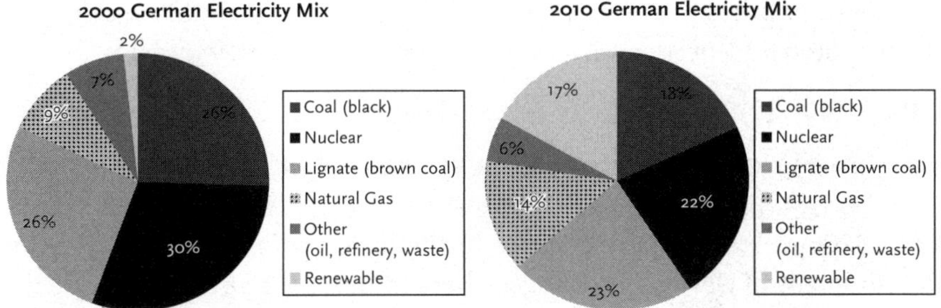

Figure 5.2. Germany's Electricity Mix
Source: Federal Environmental Agency, Germany (2012).

Greater inroads might have been possible in terms of reducing the role of coal-fired power; however, government plans to phase out nuclear power—which have been in place since the late 1990s—have necessitated continued short-term reliance on thermal coal technologies, in order to provide sufficient base-load electricity, while the nation transitions to renewable energy.

It is clear that the current government views renewable energy technology as the future core of electricity generation in Germany. As mentioned, it has declared an intention to provide at least 80% of its electricity through renewable sources by 2050. As of January 2011, around 17% of electricity, 8% of heat, and 6% of fuel utilized in Germany came from renewable sources. Yet despite modest progress to date, the renewable energy sector is an increasingly prominent employer, employing more than 350,000 people in Germany (up from 30,000 people in the 1998), with over 100,000 employed in the wind industry.[6]

In the electricity sector, wind power is being touted as the central enabling technology to meet the 2050 target. As Figure 5.3 indicates, wind power

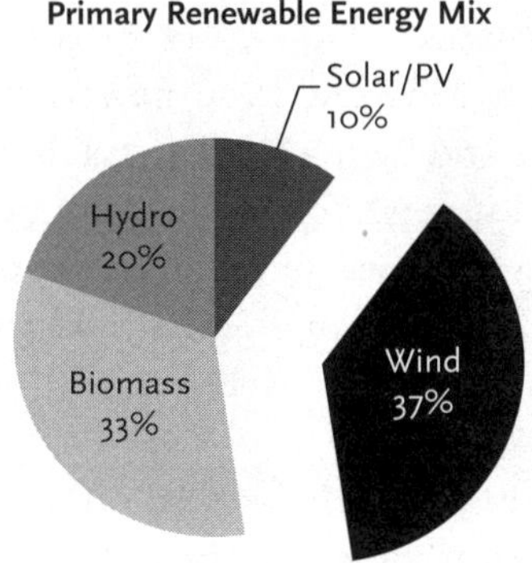

Figure 5.3. Germany's 2012 Primary Renewable Energy Mix
Source: Federal Environmental Agency, Germany (2012).

already constitutes 37% of Germany's total primary renewable energy production (and 50% of renewable contributions to the electricity grid).

5.3 HISTORY OF WIND POWER DEVELOPMENT IN GERMANY

The 1970s was a watershed decade in the history of modern wind power development in Germany. Like elsewhere in Europe, a social wave of environmental consciousness which emerged in the late 1960s began to gain momentum in the early 1970s, thanks in part to the influence of the "Limits to Growth" study and enhanced awareness stemming from the 1972 UN Conference on the Human Environment in Stockholm. By the time the first global oil crisis arose in 1973, there was already a high degree of public antipathy toward nuclear power. Therefore, plans put in place by the German government (with little public consultation) to construct a number of nuclear power plants in order to wean the nation from a dependence on oil imports were met with intense public protest.[7]

Opposition toward nuclear power in Germany in the 1970s played an important role in the rise of wind power because the antinuclear protests sired a powerful Green political party that would come to bear significant influence on national energy policy.[8] Although the first wind farm would not be constructed in Germany until 1987, the environmental ideology that emerged in the 1970s framed a path for wind power development.

As the 1980s dawned, although there was still a measure of political support for nuclear power by a faction of industrial-minded policymakers, the government perceived a clear need to diversity its policy risk by investigating alternative technologies. This precipitated an investment in wind power R&D, the bulk of which went into the development of a huge test installation called Growian—a 3MW turbine with a hub diameter of over 100 meters. The design was commissioned in 1980, and the turbine became operable in 1983. It operated between 1983 and 1986; however, the project was fraught with technical problems and by the time it was decommissioned, Growian had spent more time shut down for repairs than in operation. The project was a huge setback for wind power development in Germany because it was such a technical failure. In fact, wind power in Germany would have likely been much slower in materializing were it not for the Chernobyl nuclear disaster in April, 1986.

The nuclear disaster in Chernobyl kicked off a technological rethink in German energy circles, during which time German energy policymakers began to scan the technological horizon for other feasible energy options.[9] As described in the previous chapter, in neighboring Denmark a vibrant wind power manufacturing industry was taking root in the 1980s. Synergies in Denmark between clean energy and industrial development piqued the interest of German policymakers; subsequently, in June 1989, with failure of the Growian project still fresh in their minds, the Federal Ministry of Research returned to the drawing board, emerging with a diversified research program called the 100 MW Wind Programme.

The intent of the program was to support the design and evaluation of promising new wind power systems. All turbines covered under the program would be monitored for 10 years in order to assess technical performance. The program provided investment grants of €102 per installed kilowatt of generating capacity, up to a maximum of €46,000. Moreover, wind power operators were granted a subsidy of €0.041 per generated kilowatt hour, which would be added to the negotiated purchase price for power sold into the public electricity grid. The program attracted strong competition among wind power plant producers; consequently, in 1991 the program was expanded to support 250 MW worth of test projects. This R&D program was instrumental for reducing the cost profile of wind power, allowing the government to reduce the subsidy to €0.031 per kilowatt hour in 1991.[10]

The period between 1991 and 1995 has been dubbed the first breakthrough period for wind power.[11] The catalyst behind this breakthrough was announcement of Germany's first feed-in tariff (StrEG1991), which required Germany's grid operators to purchase wind power for 90% of the average retail price of electricity. In order to allay financial concerns of the utilities, it also established a hardship clause which allowed utilities to claim back extra costs from upstream power generators if total renewable energy contributions exceeded 5% of total system supply. Thanks to this new policy, Germany's installed wind power capacity expanded tenfold over this five-year period and utility opposition was negligible (see next page, Figure 5.4).[12]

So much wind power coming online in such a short period of time gave rise to two challenges. The first challenge was that public opposition began to escalate. Concerns over the risk of bird collisions, turbine noise, and shadow flicker escalated.[13] In response, local land-use planning authorities tightened wind farm approval standards, resulting in a backlog in siting permissions and injecting a higher degree of risk into the wind power development process. The second challenge was that wind power development in some regions was so extensive that power utilities began to express concern over grid instability caused by the stochastic nature of wind power flows. These concerns prompted the Association of German Electric Power Utilities to file a lawsuit, charging that the feed-in tariff act contravened European state aid regulations.[14] This legal challenge amplified the risk associated with wind power projects.

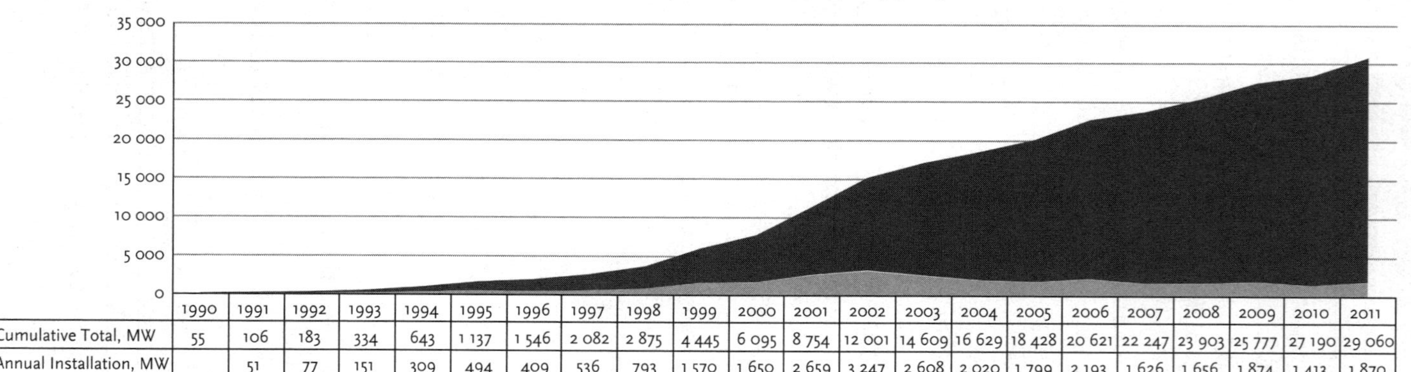

	1990	1991	1992	1993	1994	1995	1996	1997	1998	1999	2000	2001	2002	2003	2004	2005	2006	2007	2008	2009	2010	2011
■ Cumulative Total, MW	55	106	183	334	643	1 137	1 546	2 082	2 875	4 445	6 095	8 754	12 001	14 609	16 629	18 428	20 621	22 247	23 903	25 777	27 190	29 060
■ Annual Installation, MW		51	77	151	309	494	409	536	793	1 570	1 650	2 659	3 247	2 608	2 020	1 799	2 193	1 626	1 656	1 874	1 413	1 870

Figure 5.4. Evolution of Wind Power in Germany

Source: German Wind Energy Association (2012).

To compound the risks faced by wind power developers, the average retail cost of electricity declined due to a government phase-out of a coal levy and enhanced economic efficiency, which was a byproduct of the liberalization of Germany's power markets (to adhere to EU directive (96/92/EC)). This had an adverse impact on wind power development, because the 1991 feed-in act had linked wind power generation revenue to the average retail cost of electricity (the 90% clause). Consequently, although wind power capacity continued to expand in 1996 and 1997, there was sufficient market risk to slightly rein in this emerging bull market.

Two developments unfolded in 1998 to escalate the pace of wind power diffusion. In that year, the European Court of Justice ruled that the German feed-in tariff was in compliance with European state aid regulations and dismissed the utility-led legal challenge that had unsettled development prospects. In the same year, the federal building act was amended in order to speed up the wind power siting process.[15] These developments fueled a second wind boom, which saw installed capacity grow from 2082 MW at the end of 1997 to 4445 MW by the end of 1999.

This is not to say that wind power development in the late 1990s was devoid of conflict. There were still pockets of community opposition, particularly in regions with comparatively high wind power capacity. Moreover, there was still a degree of resistance from utilities. This was a time of market liberalization. Therefore, the prime focus of utilities was on consolidating market positions, not challenging new technologies.[16]

The success of wind power development in Germany emulated the Danish model in nurturing a blossoming industry featuring firms such as Enercon, Micon, Tacke, and AN Windenergie, which in aggregate employed 15,600 people by 1998.[17] Consequently, although there were concerns voiced that the feed-in tariffs were excessive, there was still broad political support for maintaining robust subsidies. This staunch commitment came to be known as the tailwind movement.

There was one other political development in the late 1990s that influenced wind power fortunes. In 1998, after 16 years of rule by a conservative-liberal coalition led by Helmut Kohl, a red/green coalition of the Social Democratic Party (SPD) and the Green party, under the chancellery of Gerhard Schroeder, assumed power. This new administration was even more supportive of wind power development than the preceding administration, announcing plans to increase renewable energy contributions to the electricity supply to 12.5% by 2010 and 50% by 2050. In April 1999, it introduced an electricity tax to provide financial support for its renewable energy development policy.[18] By the fall of 1999, a new and improved feed-in tariff had also been tabled for discussion. As opposed to the feed-in tariff of 1991, which was passed unanimously in Parliament, the new proposal faced stronger political opposition—particularly by the utilities.

Despite increased resistance, the Renewable Energy Sources Act (EEG2000) was finally passed on April 1, 2000. It boosted prospects for renewable technologies in that it provided greater cost certainty and security for developers and established a 2010 target of 10% renewable energy contributions to the electricity grid. Instead of tying feed-in rates to retail electricity costs, the new act featured fixed tariffs for renewable energy.[19] In regard to wind power, all wind power systems would receive an initial amount of DEM0.178 (approx. €0.091) per kWh for five years. Hence, systems in areas of high quality wind (exceeding a preestablished reference yield of 150%) would receive only €0.062 per kWh. For projects in areas of lower wind quality, the higher feed-in tariff rate would be prolonged by two months for every 0.75% that the yield fell under 150% of the preestablished reference yield. The effect of this policy was that it encouraged the development of inland capacity on sites that exhibited comparatively inferior wind quality.[20] Offshore projects would also receive €0.091 per kWh, but for nine years instead of five. The act also introduced a national equalization scheme in order to ensure that utilities in regions with high levels of installed renewable energy capacity would not be overburdened compared to utilities in other regions. Moreover, for the first time utilities were given permission to apply for the feed-in tariff for utility-owned projects. One other key characteristic of the new feed-in tariff was that the subsidy declined by approximately 1.5% each year for projects coming online after 2002. This regressive subsidy was designed to encourage progressive technological innovation in renewable energy technologies. One final feature of the EEG2000 was that wind power project operators were made responsible for costs associated with connecting installations to grid connection points; however, costs associated with upgrading the grids to connect new installations were to be borne by the grid operator.[21] All of these new features enhanced cost certainty and wind power development blossomed in 2001 and 2002, with 2659 MW and 3247 MW being added, respectively. By the end of 2002, installed wind power capacity exceeded 12,000 MW.

In 2001, two other developments took place that had a substantial influence on the pace of wind power diffusion. First, the federal government and the electricity industry reached a nuclear consensus, announcing an intention to phase-out nuclear energy over the next two decades.[22] The success of Germany's wind power industry arguably provided energy policymakers with a degree of confidence that utility-scale, alternative low-carbon technological solutions were feasible and that nuclear power was no longer a necessary evil. Second, in response to growing public opposition to onshore wind power developments, the federal government announced a national offshore strategy (BMU2001) that, when finally ramped up, would produce sizable results over the ensuing two decades. The aim was to foster offshore wind power development of 500 MW by 2006, 3,000 MW by 2010 and up to 25,000 MW by 2025/2030. In response to rumblings of social discontent over offshore turbines undermining the aesthetics of Germany's seaside communities, the government also announced that it would limit its offshore development to the exclusive economic zone (EEZ), meaning that sites would be situated at least 12 nautical miles from the coast.[23] Developing projects far offshore attenuated community opposition but it presented costly technical challenges that to this day are not fully resolved.

Despite the announcement of an intention to limit offshore wind power development to the EEZ, there were still ecological concerns that offshore projects might adversely impact marine life. Consequently, the government initiated a research program to investigate ecological impacts and identify mitigation measures.[24]

By the end of 2002, the future of wind power in Germany was rosy. Germany's 12,000 MW of installed wind power capacity represented 39% of global wind power capacity; annually, one in every two turbines installed around the world were being installed in Germany. Moreover, 53,200 people were working in the wind industry, an increase of almost 40,000 people in four short years. Nuclear power was on its way out and the new offshore strategy promised to triple existing wind power generation capacity within two decades.[25] In a public opinion poll undertaken in 2003, wind power was the most preferred alternative energy technology and 61% of the respondents affirmed support for the government's decision to phase-out nuclear energy (compared to 46% just three years previous).[26]

However, over the ensuing two years signs began to emerge that the further diffusion of wind power would not be without challenges. Steel, which is a critical input in the construction of wind turbine towers and many turbine components, almost doubled in price between 2002 and 2004. Copper, which is used for cabling, generators, and other electrical components, doubled in price between 2003 and 2004.[27] This unanticipated increase in factor prices eroded profit margins and forced many developers to reign in development activities. During the same period, pockets of concentrated wind power development began to engender community resistance due to the aesthetic invasiveness of wind turbine arrays.[28] Even repowering initiatives were beginning to engender community resistance, resulting in delays and rejections of permits issued for building new wind power plants that exceeded 100 m in height.

In government circles, it became apparent that the 3,247 MW of installed capacity added in 2002 represented somewhat of a developmental peak. In

2003, annual capacity expansion tailed off to 2,608 MW; in 2004, the pace of annual capacity expansion dipped further to 2020 MW; and in 2005, the pace of growth declined again, to 1,799 MW. Capacity was still increasing and in global terms, Germany's annual wind power expansion rate still led all nations; however, given Germany's ambitious development targets, it became apparent that the wind power diffusion program required new policies to refocus efforts to encourage development in new areas without further dampening momentum.

An amended Renewable Energy Sources Act (EEG2004) began to address these challenges. A new feed-in tariff was announced that paid €0.087 per kWh for the first five years to plants achieving 150% of a preestablished reference yield. In ensuing years, the rate would drop to €0.055 per kWh. The amended act also extended the policy of extending the initial 5-year €0.087 tariff by two months for every 0.75% that the yield of a given project fell under 150% of the preestablished reference yield, to continue to encourage development at less attractive sites. The act also established an offshore feed-in tariff of €0.0619 per kWh for the first 12 years of operation with bonus payments for early development (€0.0291 per kWh extra for projects commissioned by the end of 2010). This increased tariff was extended beyond 12 years for projects that were developed further than 12kms from shore or deeper than 20 m. The EEG2004 also reaffirmed the nationwide settlement system that rectified regional disparities through transfer payments between utilities.[29] In the same year, the government passed an amendment to the spatial planning act of 1965 in order to mandate special planning procedures for offshore wind turbines.[30] In sum, it was clear from these policies that the government was not going to let the wind power program lose its tailwind.

Politically, however, the discussions leading to the passage of the EEG2004 differed from past experiences in that universal political support for wind power exhibited signs of erosion. Support for the new set of wind power policies were driven by members of the governing coalition and the farmers association. Meanwhile, opposition, albeit lukewarm, emerged from factions within the conservative (CDU-CSU) and liberal (FDP) parties.[31]

In 2005, infrastructure challenges emerged. The German Energy Agency published the results of a study which concluded that 850 km of grid enhancement would be required if the government was to be successful in reaching the target of supplying 19.6% of all energy by renewable sources by 2020. Furthermore, it became apparent that the planned offshore developments would necessitate sizable infrastructure investments. In response, the government announced the creation of an Offshore Wind Energy Foundation comprised of representatives from the federal Ministry of

Environment, other state ministries, German manufacturers, wind energy interest groups, environmental associations, and large energy suppliers.[32] The goal of the foundation was to identify barriers to offshore development and smooth the development path.

Despite these emergent challenges, it was clear that wind power expansion was desirable for economic and political reasons. According to an analysis published by the German Ministry of Environment, wind power was responsible for energy cost savings of about €5 billion dollars in 2006.[33] Moreover, with nuclear power on the way out, progressive expansion of wind power was deemed essential to facilitate the ambitious goals of Germany's Climate Protection Program.[34] Consequently, in 2006, the federal government passed an infrastructure planning acceleration law, which obliged network operators to assume the cost of connecting offshore wind farms to the power grid. However, the law enabled network operators to pass on these connection costs to the consumers. Therefore, in one fell swoop, this law stemmed the uncertainty that was deterring investment in offshore wind power development and allayed lingering concerns on the part of network operators that they would once again be responsible for the financial burden of supporting government wind power development initiatives.[35]

The following year, the government advanced initiatives to shore up its plan for expanding offshore wind energy development. It announced a plan to commit €34.6 million to new wind energy R&D initiatives, with two-thirds of the allocation going to offshore wind energy research.[36] In the same year, it also signaled its resolve to bolster development by declaring an intention to increase renewable energy in the power supply to 25%-30% by 2020.[37]

In December of the following year, an EU Renewable Energy Directive (2009/28/EC) was published that estimated that renewable energy contributions to electricity demand would need to reach approximately 35% by 2020 in order to achieve the overall EU energy objective of generating 20% of primary energy with renewable energy by 2020.[38] Despite Germany's stellar performance in facilitating wind power development, the directive made it clear that Germany would still be challenged to live up to its obligations. In the subsequent National Action Plan that was designed to support this initiative, Germany announced new 2020 targets of 19.6% contribution from renewable energy to total primary energy consumption and 38.6% contribution of renewable energy to electricity consumption. If achieved, Germany would exceed its binding target of 18% contribution from renewable energy to total primary energy consumption, originally mandated by the directive.[39]

With this target in place, in January 2009 the government amended the Renewable Energy Sources Act (EEG 2009) in order to improve remuneration conditions and revive wind power development.[40] The tariff for onshore wind energy was increased to €0.0911 per kWh for the first five years (up from €0.087 per kWh), declining to €0.0497 per kWh in ensuing years. The tariff for offshore wind energy was increased to €0.13 per kWh for the first 12 years, falling to €0.035 per kWh for each year thereafter. For installations that were located in deep water, operators could apply for prolongation of the initial €0.13 per kWh subsidy. An additional €0.02 per kWh was also offered for offshore installations that were commissioned before December 31, 2005, in order to jump start development. Finally, to ensure a higher degree of responsiveness to the changing economics of energy generation, this new act stipulated that the tariffs would be revisited every three years instead of every four years, as initially established under the 1991 feed-in tariff act.

A number of technological upgrading incentives were also introduced. For example, onshore wind power installations could receive an additional €0.005 per kWh, if they were equipped with advanced grid integration technology. Furthermore, for turbines that were more than 10 years old, a special repowering bonus was announced for upgrades that doubled the rated capacity of existing systems. In order to incentivize progressive innovation, the tariff for new onshore installations would be reduced each year by 1%. For offshore installations, there was a period of grace prior to digression of the feed-in tariff—there would be no annual reduction in the feed-in tariff until 2015. Thereafter, the feed-in tariff of €0.13 per kWh for new installations would decline by 5% per year. Finally, the new renewable energy act also compelled grid operators to expand and optimize the grid in order to accommodate wind power connections. Failure to do so would leave the grid operator libel for damage claims by any renewable power producer that was willing but unable to feed electricity into the grid.

The government also introduced two new initiatives in 2009 intended to attenuate community opposition to onshore wind power and enhance local participation. The first initiative was a new tax law introduced on January 1, 2009 that ensured 70% of trade tax income coming from wind energy generation would go back to the local community where the energy was generated. The second initiative was the establishment of repowering consulting agencies that were tasked with the responsibility of liaising with districts, municipalities and local authorities in order to enhance support for the development and implementation of turbine upgrade projects.[41]

On August 12, 2009, turbines at Germany's first offshore wind farm started to feed electricity into the German power grid. This was a test field known as Alpha Ventus, situated 45 km north of Borkum Island. Federal investment in this research initiative was purported to be in the neighborhood of €50 million.[42] The raison d'être of the facility was to provide researchers with insights into optimizing offshore wind park performance.

Thanks to positive investor reception to these initiatives, 2009 proved to be a stellar year for wind power development, with 1,874 MW of new capacity added—reversing a three-year trend of declining new capacity. By the end of 2009, there were 21,164 wind turbines installed in Germany with a total installed generation capacity of 25,777 MW.

Unfortunately, unanticipated events in 2010 reaffirmed once again the types of disparate influences that can influence wind power development. The combination of a global economic recession and domestic investor insecurity arising from government decisions to prolong the lives of nuclear plants in Germany resulted in only 1,413 MW of new capacity being installed.[43] Nevertheless, by the end of 2010, Germany had crushed its 12.5% target under the EU Renewable Energy Directive, with 17.5% of primary energy coming from renewable sources. It was the only major economy in the European Union to exceed its target.[44]

As 2011 dawned, it was looking like it would be another year of progressive wind power capacity development, accompanied by the typical trials and tribulations brought about by ever-increasing contributions of wind power to the nations' electricity supply. But on March 11, 2011, an earthquake off the coast of Japan triggered a tsunami which rammed into a complex of aging nuclear reactors located in Fukushima, Japan. The ensuing nuclear disaster had an enormous impact on German nuclear power policy. Almost immediately after the disaster in Japan, Chancellor Angela Merkel, who herself is a physical chemist, announced plans to expedite a phase-out of all 17 of Germany's nuclear reactors and replace them with renewable energy.[45] As a first step, the government announced immediate closure of eight older nuclear power plants.[46] In response, orders for new wind power projects have increased and new capacity added in 2011 climbed to 1,870 MW.

Since then, a three-year revision to the Renewable Energy Sources Act that was passed on July 8, 2011 reflects the intent of the German government to accelerate renewable energy development, in general, and offshore wind energy specifically. Keeping in line with the EU Renewable Energy Directive for 2020, it sets a minimum goal of 35% renewable energy contribution to the electricity supply by 2020, rising to at least 50% by 2030, 65% by 2040, and 80% by 2050. In order to meet these targets, the offshore wind turbine tariff has been increased from €0.13 per kWh to €0.15 per kWh and an early starter bonus, for offshore wind projects that are commissioned by the end of 2015, has been increased to €0.04 per kWh. Meanwhile tariffs for onshore wind energy will remain the same.[47] In order to support offshore wind power initiatives, the government is planning a special €5.5 billion program to finance wind park expansion in the North and Baltic Seas that commenced in late 2011.[48]

In March 2012, the government intensified its commitment to meeting its ambitious renewable energy goals by announcing a program to invest approximately €200 billion in new renewable energy projects. This equates to approximately 8% of the nation's annual GDP.[49] It appears that China and the United States will receive a run for their money in regard to new wind power capacity development over the next couple of decades.

5.4 UNDERSTANDING THE GENERAL FORCES FOR CHANGE

5.4.1 Sociocultural Landscape

As previously detailed, both political and grassroots support for Germany's wind power program sprung from anti-nuclear power / environmental movements of the 1970s.[50] In the ensuing 40-year period, support for enhanced environmental governance in Germany has escalated to a stage where Germany's Green party now plays an influential role in German politics. To this day, Germany is considered to be a leading nation in environmental governance and the prevalence of a strong environmental ethic continues to serve as the basis of popular support for wind power development.

One other sociocultural characteristic that has aided the development of wind power in Germany is the proclivity on the part of Germans to invest in community initiatives, either individually or through cooperative investment. Wind power developers have successfully exploited this. According to wind power expert Paul Gipe, 51% of installed renewable energy capacity is owned by German citizens (40% individuals and 11% farmers), not corporations or utilities. In some regions such as Northern Friesland, citizen ownership of wind farms exceeds 90%. In total, there is about US$100 billion of private investment tied up in renewable energy projects with about 30% of this invested in wind power projects. As Stefan Gsanger, director of the World Wind Energy Association, points out, "if we want to reach 100% renewable energy supply, we have to ensure that local communities benefit from renewable energy development and support projects in their vicinity. Community and citizen-ownership models have a proven track record in achieving this objective."[51] Interestingly, some analysts have asserted that community ownership in renewable energy projects in the late 1980s and early 1990s fueled a virtuous cycle—community ownership enhanced public sensitivity toward climate change, which in turn catalyzed support for further renewable energy development.[52]

Despite broad public support for wind power, it cannot be said that all members of the public have a shared affinity for the technology. It has been reported that in some regions proliferation of wind power installations have engendered local resistance, particularly in communities with wind farms that generate energy for export to other German states or other nations.[53] As one team of researchers summarized it, "Germans may love their green energy, but they also have a growing proclivity toward not in my backyard (NIMBY) lawsuits and referenda."[54] Nevertheless, there still appears to be widespread support for wind energy. According to a 2011 survey commissioned by Germany's Renewable Energy Agency, two-thirds of Germans still support increasing wind power capacity. However, 71% of these respondents would prefer to see policy directed at encouraging more offshore wind power projects.[55]

The government has recognized that continued development onshore may erode public support for wind power; and as outlined in the previous section, it has responded by elevating offshore wind power development plans. Ominously, there is also evidence that many Germans hold deep convictions of the sea as a natural space and might not respond favorably to technological defilement of the aesthetic bliss associated with a day at the beach.[56] In short, there is evidence that civic support for wind power has limits.

5.4.2 Economic Landscape

The economic landscape influencing wind power development in Germany is really a story of clashing interests. There has been opposition from industries and large utilities. Industry opposition to higher contributions from wind power stem from the higher costs exhibited in the early days of wind power expansion. Larger utilities, which in Germany are mostly private owned, have exhibited a resistance to higher contributions of wind power due to the technical complications of managing stochastic power supplies and due to connection costs that utilities were responsible for under the renewable energy acts.[57] So generally, the stakeholders who have been opposed to wind power in Germany do not differ much from the industrial stakeholders that oppose wind power in other nations.

However, opposition to wind power from German utilities in the early days was mild and ineffective, compared to opposition encountered in most other nations. Industries in Germany were so hard-hit by the inflated oil prices of the 1970s that there was widespread industrial support for diversifying the energy mix even if it came at a slightly elevated cost.[58] For this reason, industries were far from threatened by the prospect of 5% or 10% of Germany's electricity supply coming from wind power—a domestic source which exhibited low variable costs. To the contrary, there was general support for limited diversification. Similarly, although there was always a degree of opposition to wind power stemming from technological concerns, government policies allowing additional costs to be passed through to consumers tended to dampen what would normally be higher levels of utility opposition. Additionally, many of Germany's smaller local utilities were community-owned, and as such they were much more receptive to environmentally favorable initiatives (such as renewable energy development) that were in the public interest.[59]

In striking contrast to the tepid opposition to wind power shown by energy-intensive industries and large utilities, there were two stakeholder groups that benefited financially from wind power development and were, therefore, ebullient advocates of wind power. First, there were a myriad of citizens that individually or in small limited partnerships were investing in wind power developments. Not only were the returns attractive for these investors, the investment was tax deductible. Therefore, investors were well-motivated to proactively campaign for sustained political support.[60] Second, German wind turbine manufacturers were represented by the Federal Wind Association, which has proven to be exacting and effective in political circles.[61]

In the early days, the blossoming wind power market bore no real drama. Proactive, proponents of wind power fueled public and political support, while opposition from major industries and large utilities remained muted. As outlined earlier, it wasn't until about 1997, when installed national wind power capacity reached 2000 MW, that the utilities banded together to publicly oppose further wind power development by challenging the government's renewable energy act as contradicting EU market liberalization regulations. By then, a major justification for opposition (higher costs) had disappeared. Between 1990 and 2000, competition in Germany's vibrant wind power market had stimulated technological progress;[62] consequently, the cost of wind power decreased by about 30%, significantly attenuating cost concerns that industries harbored.[63] With wind power prices reaching a level of competitive viability (thanks in part to the presence of green taxes), the only hope that utilities had for successful opposition was that public dissonance associated with aesthetic impairments to Germany's landscape would prompt a reversal of public support. This has yet to materialize, and so utilities have had to reluctantly accept the challenge of trying to incorporate ever increasing grid contributions from wind power.

Meanwhile, Germany's wind turbine manufacturers have been a success both domestically and overseas. Two-thirds of the total funds provided under the initial 100 MW and 250 MW projects went to German manufacturers.[64] As Germany sets its sights on offshore development, some of its larger domestic wind power turbine manufacturers are beginning to specialize in that area. Furthermore, the link between wind power and hydrogen production is becoming an important area of technical research in Germany. In short, Germany's wind turbine industry is seen as an increasingly important contributor to employment and economic development.

5.4.3 Technological Landscape

There are two factors that render Germany an ideal nation for wind power development. The first factor is that Germany's geographic location bordering the North and Baltic seas features wind conditions that enable comparatively favorable commercialization of wind power. In particular, Germany's northern coastal region is characterized by high quality wind conditions.[65] In 2000, a study commissioned by the Federal Ministry of the Environment concluded that commercially exploitable wind power potential was in the neighborhood of 250 TW hours of electricity per year. If fully realized, this would fulfill approximately 40% of German electricity consumption at the time.[66] Since 2000, wind power technology has improved significantly, increasing realizable potential.

The second favorable factor is that economic success has been predicated on innovations in industrial engineering technology that have arisen not just from German industrial powerhouses such as Siemens, but also from a host of small-scale manufacturers that specialize in high quality custom engineering. In fact, it has been said that small-scale manufacturers, who were ideologically committed to wind power, spearheaded the technological innovations that allowed German wind turbine manufacturers to excel both in domestic and overseas markets.[67]

These two technological support factors—a wealth of wind power potential and motivated engineers—melded seamlessly with popular support for wind power and Germany's cooperative investment culture to establish the foundations for success in wind power in Germany.

However, there is another often ignored factor that should not be overlooked when discussing technological conditions for supporting wind power. Throughout most of the twentieth century, the structure of Germany's power industry was predicated on the Energy Industry Act of 1935. This act created small monopolies in power generation, transmission, distribution and supply. However, as part of a EU directive to liberalize electricity markets in the mid-1990s, Germany's electricity sector was liberalized. Prior to market liberalization, it is estimated that there were about 1000 electric utilities, about 80 firms engaged in regional power distribution (and some generation) and more than 900 firms engaged in local distribution. In the ensuing shakeup, rather than the subsectors fracturing into further pieces, four major utilities emerged as dominant, claiming over 70% market share.[68] The scale of these utilities ensured that sufficient spare grid capacity existed within each utility's grid structure in order to accommodate contributions from wind power in a stable manner. Economies of scale also allowed utilities to absorb connection costs without having to pass on sizable rate hikes to end-consumers.

5.5. INFLUENCES ON GOVERNMENT POLICY

5.5.1 Sociocultural → Political

As outlined earlier, the emergence of Germany's environmental movement was largely predicated upon opposition to nuclear power development. Although anti-nuclear power activists represented a very small proportion of Germany's population, media coverage and public exposure to impassioned anti-nuclear power protests sowed the seeds of public skepticism regarding the wisdom of embracing nuclear power. Therefore, when the Chernobyl catastrophe occurred in 1986, it amplified active public opposition to nuclear power, putting Germany on the road to the phase-out of its nuclear power program.[69]

Germany's environmental lobby played an indirect role in elevating the prospects of wind power development by dampening the prospects of nuclear power development, over the past two decades; however, this same environmental lobby has had a direct contrapositive influence on wind power perception. Starting in the mid-1990s, the rapid rise of wind power installations engendered ecological concerns over bird mortality and fueled community concerns over noise pollution and the aesthetic impairment of rural landscapes. Initially, in order to address these concerns, municipal planning authorities tightened up siting standards and undertook more rigorous vetting of wind power project permit requests, creating a backlog in processing such permits.[70]

There is still public support for wind power development, though NIMBY (not-in-my-backyard) opposition has become progressively evident.[71] In fact, one researcher documented over 70 recent wind power protest campaigns in Germany that have been organized to derail specific projects.[72] As an example of the growing strength of opposition, in 2008, a public initiative condemning the overexploitation of Brandenburg for wind power generation garnered 27,000 signatures, representing 1.1% of the Brandenburg population. In this case, the municipal government responded in a manner that typifies the reactive nature of wind power governance by issuing a decree that wind turbines be located a minimum of 1000 m from residential areas.[73]

Although the government strategy to attenuate NIMBY opposition has generally been to reactively address emergent concerns on a project-by-project basis, there have also been some proactive initiatives adopted to temper what is becoming a progressively serious barrier to Germany's ambitious wind power diffusion goals. First and foremost of the initiatives is Germany's reempowerment scheme. Since 2003, the government has been pursuing a policy of encouraging the replacement of older turbines with turbines of higher capacity in order to expand capacity while limiting the need for new sites.[74] The government also designs schemes to encourage local investment in wind power projects in order to enhance community support.[75]

There is evidence that the NIMBY lessons learned by the government in regard to onshore wind power projects have not been forgotten when it comes to offshore wind farm planning. Although Germany's first offshore wind farm did not come online until 2009, as early as 2002 federal authorities had initiated a research program designed to investigate ecological impacts on the marine environment.[76] Moreover, in 2005, the Ministry of the Environment announced an intention to provide €50 million over a period of five years for R&D and environmental monitoring of a trial wind farm located 45 km off Borkum.[77] The intent of these initiatives is to understand the unique environmental challenges that might arise as the focus shifts from onshore to offshore wind power development.

Despite social challenges to select wind power projects, Germany's environmentally aware public still harbors a high degree of support for wind power development. In fact, opinion polls have demonstrated that 20% to 35% of electricity consumers have a positive willingness to pay for renewable energy of any type.[78]

5.5.2 Economic → Political

In Germany, wind turbine manufacturing activities meshed perfectly with Germany's competencies in engineering and machinery sectors, making the wind power sector an employer of considerable promise. This changes the basis upon which energy policy is decided because the employment potential of the wind power sector offsets some of the economic advantages of coal-fired power. In 1998 when wind power began its ascent, approximately 15,600 people were employed in the industry; by 2002, installed capacity had grown fourfold and the number of jobs had increased at a similar pace to 53,200.[79] According to a study in 2009, renewable energy was responsible for 157,000 jobs, with wind providing the bulk of employment. The same study quotes a representative from German wind power manufacturer Enercon as estimating that by 2030, renewable energy will be responsible for 710,000 jobs in Germany.[80]

The other economic element that significantly influenced wind power development policy in the early days of commercial development was the financial benefit that such projects provided to rural communities. Many of the early wind power projects were small-scale enterprises owned by individual farmers or farming cooperatives. Wind power projects helped rural communities to bolster income levels; and consequently, garnered high levels of local acceptance.[81] As larger wind farms came online and project investors broadened, community support waned. This taught the German government a valuable lesson about how to attenuate public opposition to wind power projects—encourage local involvement.

5.5.3 Technological → Political

Technological advances in wind power systems have muted the concerns that have historically underpinned both utility and civic opposition to the technology. In regard to utility opposition, the common criticism has been that heavy contributions by wind power systems destabilize the electricity grid due to the stochastic nature of wind power flows. In most nations, such criticism can still be heard from wind power opponents. However, in the states of Saxony-Anhalt, Brandenburg, Schleswig-Holstein, and Mecklenburg-Vorpommern, wind power satisfies more than 40% of total electricity consumption. It is difficult for utilities in other German states to put forth opposition to wind power out of concerns of destabilizing the grid when such precedents exist. In regard to civic opposition, concerns over aesthetic impairment, noise, bird mortality, impact on biodiversity, and shadow flicker have dominated the agenda.[82] However, improved site planning, noise-dampening features on modern turbines, and improved tower and blade designs have gone a long way to alleviating these concerns. It is for this reason that members of communities that host wind power farms tend to hold more positive impressions of such installations compared to members of communities that have no experience with wind farms.[83] Alleviating stakeholder concerns through technological advances has emboldened German energy policymakers to support enhanced levels of wind power development.

On the other hand, the cost of delivering wind power from supply centers (predominantly located in northern Germany and in remote coastal areas) to demand centers found in south and west Germany presents a financial challenge that the government is still trying to resolve. In particular, the government has recognized that connecting remote offshore wind parks represents a bottleneck that inhibits the pace and scale of development.[84] The government initially responded to this challenge by revising the Renewable Energy Sources Act of 2004 to permit higher feed-in tariffs and to extend the period of maximum reimbursement for offshore projects.[85] However, there is still considerable disagreement over how grid connection costs should be shared.

The government seems to have recognized that technological understanding leads to better policymaking. In 2005, it established the Offshore Wind Energy Foundation, which has a remit to support technological research, environmental monitoring of the effects of construction, research on the suitability and effectiveness of state instruments for the promotion of offshore wind energy, and the exchange and transfer of knowledge. This foundation is guided by a board of trustees that include offshore wind power plants manufacturers, developers, electricity utilities, insurance companies, financial institutions, NGOs, engineering offices, representatives from the construction industry, subcontractors, northern German coastal states, and federal ministries.[86]

5.6 POLITICAL INFLUENCES ON POLICY

5.6.1 National Political Structure

A political desire to demonstrate leadership within the European Union was a factor underpinning national support for wind power development. In the late 1990s, an EU directive on the promotion of renewable energies was announced. In support of this directive, EU nations were asked to commit themselves to the liberalization of national electricity markets. In Germany, this led to an amendment of the Energy Industry Act in 1998, which liberalized electricity generation in Germany and laid the structural foundation for enhanced contributions from renewable energy providers.

Aside from international motives, domestically there has been broad political support since the late 1980s for wind power development. As a testament to this, the initial feed-in law of 1991 was supported by a plurality of political parties. In fact, it has been asserted that the only major political party that was not in support was the Liberal party (FDP).[87] Furthermore, as the summary of the evolution of wind power development in Germany illustrated, not only did the political support base hold together when utilities and communities began to contest wind power projects, a prevalent tailwind ideology engendered entrenched support.[88]

With that said, wind power developments are implemented at regional if not municipal levels; accordingly, national political consensus to support wind power development is not necessarily sufficient to guarantee the type of success that Germany has enjoyed. This is particularly true when one recognizes that Germany is a federal nation, and as such its states hold considerable sway over land-use. As opposed to Canada, which is a federal nation that has seen wind power diffusion undermined by lack of provincial cohesiveness,[89] German state policy was largely aligned with federal wind power development policy.

Recently, some of the states that host higher levels of wind power installation have purportedly begun to tighten clearance decrees for wind turbines.[90] However, state-level support is largely sustained by broad public support for wind power. Therefore, as long as NIMBY opposition can be held in check through some of the methods described in this chapter, it is probable that both national and state-level support for wind power will continue.

5.6.2 Governing Party Ideology

Interestingly, although different governing coalitions have come and gone since the evolution of Germany's wind power program in the early 1990s, each successive ruling coalition harbored varied and unique justifications for supporting wind power development. In the early 1990s, the nation was governed by a Christian Democratic Union (CDU) and Christian Social Union (CSU) coalition led by Chancellor Helmut Kohl. The CDU-CSU coalition embraced a social market economy platform that viewed the technological and vocational skills of Germany's workforce as central to economic development and sought out initiatives to support economic development in rural West Germany and in comparatively impoverished East Germany. Wind power development proved to be a natural fit with this vision. Supporting the creation and maintenance of wind systems was compatible with national competencies in engineering, equipment manufacturing and precision instrument design.

In 1998, Kohl's coalition was toppled by a Social Democratic Party (SPD)–Green Party alliance. The importance of the Green Party in keeping the SPD in power engendered economic development policy that favored clean technologies, such as wind power. It is no coincidence that wind power installations in Germany took off in 1998. In fact, during the seven years that the SPD—Green Party alliance governed Germany, total installed wind power capacity grew from 2,875 MW to 18,428 MW. Approximately 50% of all wind power capacity installed in Germany up until 2012 was installed during the period of SPD–Green Party rule.

In the 2005 election, no party received enough votes to form a government; in the ensuing negotiations, a grand coalition between the CDU/CSU and SPD was formed under the leadership of the CDU's Angela Merkel. Merkel has a doctorate in quantum chemistry and was the Minister for the Environment and Nuclear Safety in 1994 under Helmut Kohl's administration. During her tenure as minister, she took the lead in designing the nation's strategy for phasing out nuclear power. When she became chancellor, the only adjustment she enacted in regard to nuclear power was to initiate a slowdown of the phase-out in order to permit a more orderly and less economically disruptive transition. Ideology aside, enacting any sort of major deviation from existing in energy policy during the first term of Merkel's tenure in power would have been difficult given the need to hold the grand coalition together. Similarly, after the 2009 elections when Merkel returned to power as head of a CDU-CSU-FDP coalition, the need for policy temperance to appease disparate political interests prevented any major deviation from the ongoing strategy of phasing out nuclear power and bolstering renewable energy capacity. Moreover, public will is so strongly in favor of such a transition that major energy policy change is improbable.

Policymaking through consensus has been evident in all governing administrations since the early 1990s. It is typical for policies in Germany to be developed through an inclusive consultative process involving numerous stakeholder groups. For example, the offshore wind power foundation that was launched in 2005 to study offshore wind power diffusion enlisted the participation of numerous stakeholder groups including officials from the Ministry for the Environment, Nature Conservation and Nuclear Safety, state government officials, representatives from key German manufacturers, wind energy interest groups, environmental associations, and major energy suppliers.[91] Even in the preparation of government studies, attempts are made to comprehensively consult with affected stakeholders. For example, a study commissioned in 2005 to investigate grid expansion requirements to support further wind power diffusion was prepared under the guidance of the Institute of Energy Economics, the German Wind Energy Institute, the Fraunhofer Institute for Wind Energy and Energy System Technology, the Ministry for the Environment, Nature Conservation and Nuclear Safety, the Ministry of Economics and Technology, state planning agencies, and a number of members from the academic community.[92] Not only does this inclusive process permit more stakeholders to have a voice in policymaking, but the comprehensive nature of input into these studies equipped policymakers with a comparatively sophisticated understanding of the pros and cons of different renewable energy technologies.[93]

5.6.3 Fiscal Health

Throughout the history of Germany's wind power program, the government has been struggling with financial fetters imposed by an ever-increasing amount of public debt. In 1998, when the wind power program really took off, government public debt amounted to €1.185 trillion, equating to approximately 61% of Germany's 1998 GDP. By 2005 public debt had risen to €1.524 trillion; and by 2011 German public debt had risen to €2.088 trillion, equal to 81% of Germany's 2011 GDP. This debt load continues to adversely influence government expenditures because servicing the debt consumes funds that could be used for other projects. For example, in 2005, €40 billion was earmarked to service the outstanding public debt. This represented 16% of total government expenditures for the year.[94]

In order to attenuate the rise of public debt, the government enacted a debt brake policy in 2009, which was embedded in the constitution. The debt brake places a limit on the size of planned annual structural budget deficits. The limit fluctuates to interface with Germany's countercyclical taxes and reserves. In 2011, this debt brake amounted to approximately 0.35% of GDP at the federal level and 0% at the state level. Even before the debt brake was institutionalized, for the past decade, the German government exhibited proclivity toward trying to reduce budget deficits by shifting the burden of program financing to private enterprise or passing along program costs to end-consumers whenever possible. In essence this has been the case with Germany's electricity feed-in tariff; the costs of this tariff have progressively been passed along to electricity end-consumers in the form of rate hikes.

5.6.4 Policy Regime

As the history of wind power development described earlier suggests, direct government involvement in shaping the future of renewable energy technologies was evident and influential during all stages of development.[95]

In the heady early days of the wind power program before government alarm over the increasing size of public debt, there was a tendency for

German government policy to favor direct financial subsidies to support desirable economic development. In these early days, the Kohl administration was well-entrenched and had established a political culture that was supportive of proactive government intervention.[96] The 100 MW program that was established in 1989 to kick off commercial development of wind power exemplifies the early proclivity toward supporting industry development through financial infusion.[97] Moreover, the German government provided generous funding for basic R&D of demonstration technologies such as the ill-fated Growian project. Finally, in early years the government also provided support for favorable loans to project developers.[98]

As the government began to undertake austerity measures to reduce the size of the public debt, the government's approach to supporting wind power diffusion has increasingly favored a "users pay" strategy. Currently, part of the financial support for wind power development is provided through a green tax levied on electricity consumers, while additional financial support for wind power development is provided by passing costs on to end-consumers in the form of rate hikes.

5.7 THE CULMINATION OF INFLUENCES

Political advocacy of wind power and civic support for wind power have enjoyed a symbiotic relationship. As outlined earlier, civic support for wind power has been unequivocal enough to enable bold wind power development policies. On the other hand, the government has been careful not to abuse public support by neglecting emergent community concerns.

In fact, a number of mechanisms were evident during the evolution of wind power in Germany that suggests the government specifically designed policies to entrench public support. For example, the government sponsored a number of investment schemes to promote community investment in wind parks, which the government refers to as citizens wind farms.[99] In order to be responsive to the concerns of community members that are not financially invested in wind farms, the government has been quick to develop and refine siting standards to address specific concerns. For example, an environmental impact assessment is required for all larger wind farms (with more than 20 turbines that are over 50 m in height) in order to allay concerns over the impact of such large projects on biodiversity and community aesthetics.[100] This is not to say that smaller projects are not vetted to ensure community approval. In fact the opposite is true. Siting standards are often set at the municipal level, which allows local authorities to ensure that projects conform to community expectations and norms. Finally, it should be noted that the German government has been very proactive in financially supporting R&D initiatives designed to identify and respond to threats and barriers to wind power development.

Such attentiveness to social concerns has made it possible for the government to pass on energy costs to the end-consumer without experiencing the type of emotive opposition that other nations have experienced.[101] In fact, civic support for wind power has also given rise to green energy marketing strategies whereby energy generators differentiate their product offerings by providing "green power" packages.[102]

Government policy has been instrumental in enabling the development of internationally competitive German wind turbine manufacturers. After an initial foray into providing R&D subsidies to encourage the development of wind power technology, the government shifted course by announcing a feed-in tariff which fueled the market growth necessary for Germany's small and medium-sized wind turbine manufacturers to compete for portions of an ever-increasing revenue pool.[103] This engendered competitive conditions that encouraged innovation while at the same time providing ever expanding revenue flows that German firms could leverage for enhancing R&D. In this way, Germany's feed-in tariff was instrumental in establishing a pool of highly competitive wind system manufacturers.

Similarly, when it came to the development of offshore wind power, government policy played an active role in equipping German wind farm developers with knowledge that few other competing developers possess. Germany specifically limited its offshore development plan to exclusive economic zones (EEZ) located at least 12 nautical miles from the coast. Although the technical challenge was greater because developers had to devise ways to construct projects in rougher seas with water depths of up to 40 m,[104] the operational insights and innovations derived from responding to this policy equips German wind system manufacturers with competitive advantages in remote offshore wind farm development—an emerging niche that many nations are just now beginning to try and exploit. For example, this offshore strategy has prompted utilities such as RWE to develop specialized sea going vessels for the efficient installation of wind turbines.[105] The decision to initiate development of offshore wind projects in deeper water has also encouraged German companies to begin to specialize in the development of 5 MW or greater turbines for offshore use, and in the process they have become technology leaders in this energy niche.[106]

Government policy has also been instrumental in ensuring that these offshore wind installations are being developed to very high environmental and social standards. Offshore wind projects are subject to approval by the federal maritime and hydrographic agency. Approvals for offshore wind development are also subject to existing laws and statutes such as the SeeAnIV of 1997 (which mandates an EIA for proposed offshore wind energy facilities) and the Federal Maritime Responsibilities Act of 2002.[107] Establishing high standards have forced manufacturers to innovate and improve German offshore wind system technology. This also ensures that German wind system technology can readily meet the most stringent standards in international markets.

In the early days of wind power development, the government experimented with a number of different policy tools designed to enhance the commercial prospects of wind power. Three policies in particular have been highlighted as key catalysts for the wind energy boom of the early 1990s.[108] The first of these three was the 100 MW program announced in 1989 which was in essence a renewable portfolio standard that spurred the development of 100 MW of wind power by offering a €0.04/kWh premium above wholesale market prices for electricity generated by qualifying wind power projects. The second policy initiative was an upgrade of this renewable portfolio standard to a 250 MW program in 1991.[109] In combination, these two programs made it clear to prospective investors that the government was intent on supporting progressive wind power development. The third policy initiative was the announcement of a feed-in tariff in 1991. The feed-in tariff was an attempt on the part of the government to reward actual generation rather than just installed capacity. This new tariff was accompanied by a number of other federal, regional and municipal support programs that encouraged investment in wind power through subsidies, tax incentives and soft loans.[110]

In the mid-1990s, the government liberalized the electricity generation market resulting in enhanced market opportunities for renewable energy generators. Although market liberalization gave rise to resistance from power utilities, it provided the market opportunities that provided the economies of scale needed to reduce the cost of wind power and close the competitive gap with coal-fired power.

"Progressive escalation of commitment" is perhaps the most suitable phrase for describing why German energy policy was so successful in encouraging wind power development. One example is the upgrading of the renewable portfolio standard from 100 MW to 250 MW only two years after the initial program was announced. Another example is the Renewable Energy Sources Act that was passed in 2000. This act improved upon the feed-in tariff on 1996, and obligated utilities to connect wind power projects to the grid and defused market risk by giving renewables feed-in priority at a purchase price that was guaranteed over a 20-year period.[111]

When the government shifted its sights to offshore wind power development, it was equally as committed. In 2002, the government released a strategy paper that declared intentions to develop offshore capacity of 500

MW by 2006, 3000 MW by 2010, and up to 25,000 MW by 2025/2030.[112] In order to support these goals, the government has recently announced a revised feed-in tariff with higher rates for offshore development and has been working on strategies for financing grid connections and the installation of overhead transmission cables to transmit the wind power generated in the North to demand centers in the South and West.[113]

To summarize, the government has influenced the economics of wind power development through a two-pronged approach. First, it seeded expanding markets through unwavering subsidization;[114] second, it encouraged infrastructure development and competition through structural reform. Moreover, to support Germany's wind turbine manufacturers, the government has reportedly encouraged international market development by offering export credit assistance and tying the purchase of domestic wind power technology to development aid loans.[115]

5.8 WHAT TO EXPECT GOING FORWARD

Wind power capacity will continue to expand in Germany for at least the next decade. This is because under the European Union's renewable energy directive, Germany has committed to a 2020 target of 18% renewable energy as a proportion of total primary energy consumption. For the electricity sector, the government's goal is to facilitate a 38.6% contribution from renewable sources by 2020.[116] Given the Merkel administration's decision to escalate phase-out of Germany's nuclear power reactors, low growth potential of hydropower and the high cost of alternative energy technologies, wind power capacity will have to increase considerably in order to allow Germany to meet the Renewable Energy Directive obligations. According to the German Wind Energy Association (BWE), the country appears to be on track to add 15,000 MW of onshore wind power capacity and 10,000 MW of offshore wind power capacity by 2020. If achieved, Germany's wind power capacity would reach about 45,000 MW, enough to supply about 25% of German electricity consumption.[117] In short, more market stimulus measures may be forthcoming to facilitate government goals.

Financial incentives for spurring on new development already exist. In 2009, the government published revised feed-in tariffs. Onshore turbines will receive €0.0911 per kWh in the first five years of operation and €0.0497 per kWh for a subsequent 15-year period. Given the fact that by 2015, 6000 MW of onshore capacity will be at least 15 years old and ready for repowering,[118] this new tariff should ensure that repowering does indeed take place. Under the new tariffs, offshore turbines will receive €0.13 per kWh for the first 12 years and €0.035 per kWh for a subsequent eight-year period. Moreover, offshore turbines that are commissioned by the end of 2015 will receive an additional €0.02 per kWh.[119]

Regarding onshore development, two challenges stand out. First, there are concerns that higher levels of installed capacity will threaten the technological integrity of Germany's electricity grid. As a response, the government offers an additional €0.005 per kWh for all turbines equipped with advanced grid integration technology. Second, although Germans are still largely supportive of wind power capacity expansion, NIMBY opposition is occasionally evident, particularly in states that currently host high amounts of wind power capacity, such as Saxony and Brandenburg. For example, in January 2011, the state parliament building in Pottstown, the regional capital of Brandenburg, was besieged by citizen protest over the mass construction of wind turbines in the state."[120] As one team of researchers have observed, "some fear that the zeal to install wind turbines mirrors the drive in the 1960s to build motorways" linking towns in West Germany. The initiative "was regarded as ultra modern at the time, but it created massive, irreversible eyesores."[121]

To an extent, the NIMBY opposition to wind power is fueling investment in solar PV as an alternative technology for advancing Germany toward its 2020 goal of generating 38.6% its electricity through renewable energy. Thanks to generous feed-in tariffs, as of 2012, 5.3% of the net electricity produced in Germany was produced by solar PV. The consequence of this burgeoning market is that the cost of solar PV decreased considerably over the past decade. However, the threat posed to wind power by solar is still a distant threat that will require significant advances in solar PV technology. In 2013, the feed-in tariff for solar PV was set at €0.39 per kWh (compared to the current feed-in tariff for wind power of €0.091 per kWh).[122] In the interim, the solar PV capacity that does exist has been proven to be a complement to wind capacity in attenuating power fluctuations.[123]

Regarding offshore development, the most serious challenge appears to be economic. The decision to construct offshore wind parks far offshore significantly inflates the cost of connecting these projects to the electricity grid.[124] According to the Global Wind Energy Council, the financial responsibility associated with connecting offshore wind farms to the mainland grid currently rests with the transmission system operators. Although three 400 MW HDVC light lines have already been completed, infrastructure development to date has only scratched the surface in comparison to what is needed.[125]

Heavy investment will be required for transmission lines to deliver offshore wind power from Baltic and North Sea sites to demand centers in the South and in the West.[126] Germany's power transmission companies have put forth plans to build electricity autobahns for this purpose. The cost of this plan, calling for 3800 km of high-voltage electricity lines, has been esti-mated at €20 billion dollars. On top of this expense, it is estimated that 1700 km of alternating current lines will be required and will have to enter service between 2017 and 2020 in order to meet capacity targets.[127]

The cost of developing offshore wind projects is so high that the coopera-tive investment model which drove onshore wind power development is not viable. A typical size for planned offshore wind parks is 80 turbines with a total capacity of 400 MW.[128] Such mega-offshore developments require significantly more investment, suggesting that many of the projects in the future will be driven by larger firms. This will alter the profile of wind power investment and the nature of stakeholder interest.[129]

The scale of expansion that has been exacerbated by the decision to expe-dite phase-out of Germany's nuclear reactors poses project management challenges, even if the finances of fortifying the grids can be worked out. Peter Terium, the CEO of German utility RWE, has publicly questioned the viability of installing 10,000 MW of offshore wind power by 2020, lament-ing, "I don't see how we can get 10,000 MW done. I think maybe 6000 MW is possible."[130] In July 2012, grid operator TenneT echoed the concern by announcing that it doesn't know when it will be able to connect a planned €1 billion North Sea project to the German grid because of technical issues.

Overall, it is also becoming apparent that elevated contributions of wind power are starting to engender concerns over grid resilience and stabil-ity. One recent study contends that increased contributions of renewable energy to the grid do not pose a threat to grid resilience in the short term; however, the 2020 targets will likely enhance the need for backup capac-ity and possibly necessitate stabilizing electricity imports from neighboring nations. The same study contends that large-scale construction of gas-fired power plants may be necessary to ensure grid stability up to 2030.[131]

Up until this point, German electricity consumers have exhibited a will-ingness to accept a certain degree of aesthetic impairment to landscapes and higher electricity prices in order to transition away from carbon-intensive electricity generation technologies. However, one has to wonder where the limits of tolerance lie.[132] For example, under the German feed-in tariff sys-tem, the electricity consumer shoulders the difference between the market price for electricity and the feed-in tariff. This has resulted in a surcharge of €0.036 per kWh in 2012. According to some forecasts, this surcharge might increase by 50% or more as contributions from renewable energy increase. So far household consumers have been tolerant toward paying the surcharge; however, industrial consumers have become increasingly vocal in raising concerns over the high price of electricity undermining interna-tional competitiveness.[133] The government appears to be aware of this. On

October 11, 2012, Environment Minister Peter Altmaier outlined plans to initiate a cap on the accumulated installed generation capacities for feed-in tariffs related to wind power and biomass. The intention of this is to keep market expansion in check in order to ensure sufficient grid enhancement simultaneously takes place to preserve grid integrity.

Finally, there is a high degree of uncertainty regarding how the general public will react to proliferate arrays of offshore wind turbines. There have already been concerns over aesthetic seascape impairment and charges that the developments are damaging the habitat of the harbor porpoise.

It is apparent that wind power in Germany is set to enjoy an extended period of capacity expansion. However, the cost of providing suitable infrastructure to support expansion, the technological (and economic) challenges associated with an ever-expanding contribution from wind power and the possibility of amplified public opposition as wind power moves from community investment projects to big business suggest that the path of wind power development will not be as smooth as it was in the past. As the quote at the beginning of this chapter attributed to German Chancellor Angela Merkel emphasizes, given the global imperative to facilitate a rapid transition away from carbon-based energy systems, the will to change clearly exists in Germany; the question is, can the transition be sustained at a suitable pace?

NOTES

1. Extracted from http://yourlocalsecurity.com/blog/2011/08/15/inspiration-from-the-100-most-powerful-women-in-the-world/.
2. Mendonca, Miguel, David Jacobs, and Benjamin Sovacool. 2009. *Powering the Green Economy: The Feed-in Tariff Handbook*. Oxford: Earthscan.
3. Bruns, Elke, and Dorte Ohlhorst. 2011. "Wind Power Generation in Germany: Transdisciplinary View on the Innovation Biography." *Journal of Transdisciplinary Environmental Studies* 10 (1): 45–67.
4. Ibid.
5. Fischer, Sebastian, Florian Gathmann, Anna Riemann, and Tijs van den Boomen. 2011. "Resistance Mounts to Germany's Ambitious Renewable Energy Plans." *Spiegel*, April 13. www.spiegel.de/international/germany/green-headache-resistance-mounts-to-germany-s-ambitious-renewable-energy-plans-a-756836.html.
6. Source: German Wind Energy Association. www.wind-energie.de/en/infocenter/statistiken/print?nid=1596.
7. Bruns, Elke, and Dorte Ohlhorst. 2011. "Wind Power Generation in Germany: Transdisciplinary View on the Innovation Biography." *Journal of Transdisciplinary Environmental Studies* 10 (1): 45–67.
8. Wüstenhagen, Rolf, and Michael Bilharz. 2006. "Green Energy Market Development in Germany: Effective Public Policy and Emerging Customer Demand." *Energy Policy* 34 (13): 1681–1696.

9. Bruns, Elke, and Dorte Ohlhorst. 2011. "Wind Power Generation in Germany: Transdisciplinary View on the Innovation Biography." *Journal of Transdisciplinary Environmental Studies* 10 (1): 45–67.

10. Bruns, Elke, and Dorte Ohlhorst. 2011. "Wind Power Generation in Germany: Transdisciplinary View on the Innovation Biography." *Journal of Transdisciplinary Environmental Studies* 10 (1): 45–67; and Wüstenhagen, Rolf, and Michael Bilharz. 2006. "Green Energy Market Development in Germany: Effective Public Policy and Emerging Customer Demand." *Energy Policy* 34 (13): 1681–1696.

11. Bruns, Elke, and Dorte Ohlhorst. 2011. "Wind Power Generation in Germany: Transdisciplinary View on the Innovation Biography." *Journal of Transdisciplinary Environmental Studies* 10 (1): 45–67.

12. Ibid.

13. Ibid.

14. Ibid.

15. Ibid.

16. Wüstenhagen, Rolf, and Michael Bilharz. 2006. "Green Energy Market Development in Germany: Effective Public Policy and Emerging Customer Demand." *Energy Policy* 34 (13): 1681–1696.

17. Bruns, Elke, and Dorte Ohlhorst. 2011. "Wind Power Generation in Germany: Transdisciplinary View on the Innovation Biography." *Journal of Transdisciplinary Environmental Studies* 10 (1): 45–67.

18. Zitzer, Suzanne E. 2009. *Renewable Energy Policy and Wind Energy Development in Germany*. Leibzig: Leibniz Information Centre for Economics.

19. Mabee, Warren E., Justine Mannion, and Tom Carpenter. 2012. "Comparing the Feed-In Tariff Incentives for Renewable Electricity in Ontario and Germany." *Energy Policy* 40: 480–489.

20. Zitzer, Suzanne E. 2009. *Renewable Energy Policy and Wind Energy Development in Germany*. Leibzig: Leibniz Information Centre for Economics.

21. Renewable Energy Sources Act. 2000. An English version of this is available at www. erneuerbare-energien.de/fileadmin/ee-import/files/english/pdf/application/pdf/ eeg_2012_en_bf.pdf.

22. Wüstenhagen, Rolf, and Michael Bilharz. 2006. "Green Energy Market Development in Germany: Effective Public Policy and Emerging Customer Demand." *Energy Policy* 34 (13): 1681–1696.

23. Bruns, Elke, and Dorte Ohlhorst. 2011. "Wind Power Generation in Germany: Transdisciplinary View on the Innovation Biography." *Journal of Transdisciplinary Environmental Studies* 10 (1): 45–67.

24. Ibid.

25. Ibid.

26. Wüstenhagen, Rolf, and Michael Bilharz. 2006. "Green Energy Market Development in Germany: Effective Public Policy and Emerging Customer Demand." *Energy Policy* 34 (13): 1681–1696.

27. Kuhbier, Jorg. 2007. "Offshore Wind Power in Germany." In *EU Policy Workshop*, 22 February. Berlin. www.erneuerbare-energien.de/fileadmin/ee-import/files/ english/pdf/application/pdf/eupol_referat_kuhbier_en.pdf.

28. Bruns, Elke, and Dorte Ohlhorst. 2011. "Wind Power Generation in Germany: Transdisciplinary View on the Innovation Biography." *Journal of Transdisciplinary Environmental Studies* 10 (1): 45–67.

29. Wüstenhagen, Rolf, and Michael Bilharz. 2006. "Green Energy Market Development in Germany: Effective Public Policy and Emerging Customer Demand." *Energy Policy* 34 (13): 1681–1696.

30. Portman, Michelle E., John A. Duff, Johann Köppel, Jessica Reisert, and Megan E. Higgins. 2009. "Offshore Wind Energy Development in the Exclusive Economic Zone: Legal and Policy Supports and Impediments in Germany and the US." *Energy Policy* 37 (9): 3596–3607.

31. Wüstenhagen, Rolf, and Michael Bilharz. 2006. "Green Energy Market Development in Germany: Effective Public Policy and Emerging Customer Demand." *Energy Policy* 34 (13): 1681–1696.

32. Bruns, Elke, and Dorte Ohlhorst. 2011. "Wind Power Generation in Germany: Transdisciplinary View on the Innovation Biography." *Journal of Transdisciplinary Environmental Studies* 10 (1): 45–67.

33. Weigt, Hannes. 2009. "Germany's Wind Energy: The Potential for Fossil Capacity Replacement and Cost Saving." *Applied Energy* 86 (10): 1857–1863.

34. Bruns, Elke, and Dorte Ohlhorst. 2011. "Wind Power Generation in Germany: Transdisciplinary View on the Innovation Biography." *Journal of Transdisciplinary Environmental Studies* 10 (1): 45–67.

35. Portman, Michelle E., John A. Duff, Johann Köppel, Jessica Reisert, and Megan E. Higgins. 2009. "Offshore Wind Energy Development in the Exclusive Economic Zone: Legal and Policy Supports and Impediments in Germany and the US." *Energy Policy* 37 (9): 3596–3607.

36. Ibid.

37. Bruns, Elke, and Dorte Ohlhorst. 2011. "Wind Power Generation in Germany: Transdisciplinary View on the Innovation Biography." *Journal of Transdisciplinary Environmental Studies* 10 (1): 45–67.

38. European Wind Energy Association (EWEA). 2009. *Integrating Wind: Developing Europe's Power Market for the Large-Scale Integration of Wind Power.* edited by F. V. Hulle. Brussels: European Wind Energy Association.

39. Federal Republic of Germany. 2010. National Renewable Energy Action Plan (in accordance with Directive 2009/28/EC on the promotion of the use of energy from renewable sources). Berlin: Federal Republic of Germany.

40. Bruns, Elke, and Dorte Ohlhorst. 2011. "Wind Power Generation in Germany: Transdisciplinary View on the Innovation Biography." *Journal of Transdisciplinary Environmental Studies* 10 (1): 45–67.

41. Information taken from the website of the Federal Ministry for the Environment, Nature Conservation and Nuclear Safety: www.erneuerbare-energien.de/en/.

42. Ibid.

43. Data taken from the Global Wind Energy Council website: www.gwec.net.

44. Mabee, Warren E., Justine Mannion, and Tom Carpenter. 2012. "Comparing the Feed-In Tariff Incentives for Renewable Electricity in Ontario and Germany." *Energy Policy* 40: 480–489.

45. Lazenby, Oliver. 2012. "Germany Swaps Nuclear for Solar and Wind Power." *Yes Magazine*, June 7. truth-out.org/news/item/9932-germany-swaps-nuclear-for-solar-and-wind-power.

46. Fischer, Sebastian, Florian Gathmann, Anna Riemann, and Tijs van den Boomen. 2011. "Resistance Mounts to Germany's Ambitious Renewable Energy Plans." *Spiegel*, April 13. www.spiegel.de/international/germany/green-headache-resistance-mounts-to-germany-s-ambitious-renewable-energy-plans-a-756836.html.

47. Gipe, Paul. 2011. "Germany Passes More Aggressive Renewable Energy Law." *Renewable Energy World*, July 25.

48. Fischer, Sebastian, Florian Gathmann, Anna Riemann, and Tijs van den Boomen. 2011. "Resistance Mounts to Germany's Ambitious Renewable Energy Plans." *Spiegel*, April 13. www.spiegel.de/international/

germany/green-headache-resistance-mounts-to-germany-s-ambitious-renewable-energy-plans-a-756836.html.

49. Lazenby, Oliver. 2012. "Germany Swaps Nuclear for Solar and Wind Power." *Yes Magazine*, June 7. http://truth-out.org/news/item/9932-germany-swaps-nuclear-for-solar-and-wind-power.

50. Bruns, Elke, and Dorte Ohlhorst. 2011. "Wind Power Generation in Germany: Transdisciplinary View on the Innovation Biography." *Journal of Transdisciplinary Environmental Studies* 10 (1): 45–67.

51. As reported on Paul Gipe's website: http://www.wind-works.org/cms/.

52. Musall, Fabian David, and Onno Kuik. 2011. "Local Acceptance of Renewable Energy—A Case Study from Southeast Germany." *Energy Policy* 39 (6): 3252–3260.

53. Bruns, Elke, and Dorte Ohlhorst. 2011. "Wind Power Generation in Germany: Transdisciplinary View on the Innovation Biography." *Journal of Transdisciplinary Environmental Studies* 10 (1): 45–67.

54. Fischer, Sebastian, Florian Gathmann, Anna Riemann, and Tijs van den Boomen. 2011. "Resistance Mounts to Germany's Ambitious Renewable Energy Plans." Spiegel, April 13. www.spiegel.de/international/germany/green-headache-resistance-mounts-to-germany-s-ambitious-renewable-energy-plans-a-756836.html.

55. As reported by the German Wind Energy Association: http://www.wind-energie.de/en.

56. Gee, Kira. 2010. "Offshore Wind Power Development as Affected by Seascape Values on the German North Sea Coast." *Land Use Policy* 27 (2): 185–194.

57. Wüstenhagen, Rolf, and Michael Bilharz. 2006. "Green Energy Market Development in Germany: Effective Public Policy and Emerging Customer Demand." *Energy Policy* 34 (13): 1681–1696.

58. Bruns, Elke, and Dorte Ohlhorst. 2011. "Wind Power Generation in Germany: Transdisciplinary View on the Innovation Biography." *Journal of Transdisciplinary Environmental Studies* 10 (1): 45–67.

59. Wüstenhagen, Rolf, and Michael Bilharz. 2006. "Green Energy Market Development in Germany: Effective Public Policy and Emerging Customer Demand." *Energy Policy* 34 (13): 1681–1696.

60. Markard, Jochen, and Regula Petersen. 2009. "The Offshore Trend: Structural Changes in the Wind Power Sector." *Energy Policy* 37 (9): 3545–3556.

61. Bruns, Elke, and Dorte Ohlhorst. 2011. "Wind Power Generation in Germany: Transdisciplinary View on the Innovation Biography." *Journal of Transdisciplinary Environmental Studies* 10 (1): 45–67.

62. Ibid.

63. Wüstenhagen, Rolf, and Michael Bilharz. 2006. "Green Energy Market Development in Germany: Effective Public Policy and Emerging Customer Demand." *Energy Policy* 34 (13): 1681–1696.

64. Lewis, Joanna I., and Ryan H. Wiser. 2007. "Fostering a Renewable Energy Technology Industry: An International Comparison of Wind Industry Policy Support Mechanisms." *Energy Policy* 35 (3): 1844–1857.

65. Weigt, Hannes. 2009. "Germany's Wind Energy: The Potential for Fossil Capacity Replacement and Cost Saving." *Applied Energy* 86 (10): 1857–1863.

66. Wüstenhagen, Rolf, and Michael Bilharz. 2006. "Green Energy Market Development in Germany: Effective Public Policy and Emerging Customer Demand." *Energy Policy* 34 (13): 1681–1696.

67. Bruns, Elke, and Dorte Ohlhorst. 2011. "Wind Power Generation in Germany: Transdisciplinary View on the Innovation Biography." *Journal of Transdisciplinary Environmental Studies* 10 (1): 45–67.

68. Wüstenhagen, Rolf, and Michael Bilharz. 2006. "Green Energy Market Development in Germany: Effective Public Policy and Emerging Customer Demand." *Energy Policy* 34 (13): 1681–1696.

69. Bruns, Elke, and Dorte Ohlhorst. 2011. "Wind Power Generation in Germany: Transdisciplinary View on the Innovation Biography." *Journal of Transdisciplinary Environmental Studies* 10 (1): 45–67.

70. Ibid.

71. Meyerhoff, Jürgen, Cornelia Ohl, and Volkmar Hartje. 2010. "Landscape Externalities from Onshore Wind Power." *Energy Policy* 38 (1): 82–92.

72. Lazenby, Oliver. 2012. "Germany Swaps Nuclear for Solar and Wind Power." *Yes Magazine*, June 7. http://truth-out.org/news/item/9932-germany-swaps-nuclear-for-solar-and-wind-power.

73. Meyerhoff, Jürgen, Cornelia Ohl, and Volkmar Hartje. 2010. "Landscape Externalities from Onshore Wind Power." *Energy Policy* 38 (1): 82–92.

74. Bruns, Elke, and Dorte Ohlhorst. 2011. "Wind Power Generation in Germany: Transdisciplinary View on the Innovation Biography." *Journal of Transdisciplinary Environmental Studies* 10 (1): 45–67.

75. Toke, David, Sylvia Breukers, and Maarten Wolsink. 2008. "Wind Power Deployment Outcomes: How Can We Account for the Differences?" *Renewable and Sustainable Energy Reviews* 12 (4): 1129–1147.

76. Bruns, Elke, and Dorte Ohlhorst. 2011. "Wind Power Generation in Germany: Transdisciplinary View on the Innovation Biography." *Journal of Transdisciplinary Environmental Studies* 10 (1): 45–67.

77. Kuhbier, Jorg. 2007. "Offshore Wind Power in Germany." In *EU Policy Workshop*, 22 February 2007. Berlin. www.erneuerbare-energien.de/fileadmin/ee-import/files/english/pdf/application/pdf/eupol_referat_kuhbier_en.pdf.

78. Wüstenhagen, Rolf, and Michael Bilharz. 2006. "Green Energy Market Development in Germany: Effective Public Policy and Emerging Customer Demand." *Energy Policy* 34 (13): 1681–1696.

79. Bruns, Elke, and Dorte Ohlhorst. 2011. "Wind Power Generation in Germany: Transdisciplinary View on the Innovation Biography." *Journal of Transdisciplinary Environmental Studies* 10 (1): 45–67.

80. Sovacool, Benjamin K. 2009. "The Importance of Comprehensiveness in Renewable Electricity and Energy-Efficiency Policy." *Energy Policy* 37 (4): 1529–1541.

81. Bruns, Elke, and Dorte Ohlhorst. 2011. "Wind Power Generation in Germany: Transdisciplinary View on the Innovation Biography." *Journal of Transdisciplinary Environmental Studies* 10 (1): 45–67.

82. Bruns, Elke, and Dorte Ohlhorst. 2011. "Wind Power Generation in Germany: Transdisciplinary View on the Innovation Biography." *Journal of Transdisciplinary Environmental Studies* 10 (1): 45–67.

83. Valentine, Scott Victor. 2011. "Sheltering Wind Power Projects from Tempestuous Community Concerns." *Energy for Sustainable Development* 15 (1): 109–114.

84. Bruns, Elke, and Dorte Ohlhorst. 2011. "Wind Power Generation in Germany: Transdisciplinary View on the Innovation Biography." *Journal of Transdisciplinary Environmental Studies* 10 (1): 45–67.

85. Ibid.

86. Kuhbier, Jorg. 2007. "Offshore Wind Power in Germany." In *EU Policy Workshop*, 22 February 2007. Berlin. www.erneuerbare-energien.de/fileadmin/ee-import/files/english/pdf/application/pdf/eupol_referat_kuhbier_en.pdf.

87. Wüstenhagen, Rolf, and Michael Bilharz. 2006. "Green Energy Market Development in Germany: Effective Public Policy and Emerging Customer Demand." *Energy Policy* 34 (13): 1681–1696.

88. Bruns, Elke, and Dorte Ohlhorst. 2011. "Wind Power Generation in Germany: Transdisciplinary View on the Innovation Biography." *Journal of Transdisciplinary Environmental Studies* 10 (1): 45–67.

89. Valentine, Scott Victor. 2010. "Canada's Constitutional Separation of (Wind) Power." *Energy Policy* 38 (4): 1918–1930.

90. Bruns, Elke, and Dorte Ohlhorst. 2011. "Wind Power Generation in Germany: Transdisciplinary View on the Innovation Biography." *Journal of Transdisciplinary Environmental Studies* 10 (1): 45–67.

91. Ibid.

92. Ibid.

93. Wüstenhagen, Rolf, and Michael Bilharz. 2006. "Green Energy Market Development in Germany: Effective Public Policy and Emerging Customer Demand." *Energy Policy* 34 (13): 1681–1696.

94. Source of data: www.datosmacro.com.

95. Ohl, Cornelia, and Marcus Eichhorn. 2010. "The Mismatch between Regional Spatial Planning for Wind Power Development in Germany and National Eligibility Criteria for Feed-In Tariffs—A Case Study in West Saxony." *Land Use Policy* 27 (2): 243–254.

96. Wüstenhagen, Rolf, and Michael Bilharz. 2006. "Green Energy Market Development in Germany: Effective Public Policy and Emerging Customer Demand." *Energy Policy* 34 (13): 1681–1696.

97. Bruns, Elke, and Dorte Ohlhorst. 2011. "Wind Power Generation in Germany: Transdisciplinary View on the Innovation Biography." *Journal of Transdisciplinary Environmental Studies* 10 (1): 45–67.

98. Ibenholt, Karin. 2002. "Explaining Learning Curves for Wind Power." *Energy Policy* 30 (13): 1181–1189.

99. Toke, David, Sylvia Breukers, and Maarten Wolsink. 2008. "Wind Power Deployment Outcomes: How Can We Account for the Differences?" *Renewable and Sustainable Energy Reviews* 12 (4): 1129–1147.

100. Portman, Michelle E., John A. Duff, Johann Köppel, Jessica Reisert, and Megan E. Higgins. 2009. "Offshore Wind Energy Development in the Exclusive Economic Zone: Legal and Policy Supports and Impediments in Germany and the US." *Energy Policy* 37 (9): 3596–3607.

101. Valentine, Scott Victor. 2011. "Japanese Wind Energy Development Policy: Grand Plan or Group Think?" *Energy Policy* 39 (11): 6842–6854.

102. Wüstenhagen, Rolf, and Michael Bilharz. 2006. "Green Energy Market Development in Germany: Effective Public Policy and Emerging Customer Demand." *Energy Policy* 34 (13): 1681–1696.

103. Bruns, Elke, and Dorte Ohlhorst. 2011. "Wind Power Generation in Germany: Transdisciplinary View on the Innovation Biography." *Journal of Transdisciplinary Environmental Studies* 10 (1): 45–67.

104. Ibid.

105. Lazenby, Oliver. 2012. "Germany Swaps Nuclear for Solar and Wind Power." *Yes Magazine*, June 7. http://truth-out.org/news/item/9932-germany-swaps-nuclear-for-solar-and-wind-power.

106. Kuhbier, Jorg. 2007. "Offshore Wind Power in Germany." In *EU Policy Workshop*, 22 February 2007. Berlin. www.erneuerbare-energien.de/fileadmin/ee-import/files/english/pdf/application/pdf/eupol_referat_kuhbier_en.pdf.

107. Portman, Michelle E., John A. Duff, Johann Köppel, Jessica Reisert, and Megan E. Higgins. 2009. "Offshore Wind Energy Development in the Exclusive Economic Zone: Legal and Policy Supports and Impediments in Germany and the US." *Energy Policy* 37 (9): 3596–3607.

108. Bruns, Elke, and Dorte Ohlhorst. 2011. "Wind Power Generation in Germany: Transdisciplinary View on the Innovation Biography." *Journal of Transdisciplinary Environmental Studies* 10 (1): 45–67.

109. Wüstenhagen, Rolf, and Michael Bilharz. 2006. "Green Energy Market Development in Germany: Effective Public Policy and Emerging Customer Demand." *Energy Policy* 34 (13): 1681–1696.

110. Ibid.

111. Bruns, Elke, and Dorte Ohlhorst. 2011. "Wind Power Generation in Germany: Transdisciplinary View on the Innovation Biography." *Journal of Transdisciplinary Environmental Studies* 10 (1): 45–67.

112. Ibid.

113. Lazenby, Oliver. 2012. "Germany Swaps Nuclear for Solar and Wind Power." *Yes Magazine*, June 7. http://truth-out.org/news/item/9932-germany-swaps-nuclear-for-solar-and-wind-power.

114. Ibenholt, Karin. 2002. "Explaining Learning Curves for Wind Power." *Energy Policy* 30 (13): 1181–1189.

115. Lewis, Joanna I., and Ryan H. Wiser. 2007. "Fostering a Renewable Energy Technology Industry: An International Comparison of Wind Industry Policy Support Mechanisms." *Energy Policy* 35 (3): 1844–1857.

116. Data taken from the Global Wind Energy Council website: www.gwec.net.

117. Ibid.

118. Ibid.

119. Portman, Michelle E., John A. Duff, Johann Köppel, Jessica Reisert, and Megan E. Higgins. 2009. "Offshore Wind Energy Development in the Exclusive Economic Zone: Legal and Policy Supports and Impediments in Germany and the US." *Energy Policy* 37 (9): 3596–3607.

120. Fischer, Sebastian, Florian Gathmann, Anna Riemann, and Tijs van den Boomen. 2011. "Resistance Mounts to Germany's Ambitious Renewable Energy Plans." *Spiegel*, April 13. www.spiegel.de/international/germany/green-headache-resistance-mounts-to-germany-s-ambitious-renewable-energy-plans-a-756836.html.

121. Ibid.

122. Wirth, Harry. 2013. *Recent Facts About Photovoltaics in Germany.* Frieberg: Fraunhofer Institute for Solar Energy Systems ISE.

123. Wirth, Harry. 2013. Frieberg: Fraunhofer Institute for Solar Energy Systems ISE. The complementary relationship between solar PV and wind is presented graphically in Figure 10.4, page 86.

124. Bruns, Elke, and Dorte Ohlhorst. 2011. "Wind Power Generation in Germany: Transdisciplinary View on the Innovation Biography." *Journal of Transdisciplinary Environmental Studies* 10 (1): 45–67.

125. Data taken from the Global Wind Energy Council website: www.gwec.net.

126. Bruns, Elke, and Dorte Ohlhorst. 2011. "Wind Power Generation in Germany: Transdisciplinary View on the Innovation Biography." *Journal of Transdisciplinary Environmental Studies* 10 (1): 45–67.

127. Wiesmann, Gerritt. 2012. "Germany Plans to Build Wind Power Grid." *The Financial Times*, May 30.

128. Markard, Jochen, and Regula Petersen. 2009. "The Offshore Trend: Structural Changes in the Wind Power Sector." *Energy Policy* 37 (9): 3545–3556.

129. Ibid.

130. Lazenby, Oliver. 2012. "Germany Swaps Nuclear for Solar and Wind Power." *Yes Magazine*, June 7. http://truth-out.org/news/item/9932-germany-swaps-nuclear-for-solar-and-wind-power.

131. Grave, Katharina, Moritz Paulus, and Dietmar Lindenberger. 2012. "A Method for Estimating Security of Electricity Supply from Intermittent Sources: Scenarios for Germany Until 2030." *Energy Policy* 46: 193–202.

132. Bruns, Elke, and Dorte Ohlhorst. 2011. "Wind Power Generation in Germany: Transdisciplinary View on the Innovation Biography." *Journal of Transdisciplinary Environmental Studies* 10 (1): 45–67.

133. Koyama, Ken. 2012. *IEEJ Newsletter*. Tokyo: Institute of Energy Economics Japan.

CHAPTER 6

Wind Power in China

6.1 INTRODUCTION

In order to properly understand the big picture, everyone should fear becoming mentally clouded and obsessed with one small section of truth.

—Xun Zi

Analyzing electricity generation sector developments in China is akin to observing a person emerging from a supermarket with a shopping cart half-full with dietary products and half-full with chocolates and other sweets and trying to determine whether or not the person is going on a diet. On the one hand, in 2009 China surpassed the United States as the world's largest emitter of greenhouse gases (GHG). Not only is China the world's largest consumer (and producer) of coal; by 2030, coal consumption in China is expected to increase 41% from 2010 levels.[1] On the other hand, China boasts a burgeoning wind power market. In 2012, nearly one of every three MW of installed wind power capacity was installed in China. As of December 2012, China enjoys top global spot in aggregate installed wind power capacity with 75,324 MW installed, 20% of global capacity.[2] China is also the fastest growing nuclear power market in the world with 40,000 MW of installed nuclear power capacity expected by 2020.

Although Deng Xiaoping passed away in 1997, his famous axiom "It doesn't matter if a cat is black or white, so long as it catches mice" (In Mandarin 不管白猫黑猫, 会捉老鼠就是好猫), which was invoked to describe the underlying premise behind political and market reform embraced by the Communist Party of China (CPC), still reigns true in China today. Accordingly, one cannot help but wonder what type of mice China's laudable commitment to wind power is intended to catch. Is this a strategic initiative that will lead to

China establishing new benchmarks for wind power leadership, or is it simply a small part of an all-out effort on the part of the CPC to keep up with burgeoning demand for energy in whatever way that works?

In this chapter, sociopolitical economic influences on wind power development will be examined with an intention to try to explain China's remarkable recent achievements in wind power. The reader will find that wind power policy in China is not just about energy economics; rather, it is a tale of pragmatic planning, strategic foresight, and gradualist politics.

6.2 AN OVERVIEW OF ELECTRICITY GENERATION IN CHINA

Due to the dual distinction of being the world's most populous nation and the world's fastest-growing economy over the past decade, total primary energy (TPE) consumption in China has mushroomed. Upon first glance of Figure 6.1, it appears that China's growth in TPE consumption over the past decade has merely kept pace with global growth trends (top of Figure 6.1). However, closer examination reveals a more engaging fact—China's TPE consumption growth has largely *driven* the global increase (bottom of Figure 6.1). Between 2000 and 2010, growth in China's TPE consumption accounted for 53% of the global increase. China's TPE consumption now accounts for a whopping 20.3% of global energy consumption. Only the

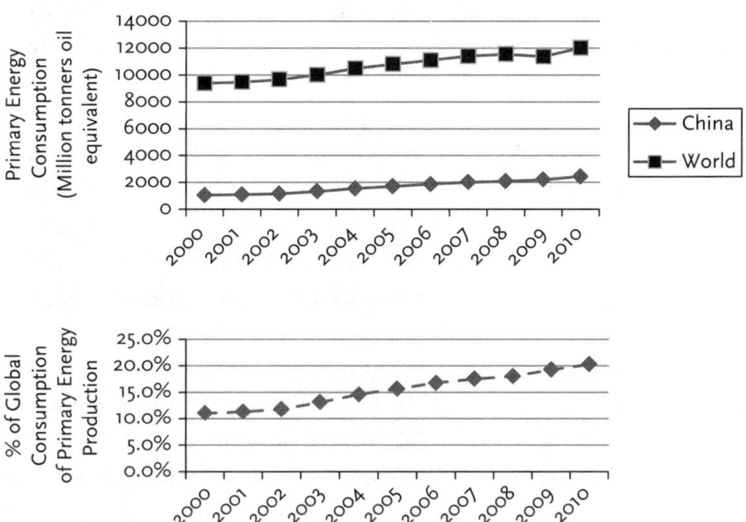

Figure 6.1. Primary Energy Consumption Trends 2000–2010
Source of data: BP (2011).

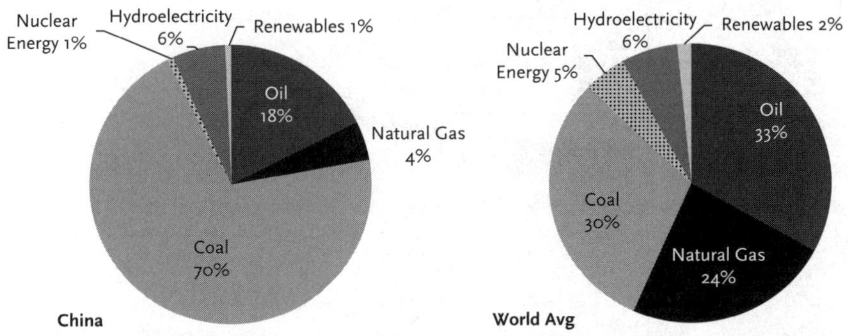

Figure 6.2. Comparison of Fuel Profile in China vs. Global Average in 2011
Source: BP (2011).

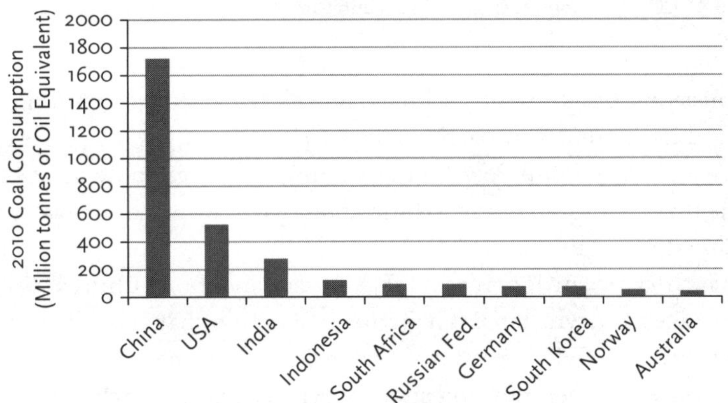

Figure 6.3. Top 10 Coal Consuming Nations in 2010
Source of data: BP (2011).

United States, with a 19% share of global energy consumption, comes near to matching China's prodigious consumption levels.

As Figure 6.2 suggests, the fact that China's immense energy appetite is currently being satiated by a CO_2-intensive energy mix (largely due to the dominance of coal-fired power generation) is of great international consternation regarding climate change mitigation efforts.

Figure 6.3 graphically illustrates the comparative scale of Chinese coal consumption. Of the top 10 coal consuming nations, China's total coal consumption in 2010 was 26% higher than the nine others combined.

It is primarily due to coal-fired power that China is now the largest national contributor of GHG emissions. As Figure 6.4 illustrates, between 1990 and 2009, CO_2 emissions in China increased threefold. China's 4.63-gigaton increase constituted 69% of the 6.74-gigaton global increase over the same period. By 2009, CO_2 emissions in China accounted for 24% of global annual

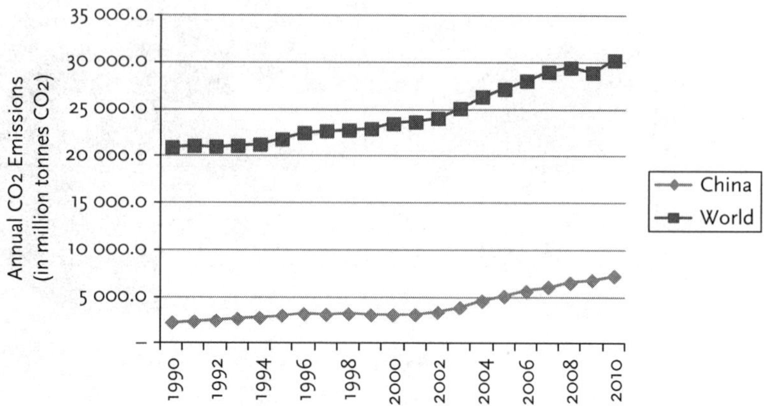

Figure 6.4. CO_2 Emissions Trends in China and Globally
Source: IEA (2012).

CO_2 emissions. Consequently, China has been criticized internationally for a perceived indifferent approach to climate change mitigation. As former US Secretary of State Colin Powell quipped in 2009, "You know what the first thing is that Hu Jintao doesn't think about when he wakes up every morning? Climate change."[3]

Yet such criticism ignores notable developments in China's electricity generation sector, which paint a picture of a nation that is far from indifferent to stemming GHG emissions. Expansion of alternative energy generation capacity has been nothing short of remarkable in terms of pace, scale, and scope of development.

As Figure 6.5 illustrates, wind power in China has flourished. China is now the world's largest wind power market, as measured in installed capacity and annual growth. According to the Global Wind Energy Council,

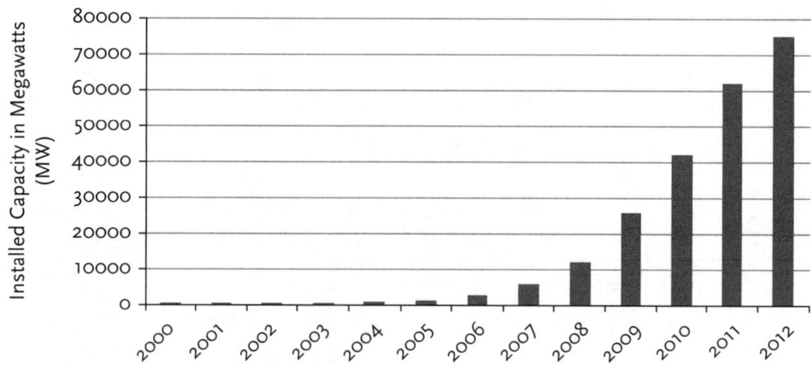

Figure 6.5. Wind Power Expansion in China
Source: Global Wind Energy Council (2013).

in 2012 China added 12,960 MW of installed capacity. The nation now hosts over 26% of all global wind power generation capacity.[4] To illustrate just how aggressive China's energy policymakers have been in supporting wind power development, China released a Mid-and-Long Term Development Plan for Renewable Energy in 2006 that set a 2020 target for installed wind power capacity of 30,000 MW.[5] As of December 2012, 75,324 MW of wind power capacity was already in place, surpassing the 2020 target by 151%. With technically exploitable wind power potential estimated to be as high as 2,548,000 MW—2.5 times current national electricity generation capacity—it appears that wind power holds great promise in China.[6]

In hydroelectric development, generation capacity in China tripled between 2000 and 2010 (Figure 6.6). As of 2010, China possessed 21% of global hydropower generation capacity. China's hydropower output now surpasses the combined hydropower production in Brazil and Canada (the nations with the second- and third-highest levels of hydropower output).

In terms of nuclear power, the nation's fourteen reactors currently provide less than 2.5% of the country's electricity. However, there are 25 reactors under construction—the most in the world—and some nuclear power analysts in China contend that there is a high probability that the government will upgrade its targets to 60,000–70,000 MW for 2020.[7] By 2035, China is projected to displace the United States as the nation with the highest amount of installed nuclear power capacity,[8] and by 2050, the CPC aims to reach 400,000 MWe of installed capacity (roughly 400 reactors).[9]

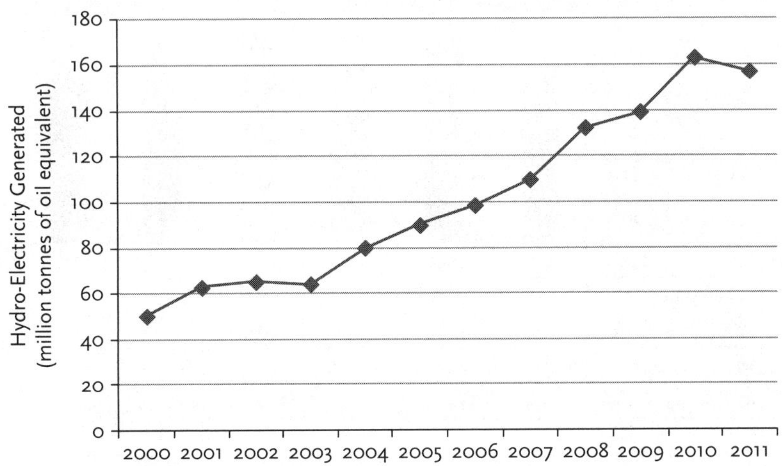

Figure 6.6. Hydropower Expansion in China
Source: BP (2011).

To support these ambitions, the CPC actively sponsors major research and development efforts to improve indigenous designs and develop technological prowess in fuel enrichment, fuel processing, and waste storage.

Even China's notorious coal-fired power sector has undergone sweeping changes, aimed in part at reducing the adverse environmental impact associated with low-tech coal-fired energy generation.[10] A 2006–2009 government initiative to replace small-scale, inefficient coal-fired power plants with more efficient coal-fired generation technologies eliminated 60,000 MW of inefficient coal-fired generation capacity.[11] Consequently, between 1993 and 2009 coal utilization efficiency in power generation improved 18%, from 417 kilograms of coal equivalent per kilowatt hour (kgce/kWh) to 342 kgce/kWh.

Overall, developments within China's electricity sector depict a nation where inroads by alternative energy development have been negated by increases in fossil fuel electricity generation capacity, as the nation strives to develop an electricity network that can effectively support unbridled industrial growth.[12] Between 1990 and 2009, when global GDP in purchasing power parity terms (GDP-PPP) grew by 92.7%, China's GDP-PPP grew 532.8%. In aggregate, China's 2009 GDP-PPP amounted to 19.3% of global GDP-PPP.[13] Over this same 20-year economic boom period, China's total primary energy (TPE) consumption increased 160% from 876.1 million tonnes of oil equivalent (mtoe) to 2,272 mtoe.[14] Figure 6.7 illustrates how expansion of coal-fired power capacity averaging 1,200 MW per week between 2005 and 2010 has negated progress in alternative energy capacity expansion and left China's energy mix virtually unchanged.

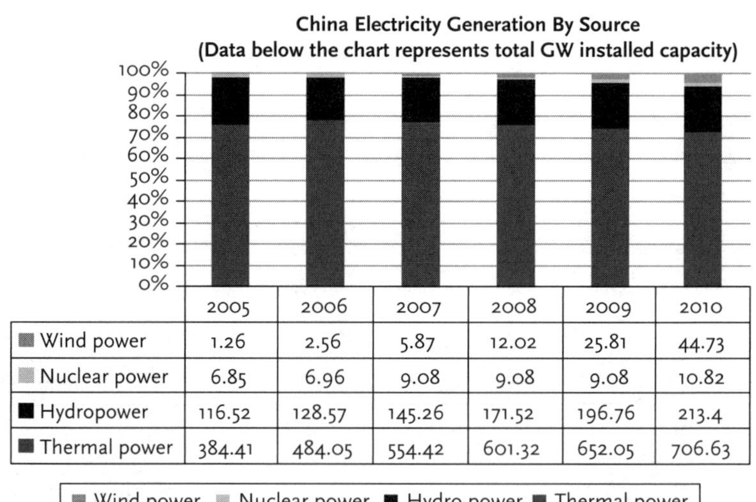

	2005	2006	2007	2008	2009	2010
Wind power	1.26	2.56	5.87	12.02	25.81	44.73
Nuclear power	6.85	6.96	9.08	9.08	9.08	10.82
Hydropower	116.52	128.57	145.26	171.52	196.76	213.4
Thermal power	384.41	484.05	554.42	601.32	652.05	706.63

Figure 6.7. China's Electricity Generation Mix
Source of data: Yuan (2011).

Perhaps surprisingly, given international criticism, China's energy efficiency ratio is comparable to many nations that are critical of China's energy policy. In 2009, China consumed 19.5% of global TPE to generate 19.3% of global GDP-PPP. By comparison, in 2009 the United States consumed 19.4% of global TPE to generate only 17.7% of global GDP-PPP, and Canada consumed 2.7% of global TPE to generate just 1.6% of global GDP-PPP.[15]

Although critics and indeed even prominent Chinese leaders would agree that China's carbon-intensive energy profile is far from desirable, efforts at decoupling energy consumption from economic growth should not be lightly dismissed. Since 1990, CO_2 emissions as a percentage of GDP (PPP) declined 51.6% in China, compared to a global average improvement of 28.2%. China's carbon intensity, which has been estimated at 0.55 kilograms of CO_2 per US$ in 2009, is commensurate with nations such as Canada (0.51 kg/US$) and Australia (0.56 kg/US$).[16] Although China's 2009 per capita CO_2 emission rate of 5.14 tonnes per person was still about 20% higher than the global average of 4.29 tonnes per person, this is considerably lower than average per capita emissions of 9.83 tonnes found in Organisation for Economic Co-operation and Development (OECD) nations.

Clearly, China has a lot to be criticized for in terms of aggregate contributions to global GHG levels; however, it also deserves recognition for endeavoring to decouple CO_2 emissions from the economic growth process. A review of recent government policies demonstrates the scale and scope of the CPC's ambitions.

In 2006, the government released its Eleventh Five-Year Plan of National Economic and Social Development (2006–2010). A key objective was to reduce energy consumption per unit of GDP by 20% from 1.22 tonnes of carbon equivalent (TCE) to 0.98 TCE. Offset by continued economic growth, this meant that TPE consumption would rise 23% to 2.7 billion TCE by 2015. These projections reaffirmed the need to decarbonize the electricity grid in order to abate CO_2 emissions. To address this, the CPC also passed a renewable energy law (REL), which came into effect on January 1, 2006.

The renewable energy law decidedly influenced the pace of renewable energy development. Prior to the passage of the law in 2005, installed capacity for wind power and hydropower were 1,260 MW and 116,500 MW, respectively.[17] By 2010, installed wind power capacity had blossomed to over 50,000 MW (about 5.1% of total electricity generation capacity) and installed hydropower capacity had surpassed 200,000 MW (about 20.5% of total electricity generation capacity). The main gist of this law was that it forced utilities to purchase all generated renewable energy from approved projects at favorable prices. It also established a foundation for renewable energy research and development (R&D), funding for projects in remote areas, access to preferred finance rates, and tax benefits for approved projects.[18]

In September 2007, the National Development and Reform Commission (NDRC)—China's central economic planning body—released its Medium and Long-Term Development Plan for Renewable Energy in China. Overall, the plan established targets for alternative energy of 10% of TPE by 2010 (up from 7.5% in 2005) and 15% of TPE by 2020. To facilitate this, the targets outlined in Table 6.1 were established. In 2007, the government's alternative energy strategy further congealed with the announcement of plans to install 40,000 MW of nuclear power capacity by 2020.

By 2009, installed hydropower capacity (197,000 MW) had already surpassed the 2010 target, while installed wind power capacity (20,000 MW) amounted to quadruple the 2010 target. Although solar PV was slower to develop, 200 MW was achieved by 2009.[19] In aggregate, actual contributions from alternative energy sources amounted to 9.1% of TPE consumption in 2009, just shy of the 2010 target.[20]

On March 5, 2011, the government released its Twelfth Five-Year Plan of National Economic and Social Development (2011–2015). According to the plan, energy consumption per unit of GDP declined by 19.06% between 2006 and 2010, falling just short of the 20% target laid out in the eleventh five-year plan. The plan outlined 2015 targets to further reduce energy consumption per unit of GDP by 16% from 2010 levels, increase the proportion of nonfossil fuels in TPE to 11.4%, and reduce CO_2 emissions per unit of GDP by 17% from 2010 levels (See Table 6.2).[21] It also earmarked, as strategic energy priorities, further development of new-generation nuclear energy, photovoltaic and photo-thermal power generation, wind power technology, intelligent power grids, and biomass energy.[22]

Table 6.1 2007 MEDIUM AND LONG-TERM DEVELOPMENT PLAN FOR RENEWABLE ENERGY IN CHINA

	2005	2010	2020
Hydropower (MW)	117,000	190,000	300,000
Biomass (MW)	2000	5500	30,000
Wind power (MW)	1260	5000	30,000
Solar PV (MW)	70	300	1,800

Source of data: China NDRC (2007).

Table 6.2 2015 RENEWABLE ENERGY DEVELOPMENT TARGETS IN CHINA			
	2009	2011–2015	2015 Projection
Hydropower (MW)	197,000	+120,000	317,000
Wind power (MW)	42,000	+70,000	112,000
Solar PV (MW)	200	+5000	5,200
Nuclear (MW)	10,800	+40,000	50,800

Author's calculations based on data from China NDRC five-year plans. China NDRC website.

There are also indications that the CPC is currently revising its 2007 Medium and Long-Term Development Plan for Renewable Energy in China. The revised plan purportedly targets 300,000 MW of hydropower, 150,000 MW of wind power, 30,000 MW of biomass power, and 20,000 MW of solar PV, for a total of 500,000 MW of renewable power capacity by 2020. If achieved, this would account for almost one-third of China's expected total power capacity (1,600,000 MW) by 2020.[23]

Certitude of meeting alternative energy development targets, progressive improvement in energy efficiency (albeit lower than planned) and a massive reforestation program intended to add 40 million hectare (ha) of forest area between 2006 and 2020, have seemingly provided the Chinese government with the confidence necessary to play a less tentative role in international climate change mitigation negotiations. Prior to the Fifteenth Conference of the Parties to the Kyoto Protocol (COP15) in 2009, the government announced an intention to reduce carbon intensity (as a percentage of GDP) to 40 to 50% of 2005 levels by 2020, effectively eclipsing the objectives set forth by the United States.[24] During the subsequent COP17 conference held in South Africa, China broke from its reactive negotiation approach by expressing a willingness to discuss adopting GHG reduction targets provided that industrialized nations were willing to commit to further GHG reduction targets.

6.3 HISTORY OF WIND POWER DEVELOPMENT IN CHINA

One of the reasons why CPC leaders feel empowered to proactively engage in international GHG reduction negotiations is the remarkable progress the nation has made in wind power development over a short period of time. Although there is evidence of wind power research initiatives as early as the late 1950s, when a number of 10 kW turbines were erected in provinces such as Jilin, Liaoning, and Xinjiang,[25] consolidated research into wind power didn't really gel until the oil shocks of the early 1970s, when China began to develop wind energy primarily to supply electricity to remote communities.[26] It wasn't until 1986 that the first wind farm was completed, in Rongcheng County of Shandong Province. This project was primarily for exploratory R&D and consisted of three imported 55 kW Vestas wind turbines.[27] In 1989, through favorable financial backing, the CPC encouraged the development of two larger-scale proof of concept projects—one in Dabancheng, Xinjiang Province consisting of thirteen Bonus (now Siemens of Germany) 150 kW wind turbines, and one in Sonid Youqi, Inner Mongolia consisting of five USW (now Kentech of the US) 100 kW wind turbines. At the time, the Dabancheng project was the largest wind farm in Asia.

The years 1991 to 1993 represent a period of consolidated R&D and learning by doing. Throughout this period, a host of projects employing different brands of foreign-made turbines and different output capacities were established. Between 1991 and 1992, the Dalankuo wind farm was constructed in Nan'ao, Guandong. This project consisted of three 130 kW and six 150 kW Vestas (Denmark) turbines. Between 1992 and 1993, the Dabancheng project was expanded with the addition of four 300 kW Vestas turbines, four 300 kW Bonus turbines, and four 500 kW Bonus turbines. The Sonid Youqi project in Inner Mongolia was also expanded in 1993 when four turbines from Dutch manufacturer HSM were added. Table 6.3 (next page) demonstrates the limited scale and scope of wind power developments in the early days of the program.

In 1994, the Ministry of Electric Power announced two initiatives—the Chengfeng Plan and the Shuangjia Plan—to move wind power development from the proof of concept stage to commercialization. In support of these programs, regulations were announced that required grid companies to facilitate grid connections to all wind farms and purchase all electricity generated. On-grid tariffs were determined on a project-by-project basis incorporating capital, financing, operation and maintenance costs, and profit margins that were in the neighborhood of 8 to 10%.[28] The allure of guaranteed profits drew private investment into the industry and lead to elevated development. As Table 6.3 illustrates, between 1994 and 1996 cumulative installed wind power capacity increased from 10.14 MW to 57.44 MW, but all of the development involved foreign-made wind turbines.

In 1996, the State Development and Planning Commission (SDPC) announced the Ride the Wind Program, which mandated that any SDPC wind power project must contain at least 40% locally made components.

Table 6.3 CHINA WIND POWER DEVELOPMENT FROM INCEPTION TO CUSP OF BOOM (1989–2007)

Year	# of Projects	# of Turbines	Capacity Added (MW)	Cumulative Capacity (MW)
1989	2	18	2.45	2.45
1991	1	3	0.39	2.84
1992	3	14	3.3	6.14
1993	3	12	4	10.14
1994	5	49	12.9	23.04
1995	5	39	9.6	32.64
1996	10	70	24.8	57.44
1997	10	151	84.65	142.09
1998	10	123	71.9	213.99
1999	12	83	50.25	264.24
2000	12	125	77.34	341.58
2001	11	91	57.21	398.79
2002	13	109	66.31	465.1
2003	18	131	98.3	563.4
2004	15	249	196.75	760.15
2005	36	595	506.91	1267.06
2006	55	1452	1335.65	2602.71
2007	121	3155	3303.65	5906.36
	342	6469		

Source: China Wind Power website.

This regulation catalyzed investment in Chinese plants by foreign wind manufacturers and stimulated a series of joint ventures between foreign wind turbine firms and Chinese manufacturing concerns.[29] It also created a market bias in favor of state-owned enterprises such as Sinovel and Goldwind, which were pursuing government-funded wind turbine development programs. The continued fair-profit guarantees granted to wind power developers along with ramped up market activity created by the Ride the Wind Program caused cumulative capacity to increase tenfold from 57.44 MW at the end of 1996 to 563.4 MW at the end of 2003.

In 2003, the NDRC incorporated lessons from a three-year power pool trial program in Shanghai city and Zhejiang and Shandong provinces into a concession system for selecting large-scale wind power projects.[30] It took a few years under the system for participants to become adept at estimating project costs and benefits; consequently, many bids were unprofitable during the first years of the program. However, in addition to catalyzing growth in large-scale wind power projects, the concession system gave the State Power

Corporation and the NDRC's Energy Pricing Bureau valuable insight into the nature of wind power costing.

In 2004, the government utilized this knowledge in developing on-grid electricity tariff benchmarks that were based on average social costs of power generation. These tariffs varied depending on the province and the technology.[31] Thanks to this new system, 196.75 MW of new wind power capacity was added in 2004—doubling the previous annual growth record set the year before. By 2005, it was becoming apparent that the CPC's support for wind power represented a significant commercial opportunity. Newly installed capacity for the year exploded to 506.91 MW, representing almost 40% of total national wind power capacity.

Despite such impressive development, 2006 marks the true dawn of the commercial age for wind power in China. In January 2006, the Renewable Energy Law of China came into effect. The main gist of this law was that it forced utilities to purchase all generated renewable energy from approved projects at favorable prices. It also established a foundation for renewable energy R&D, funding for the development of projects in remote areas, facilitated access to preferred finance rates, and instituted tax benefits for approved projects.[32] In response, 1,335.65 MW of newly installed wind power capacity was added in 2006 alone—doubling total cumulative capacity.

In September 2007, the NDRC clarified the target for wind power development in its Medium and Long-Term Development Plan for Renewable Energy in China—30,000 MW by 2020. Given such a clear declaration by the CPC regarding renewable energy development strategy, the wind power development market exploded. As Table 6.4 illustrates, since 2006 cumulative capacity has grown nearly thirtyfold, despite global recession. China

Table 6.4 CHINA'S WIND POWER BOOM

Year	Capacity Added (MW)	Cumulative Capacity (MW)	Annual Growth
2006	1339	2599	106%
2007	3311	5910	127%
2008	6110	12020	103%
2009	13785	25805	115%
2010	16482	42287	64%
2011	20013	62300	47%
2012	13024	75324	21%

Source of data: Global Wind Energy Council (2013).

now boasts the most installed wind power capacity in the world and challenges the United States as the fastest growing annual wind power market.

In the remainder of this chapter we will examine the STEP components that have catalyzed such support for wind power in China. Section 6.4 will examine the sociocultural, economic, and technological conditions that provided a nurturing national landscape for supporting China's wind power boom. In section 6.5, the governance conduits—sociocultural, technological, and economic forces that influence political behavior—will be highlighted in order to demonstrate how CPC policy evolved in response to feedback from these conduits. In section 6.6, crucial regime characteristics of China's system of governance will be examined in order to understand what engendered the types of support policies that the CPC developed for stimulating wind power development. Finally, in section 6.7, the policy conduits—the policy initiatives that influenced sociocultural, technological, and economic development—will be summarized in order to understand how the policy response meshes with these other elements. The chapter will conclude with a discussion of current trends in China's national landscape that might have the greatest impact on wind power market development.

6.4 UNDERSTANDING THE GENERAL FORCES FOR CHANGE

6.4.1 Sociocultural Landscape

Many visitors to China's major commercial centers (such as Beijing or Shanghai) remark on the level of activity that they are confronted with. There is an optimistic buzz in the air that is characteristic of nations in the midst of economic ascent. Most assuredly, the average inhabitant in these new megalopolises harbors concerns. Escalating land prices and the skyrocketing cost of living presents sizable challenges to many of the rural Chinese migrants who have come to China's mega-cities in search of employment. Traffic congestion is severe and environmental problems abound. Yet there is an underlying sense of awaking: a sense that the future holds great promise.

One of the underlying reasons for such optimism stems from the strength of China's workforce. Premier Zhou Enlai's Four Modernizations initiative, which was launched in 1963, catalyzed large-scale government investment in enhancing scientific and technological knowledge. Almost 50 years later, China is now a nation with notable workforce diversity. Although it has an enormous blue-collar workforce to support China's massive manufacturing ventures, it also boasts competencies in professions such as engineering, science, electronics and medicine.

With that said, there is also a growing poverty gap with over 300 million Chinese still living in impoverished conditions. As affluence in the commercial centers continues to mount, the disparity between rural and urban quality of life is of growing political concern.

6.4.2 Economic Landscape

Sociocultural justifications aside, there at least three key economic reasons for China to be optimistic regarding its economic future. First, it is a nation of abundant natural resources. Water resources rank sixth in the world, supporting both hydropower generation (that ranks first in the world) and widespread agricultural activity (which has established China as a major producer of wheat, corn, rice, and cash crops). China's proven deposits in tungsten, titanium, zinc, rare earths, magnesite, plaster stone, graphite, tin, mercury, asbestos, coal, nickel, lead, iron, manganese, and platinum all rank within the top five globally. Given the natural resources and diverse workforce found in China, its position as "factory to the world" is duly justified.

Second, China's level of private, public, and corporate savings eclipse that found in most other nations, giving China growing financial clout. Throughout the 2000s, gross national savings in China ranged between 39.2% to 54.3% of gross national income. On average, the nation saved one out of every two dollars that it generated. This level of financial prudence becomes particularly striking when one contrasts this with the gross national savings rates in affluent nations such as Japan (26.8%–33.2%) and the United States (12.1%–17.7%) or with savings rates in developing nations such as Mexico (18.6%–25.5%) and India (21.4%–37.6%) over the same period.[33]

Third, the influx of foreign investment in China is staggering. In 2010, China ranked second to only the United States (US$236.2 billion) in foreign direct investment attracting US$185.1 billion. To put foreign direct investment in China into proper perspective, $15 out of every $100 invested globally in foreign markets was invested in China.

6.4.3 Technological Landscape

China's diverse workforce, its natural resource wealth, and the influx of foreign investment have fostered a unique technological environment where rudimentary agricultural practices, low-tech manufacturing, and high-tech industry thrive simultaneously albeit in different regional proportions. In many impoverished rural areas, rudimentary agricultural practices undertaken by poorly educated laborers remain as a remnant from Mao's Great Leap Forward. In the 1980s, Deng Xiaoping's vision of a socialist market economy led to the emergence of a thriving state-owned enterprise sector which specialized in low-tech manufacturing. As a result of these reforms, China now leads the world in primary manufacturing of commodities such as steel, concrete, and flat glass.[34] In the 2000s, market reforms under Hu Jintao have catalyzed an upwelling of high-tech industrial capability through policies that have encouraged foreign direct investment and joint ventures.[35] Many of the world's most sophisticated manufacturers have established plants in China to take advantage of China's increasingly well-educated labor force, its lower cost base, and improved access to the blossoming Chinese consumer market.

China's evolution in terms of both economic influence and vocational competency, along with an influx of foreign investment and know-how, has engendered a technological environment in China that exhibits an ever-increasing capacity to support technological change. As the next sections will outline, the evolution of China's national landscape feeds into evolutionary changes within China's governing regime and this feedback process interfaces with regime characteristics to produce the wind power development policy that currently exists in China.

6.5 INFLUENCES ON GOVERNMENT POLICY

6.5.1 Sociocultural → Political

Despite progressive reform, the CPC has a well-documented disposition for responding to large-scale civil disobedience with force.[36] Perhaps as a result, although there have been some exceptional instances of civic activism in regard to ill-conceived large-scale energy projects (such as the Three Gorges Dam project),[37] government-led development projects tend to engender public acquiescence. On the one hand, this predilection enables China's state-owned enterprises (SOE) to pursue the development of energy projects that carry high social costs, which in other nations might not be tenable. Public acquiescence has played a role in enabling both government commitment to nuclear power and ongoing support for environmentally damaging coal-fired power plant expansion.[38] On the other hand, low levels of civic activism also implies that NIMBY (not-in-my-back-yard) opposition to wind projects, which tend to emerge as wind power capacity evolves in a given country,[39] will be weaker in China.

Amplification of engineering competency has provided the skill base necessary for policymakers to begin thinking of alternative energy in strategic economic development terms. Since Zhou Enlai's Four Modernizations initiative in the early 1960s, China has long exhibited a commitment to nurturing engineers.[40] In fact, many of China's ruling elite received training as engineers.[41] The scale and scope of engineering programs in China has progressively grown. By 2006, higher education enrollments reached 18.85 million students, with one-third enrolled in engineering programs. To put this into perspective, there were 50% *more* students enrolled in engineering programs in 2006 in China than there were enrolled in all higher education programs in 1990.[42]

Within China's burgeoning wind power industry, there are now more than 60 domestic turbine manufacturers capable of competing at international levels and a plethora of smaller enterprises that engineer affordable components in support of these manufacturers.[43] The Chinese nuclear power industry enjoys similar human capital benefits.[44] Conversely, due to advances in education, the numbers of unskilled laborers are also diminishing which debases the importance of coal mining and coal-fired power as sources of jobs for unskilled laborers. One recent study in China estimated that between 2006 and 2010, there were 472,000 net job gains from alternative energy projects that displaced coal-fired plant projects.[45]

In terms of community expectations, there is evidence of public dissonance and escalating public pressure to minimize environmental degradation.[46] In the first decade of the twenty-first century, 8% of China's population was purportedly adversely affected by environmental disasters caused in part by inefficient development planning.[47] Those who took in the splendor of the Beijing Olympics would have also noticed the shroud of smog enveloping many of the venues. International experience suggests that growing affluence will progressively amplify public demand for less environmentally invasive industrial practices and this engenders support for alternative energy.[48]

Modern Chinese society is globally connected and has aspirations to match the economic well-being of industrialized nations. As society evolves economically, perceptions of what constitutes acceptable governance also evolve. In energy policy, both political and public perspectives on good energy governance now reflect perspectives found in most industrialized nations. Energy policymakers are now expected to establish a balance in terms of energy affordability, accessibility, minimal ecological impact, and energy supply security.[49] This means that the 1990s trend of building one or two coal-fired power plants each week is losing political favor.

Nevertheless, the growing poverty gap that was alluded to in the previous section makes the government sensitive to the price of commodities deemed to be social goods. Consequently, there is pressure on the government to manage energy costs in a way that does not create undue hardship for impoverished rural communities.

Aside from increasing affluence, one key driver of changing perceptions on good energy governance arises from public health concerns. In China, respiratory and cardiovascular problems stemming from pollution has become the primary cause of poor health.[50] Heightened public concern over the adverse health impact of fossil fuel electricity generation technology is increasingly fueling public support for alternative energy technologies, which until recently included nuclear power.[51]

In the 1980s, nuclear power development in China was seen as a symbol of national pride.[52] However, following the nuclear crisis in Fukushima Japan, public confidence in nuclear power development has deteriorated. Although many analysts believe that plans for further nuclear power expansion will not be derailed by the Fukushima disaster, the media and the public are now beginning to question the safety of overly ambitious nuclear power development.[53] The convergence of these two trends—health concerns stemming from coal-fired power and skepticism over the safety of nuclear power—is engendering increasing public support for renewable energy solutions in general and wind power in particular.

6.5.2 Economic → Political

The progressive escalation of fossil fuel prices over the last five years has had some predictable consequences in China. High fossil fuel prices have heightened the economic contribution of fossil fuel resources in resource abundant regions. This is particularly salient for coal-rich rural provinces such as Shanxi. However, the escalation of coal prices has also diminished the national incentive to utilize coal for generating power. Moreover, as coal prices escalate, alternative energy technologies such as wind power and mini-hydro have become increasingly cost competitive. The evolving competitive dynamics partly explains aggressive development aspirations for wind power, hydropower, and nuclear power in China.

In addition to price inflation, fossil fuel commodity prices have been extremely capricious in recent years. Price volatility is particularly damaging to utilities and coal-fired power generators. The NDRC's Price Department has a practice of setting wholesale prices based on province-specific average social costs, which tends to push inefficient generators to improve operating costs. Therefore, coal price volatility (which accounts for between 50 to 70% of coal-fired generation costs) complicates the process of rate setting. Moreover, when coal prices exceed benchmark prices for an extended period, it can create enormous financial hardship on power generation firms because the NDRC only allows 70% of coal price increases to be passed along to end-users through tariff increases. Although it is true that sudden dips in the cost of coal can cause windfalls for these same firms, coal price volatility is occurring during an inflationary trend, which suggests that there will be more losers than winners.[54] Indeed, a recent study of coal-fired power plants in 10 Chinese provinces found that all the coal-fired plants examined posted financial losses during the first three quarters of 2010.[55] In short, price volatility incentivizes utilities to embrace forms of electricity generation that possess more predictable cost profiles. This enhances the appeal of alternative technologies (particularly wind power and hydropower) that exhibit more stable variable cost profiles.

The economic cost of externalities associated with electricity technology also influences energy market dynamics. Chiefly, reliance on coal-fired electricity production has contributed to undesirable levels of atmospheric pollution. Additionally, as mentioned earlier, transporting coal from supply to demand centers in China congests rail networks, resulting in transport delays and increased costs for other transportable products. In 2007, 1220.8 million tons of coal was shipped an average of 607 km/ton along China's rail arteries, comprising 46.7% of all goods shipped by rail in China.[56] However, coal is not the only villain in the electricity sector. In hydropower too, mega projects such as the Three Gorges Dam have produced extensive environmental and social costs that the government is keen to avoid.[57] When there were no economically viable technological alternatives to coal-fired electricity and large-scale hydropower, bearing the cost of negative externalities was considered unavoidable; however, this is all changing as costs for renewable and conventional energy converge.

Increasingly, international financial incentives and disincentives also influence China's electricity market. The Clean Development Mechanism (CDM), which allows fungible Certified Emission Reduction (CER) credits to be generated for approved alternative energy projects in developing nations, has played an enabling role in the development of alternative energy projects in China. CDM projects in China have accounted for 58.8% of global CERs issued since the program began.[58] Furthermore, an agreement made during the COP17 Conference in South Africa will give developing nations (such as China) access to a US$100 billion Green Climate Fund. Therefore, despite the possibility that the Kyoto Protocol (and the CDM) might collapse, international financial incentives for China to support renewable energy expansion remains. In terms of disincentives, international support appears to be amassing for border taxes—a carbon tax applied onto imported goods based on the energy profile of the exporting nation.[59]

This is a disturbing development for Chinese firms, because many of their exports are energy intensive. In 2008, CO_2 emissions associated with the production of goods for export accounted for approximately 3.1 gigatons of China's aggregate CO_2 emissions of 6.5 gigatons.[60] Any border taxes applied to Chinese exports could have a significant economic impact on industrial profitability. Consequently, it is in the interests of the CPC to ensure that China's electricity infrastructure is flexible enough to mitigate the impact of international disincentives of this type.

The economic security justification for coal-fired power is also crumbling, and this will alter market dynamics. Historically, the latent economic value of China's coal reserves has underpinned a coal-dominated electricity generation sector. However, given that the consumption of coal-fired electricity is expected to double in China by 2035, coal will only be an abundant domestic resource for a few more decades.[61] Meanwhile, wind power, solar PV, and nuclear power are increasingly seen in political circles as technologies that will enhance economic security.[62]

6.5.3 Technological → Political

Grid coverage and resiliency in China is a concern. In response to unprecedented growth in demand for electricity, the government has initiated numerous grid expansion projects to connect new power sources and interconnectivity projects between China's regional power grids to improve grid resilience. Between 2005 and 2009, grid investment rose 150% from RMB153 billion (approx. US$25 billion) to RMB385 billion (approx. US$63 billion).[63] Nevertheless, as Figure 6.8 (next page) indicates, since 1978 investment in grid infrastructure has lagged behind investment in generation capacity expansion.[64] Consequently, the power grid is no longer capable of effectively supporting installed capacity. To illustrate, in 2010 it has been purported that only 70% of installed wind power capacity was actually connected to the grid.[65] Of greater consternation, this also suggests that requisite funding for R&D to enhance smart grid technology and improve network resilience to incorporate higher contributions from renewable sources has been insufficient to keep pace with renewable energy supply expansion.

Although insufficient infrastructure partly explains why 30% of total installed wind power capacity was not connected to the grid in 2010, another impediment is the inherent rigidity of China's electricity mix. Coal-fired power dominates China's electricity mix. The current system lacks sufficient peak-load capacity to dampen power fluctuations from stochastic renewable energy technologies, such as wind power. Consequently, although China's

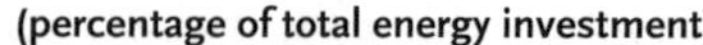

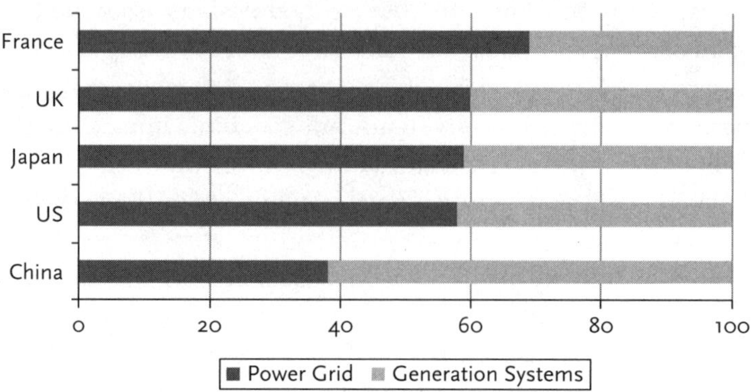

Figure 6.8. National Comparisons of Investment in Energy Infrastructure vs. Generation Systems Since 1978
Source of data: Li (2012).

renewable energy law requires utilities to purchase all available renewable energy at prescribed prices, there is evidence that many utilities have been slow to accommodate new projects because the absence of sufficient peak-load technologies forces utilities to run coal-fired power plants at suboptimal levels to provide peak-load support.[66] This inflates generation costs and has exacerbated utility resistance to wind energy.[67]

The geographic separation of renewable energy supply centers from key demand centers also poses logistical challenges and further inflates costs. Seventy-eight percent of China's hydropower resources are concentrated in the sparsely populated west. Meanwhile, the 11 provinces with the highest population concentrations possess only 6% of total hydropower resources.[68] The wind power story is similar. The areas of greatest onshore wind power potential lie in China's north—in regions such as Xinjiang, Gansu, and Inner Mongolia. Conversely, aside from Hebei and Jiangsu Provinces, the major electricity demand centers in eastern and southern China possess much lower onshore wind power potential. In combination, this also suggests that offsetting stochastic wind power flows is problematic because hydropower resources predominate in the west while wind power potential predominates in the north.

All of this points to the foundations of a logistical dilemma. In order to tap China's abundant renewable energy potential, grid connections must be established to deliver electricity from supply centers to geographically distant demand centers. As the next section describes, the CPC is striving to address this challenge but it bears noting that the geographic challenge of getting energy to demand centers is not new in China. Many of China's coal mines are located in China's heartland, far from the major coastal demand centers. Consequently, coal transport uses almost half of all rail transport capacity in China.[69] Therefore, to state that renewable energy in China is constrained by geographic barriers (as some renewable energy critics suggest) contradicts China's historical propensity to resolve such challenges.

Energy resource management has become an added concern for China's energy policymakers. Up until the first half of the 1990s, China was self-sufficient in its TPE supply. However, consumption growth rates ranging between 5 to 8% per year resulted in China becoming a net TPE importer in 1997.[70] In coal, rich reserves have permitted China to maintain a net export balance. However, since 2000 China's export surplus in coal has declined from 52.9 million tons to 5 million tons in 2008.[71] In petroleum, China imported only 7.6% of its supplies in 1995, but by 2007, petroleum imports exceeded 50% of total oil consumption.

China's voracious consumption of energy has also significantly diminished fossil fuel reserve-to-production ratios. Table 6.5 provides a comparison of fossil fuel reserve-to-production ratios for China, the United States, India, and the global average. At current rates of consumption, China may have less than 40 years remaining before it becomes almost entirely dependent on imported fossil fuel.[72] Alarm over fossil fuel resource depletion has catalyzed a number of initiatives designed to improve energy efficiency and stimulated political support for alternative energy development.[73]

The broadening of technological expertise provides policymakers with electricity generation diversification options that did not always exist. Prior to the 1970s, China's two core electricity generation technologies were coal-fired power and hydropower. Yet even so, engineering quality was desultory, as evidenced by the tragedy of the Banqiao Dam in Henan Province, which collapsed in 1975 causing deaths in excess of 100,000. However, in the 1970s, technological expertise began to gel in new energy areas. China's commercial nuclear program formally began in 1972 with the 728 Project,

Table 6.5 COMPARATIVE RESERVES TO PRODUCTION RATION FOR COAL, OIL AND NATURAL GAS

(Years)	China	World Average	United States	India
Coal	41	122	224	114
Oil	11.1	42	12.4	18.7
Natural Gas	32.3	60.4	11.6	35.6

Data from Zhang (2011).

an initiative to develop submarine reactors.[74] Since then, Chinese engineers have completed 15 reactors in four locations, totaling 11,881 MWe of installed capacity.[75] One of the government's key objectives in the early days of the program was to amass requisite knowledge and capability in all stages of the fuel cycle (uranium extraction and refining, fuel processing, plant construction, operation, and decommissioning) so that China could become self-sufficient by the end of the 1990s. As a consequence, Chinese policymakers now consider nuclear power expansion as a strategically attractive alternative to coal-fired power.[76] Similarly, although the first wind farm that was completed in 1986 in Shandong used imported turbines,[77] technological transfer has resulted in four Chinese manufacturers within the top 10 wind power turbine manufacturers in the world, and scores of component suppliers.[78]

The breadth of expertise that now exists within power engineering circles in China undermines the already wavering strength of coal-fired power interests. Historically, the coal industry has been an important employer of China's rural, unskilled laborers—employing nearly 4.9 million workers in 64,000 coal mines in 1997. Coal revenues have also been of high economic importance to coal-rich provinces such as Shanxi, Inner Mongolia, Liaoning, Heilongjiang, Shandong, and Henan. Consequently, a network of political support has developed around the coal sector.[79] However, since the late 1990s, government initiatives to centralize coal production (to enhance safety and improve economies of scale) reduced the number of coal mines to 3,215 in 2008. As of 2006, employment in coal mining had fallen to 2.6 million, a 47% decline from 1997 levels.[80] Although the economic importance of coal as a commodity remains high in many of the coal-rich rural provinces, the political importance of the coal mining sector has diminished as employment numbers fall and educational levels improve, thereby broadening employment prospects for workers. Furthermore, with exports now comprising almost 50% of all Chinese coal production, pressure to sustain the coal mining industry through a commitment to coal-fired power plants has lessened.[81] Overall, coal-fired power plant advocates enjoy far less political leverage nowadays than they did a mere decade ago.

6.6 POLITICAL INFLUENCES ON POLICY

6.6.1 National Political Structure

China has been described as a single-party, socialist state, governed by the Communist Party of China (CPC) since 1949. The CPC describes its governing structure as democratic centralism, wherein political power is centralized within a tight network comprised of the National People's Congress (NPC), the president, and the State Council (the centralism feature). In theory, within this centralized political network matters of policy are debated and alternative perspectives are aired (the democratic feature).

The NPC is China's highest state body, the only legislative house in China and in theory the nation's political core, as it is responsible for electing the president and the premiere of the State Council. Members of the NPC are elected by the Provincial People's Congress, which is in turn consists of elected members of the People's Congress at the county level. Members of the People's Congress are elected by voters, so there is a degree of theoretical grassroots power as voters have some say over who represents their interests within the CPC.

In practice, the true core of political power resides within the Central Committee of the Communist Party of China, where the General Secretary is the highest ranking official, head of the Secretariat, a standing member of the Politburo and President of China. It is within the Central Committee of the CPC that policy is formulated. Furthermore, the General Secretary of the Central Committee nominates China's Premier, who is responsible for organizing and supervising China's civil bureaucracy.

In order to maintain centralist control, governance in China is based on a dual leadership system whereby the heads of local bureaus or offices report to both a local leader and the leader of the corresponding office, bureau, or ministry that is at the next highest functional level. Whereas in the past the CPC managed to preserve authoritarian control over regional politics via control over finances, reforms beginning in the Deng Xiaoping era have gradually resulted in a diffused power base, giving local or regional bureaucrats greater decision-making autonomy. In fact, there are well-documented cases of leaders in provinces such as Guangdong and Zhejiang and municipalities such as Shanghai and Tianjin pursuing initiatives that are at odds with CPC policies. The Special Administrative Regions of Hong Kong and Macau are also theoretically granted a high degree of political independence, provided the policies enacted by regional leaders do not conflict with central government dictates. However, when regional initiatives clash with central policies, there have also been well-documented incidents of Beijing reasserting control.

In regard to energy policy, the devolution of national political structure from tight control at the core to ever-increasing economic autonomy at regional and local levels has engendered a policy regime within which regional disparities are beginning to emerge. This fragmentation of power comes with pros and cons. On the one hand, regions that are rich in particular energy resources embrace policies to exploit these resources. Therefore, as opposed to the days of stronger central government control where local apathy led to poor implementation of central policy, the economic autonomy granted to provinces and municipalities engender higher levels of local initiative. On the other hand, it is becoming increasingly problematic to create a unified energy policy with different regions pursuing different development agendas. Going forward, creating a unified national energy strategy in order to provide the integrated efficiencies necessary to support higher levels of wind power under a system of increasingly decentralized administrative control represents a key challenge for the CPC.

6.6.2 Governing Party Ideology

Since the 1970s, the CPC has been absorbed in an ongoing experiment involving a dual track governance system that seeks to establish a balance between socialist and market economy principles. In the 1970s, agriculture was decollectivized in order to improve productivity and special economic zones were established in order to encourage international trade. In the 1980s and 1990s, many government-controlled industries were either privatized or control was delegated to local levels of government. In the 1990s and 2000s, policies were initiated to encourage foreign direct investment and reduced tariff barriers in order to make Chinese industry more competitive.

Throughout this reform process, a few ideological principles have emerged as dominant. First, there is the principle of pragmatism as espoused by Deng Xiaoping's quote referenced at the beginning of this chapter that the color of the cat is less important than its ability to catch mice. Thanks to this ideological undercurrent, the CPC's approach to policymaking is malleable and subject to change based on performance. Second, China's development policy is predicated on the principle of scientific development, which advocates design, implementation, and evaluation of policy based on multifaceted goals that integrate economic, social, and environmental objectives. Third, the CPC clearly understands that it is imperative to avoid "building a cart behind closed doors" (in Mandarin 閉門造車). This idiom refers to recognition within the CPC that for Chinese industry to succeed, Chinese businesses must be exposed to world-class competition and incorporated upon a base of world-class technology.

In combination, these concepts of pragmatism, scientific development, and progressive exposure to competition have produced an electricity generation sector that is dynamically evolving. Although coal-fired power still forms the core of China's electricity generation sector, the government is well aware of the need to diversify for the economic, environmental and resource security reasons stated earlier. In attempting to encourage such a transition, it has enacted a number of policies to nurture local wind turbine and component manufacturers and is striving to provide the electricity grid infrastructure and economic climate necessary to encourage further expansion of wind power capacity.

On the other hand, a pragmatic approach to development planning, support for scientific development and a belief that China's industrial competitiveness must be honed through progressive exposure to world class competition also lends support to the promotion of nuclear power and carbon capture and sequestration (CCS) research. As mentioned earlier, the CPC still intends to commission 60,000 MW of nuclear power by 2020. It is also well down the road in piloting CCS technology and has already collaborated with Australian partners on domestic trial projects. For example, the Xian Thermal Power Research Institute has already developed a CCS system that is purportedly capable of recovering more than 85% of CO_2 emissions, using equipment that has been entirely designed and manufactured in China.[82]

6.6.3 Fiscal Health

The financial health of the CPC plays an enormous role in influencing the development of energy policy, because many of China's electricity generation firms are still government-owned and this means that the ability to commission new power projects is directly linked to fiscal financial health. This is particularly true at the provincial level. Since the 1980s, provincial-level control has grown to the point where 45% of generation assets were owned by provincial-level companies in 2006.[83]

The financial ability of the CPC to facilitate technological transition within the electricity sector is enhanced thanks to an expanding economy and a high level of government savings. Despite extensive investment demands to bolster infrastructure and enhance social services, the government has managed to maintain fiscal surpluses since 1992.[84] Table 6.6 outlines the extent of savings accrued. Between 2002 and 2008, the CPC amassed over US$922.7 billion in savings. Overall, the combination of a high degree of government savings and a proclivity on the part of the CPC to strategically guide economic development means that policymakers are able to think and act strategically in regard to electricity policy. In testament to this, as mentioned earlier, China's eleventh national development plan (2006–2010) established robust targets for hydropower, wind power, solar PV, biomass,

Table 6.6 FISCAL SURPLUSES IN CHINA (AS A PERCENTAGE OF GDP)							
	2002	2003	2004	2005	2006	2007	2008
Govt Savings Rate	3.2%	3.3%	4.1%	4.2%	5.0%	6.4%	6.4%
China GDP (2011 US$trillion)	1.45	1.64	1.93	2.26	2.71	3.49	4.52
Govt Savings (2011 US$billion)	46.4	54.1	79.1	94.9	135.5	223.4	289.3

Source of data: Govt savings (Ma, 2010); China GDP (World Bank); GDP Savings (author's calculations).

and energy efficiency. In 2009, China purportedly invested US$34.6 billion in renewable energy development, far exceeding the United States, which posted the second highest level of renewable energy investment of US$18.6 billion.[85]

6.6.4 Policy Regime

The historical role of the CPC in setting broad economic development policy has enhanced local control over specific initiatives. An ideology that is pragmatically focused on developing an electricity system that is cost competitive, efficient, and supportive of scientific development principles has led to China tapping into its national savings to catalyze massive development in wind power in some regions. As the next section will demonstrate, it has done so through robust but gradualist policies aimed at influencing behavior in the sociocultural, technological, and economic realms.

Before moving onto the next section, the topic of gradualism merits attention because gradualism defines the CPC's approach to policymaking in wind power development (and indeed, in most other areas). In Mandarin there is an expression favored by Deng Xiaoping to explain the CPC's approach to policy making—"crossing the river by feeling for stones" (in Mandarin 摸著石頭過河).

Decision making within the CPC has been the heavily influenced by the Great Leap Forward, which was a policy between 1958 and 1961 that was intended to catalyze a rapid transition from agrarian to industrial practice in China but instead resulted in famine and the deaths of millions.[86] Accordingly, since Deng's era of leadership China's leaders have enacted policy in small steps, testing out the impact on a small scale and adapting policy based on stakeholder feedback, before launching more consorted efforts which are further monitored and revised as required. In the next section, which reviews the policy response of the CPC, the presence of a gradualist approach in relation to wind power development will be evident.

6.7 THE CULMINATION OF INFLUENCES

Previous sections provided the contextual background necessary to understand why the CPC has adopted a rigorous wind energy development strategy. The social, technical, and economic influences described have all incentivized the government to react and change its strategy regarding electricity generation governance, which historically emphasized coal-fired electricity as the dominant base-load technology supported by hydropower to provide peak-load support. This section summarizes the main cause-and-effect relationships that have catalyzed change.

First, from a social perspective, there is increasing awareness that emissions from coal-fired power plants are largely responsible for the hazes, which make Chinese urban centers the most polluted in the world. Combined with the health costs associated with urban pollution, the government had both social and economic incentives to take action to abate coal-fired power plant emissions. The first step was to phase out all smaller, inefficient coal-fired power plants. Over the past decade, over 60,000 MW of highly pollutive technology has been removed from the electricity grid. The next step was a gradual move to diversify the energy mix by supporting expansion of wind power, mini-hydro and nuclear power. While this progresses, the government continues to fund research into carbon capture and sequestration. Throughout this phase-out process, the gradualist approach of the CPC is evident.

Second, China's ever-increasing base of engineering competency, along with applied know-how gained through joint ventures with foreign wind turbine manufacturers, has given the government the confidence necessary to encourage indigenous development of wind power; however, the government also deserves praise for its policies in support of amassing such competency. The process of enhancing engineering prowess in electricity generation technology has been methodically planned. The CPC has even gone as far as to establish a specialized university—the North China Electric Power University in Beijing—with separate departments solely dedicated to advancing renewable energy and nuclear power science. To ensure China's new engineers developed world-class applied experience, the government has pursued a policy that encouraged joint ventures with foreign nuclear power and wind power firms.[87] For example, China's first nuclear power plants were joint ventures that helped China's nuclear power engineers acquire applied technical knowledge. As China's nuclear power engineers gained applied experience, the government adopted a policy of allowing selective competition in order to push domestic entities to higher levels of performance.[88] Similarly, in wind power development, the Chinese government initially established local content regulations in order to ensure that Chinese firms were given the chance to develop manufacturing competency and scale. Once Chinese wind power manufacturers became established, these regulations were relaxed.[89]

Third, it has become increasingly apparent to CPC energy planners that the transition to wind power depends on improving the capacity to deliver wind power from geographically remote supply regions to coastal demand centers. Consequently, the CPC is currently pursuing plans to spend billions over the next 20 years to enhance grid coverage and resiliency. For example in 2007, the government awarded a contract to Siemens to construct a 1400 km high-voltage 800 kV DC transmission system with 5000 MW of capacity to connect Yunnan to Guangdong. Another project is now underway to connect Xinjiang's wind power riches to the rest of China. There is also a 1100 km long power line being constructed between Qinghai and Tibet that will cost roughly US$850 million (about US$772,000 per kilometer) due to the complexity of the terrain to traverse. Overall, it has been estimated that on average a 500 kV, 250,000 kV amp transmission line costs about $220,000 per kilometer in China.[90] In other words, although there is awareness that grid infrastructure needs to be improved, such projects take time and the sheer cost of expanding coverage and enhancing grid resilience forces policymakers to make some delicate fiscal trade-offs.[91] In the short term, this dilemma enhances the attractiveness of larger nuclear and coal-fired generation plants that can be located near existing grid connections and partly explains the government policy of encouraging the development of wind farms of 100 MW or more.[92]

As the grid fortifies, the technological barriers to connecting renewable energy installations begin to diminish; however, in the meantime, China's pragmatic political ideology sees the government directing its state-owned utilities to commission increased capacity in coal-fired power plants in order to prevent disruption to the electricity supply. Nevertheless, just because China is still building coal-fired power plants at a remarkable pace, this should not be misconstrued to infer that China is not interested in weaning itself from a dependence on coal-fired power. The social, technological, and economic signs that China needs to transition away from coal-fired power are clear and well understood in political circles. Pollution, dwindling coal reserves, coal price inflation, the capricious behavior of coal markets, and the congestion of China's rail networks caused by coal transportation are all influencing CPC energy policymakers to endorse wind power development.

Finally, within the CPC's own political framework there are forces that are encouraging the development of wind power. The decentralization of administrative governance as part of the CPC's ongoing political reform process designed to recognize regional strengths and weaknesses to facilitate scientific development has encouraged political leaders in provinces of abundant wind power potential to leverage central government financial support for local economic benefit. The CPC's philosophy of encouraging more competitive domestic firms by allowing a gradual easing of entry barriers for foreign firms is yet another example of gradualist ideology influencing policy design. Last but not least, the fiscal health of the CPC allows it to continue to pursue its strategy of driving change in China's energy sector by subsidizing alternative energy and by pursuing its own development projects through state owned enterprises.

6.8 WHAT TO EXPECT GOING FORWARD

The dominant trends in China's national landscape suggest that support for wind power will continue to escalate. Economics that tends to have the greatest influence on electricity system development favors wind power over coal-fired power. As outlined in Chapter 1, in December 2011 coal was trading at three times what it was in December 2000. Given IEA estimates that global coal consumption will increase by 53% between 2007 and 2030,[93] it would be reasonable to expect coal prices to be at least as high as—if not higher than—current levels. Conversely, wind power costs have exhibited a gradual downward trend as technological innovations offset inflated costs of key wind power system components (i.e., steel), labor, and siting.[94] Moreover, there is currently a glut of wind power suppliers in China.[95] This suggests that competition for business will ensure that China remains a cost-effective wind power market over the next decade.

With that said, there is no guarantee that wind power will be the preferred technology going forward. There is still pervasive support for nuclear power expansion in the nation. Many of China's political leaders possess training as engineers and an undercurrent of technocratic optimism is evident when discussions turn to the future of nuclear power.[96]

From a sociocultural perspective, growing affluence is a key element that should fuel elevated support for wind power. History demonstrates that as societies become more affluent, tolerance toward environmentally invasive industrial practices wanes, willingness to accept higher living costs in return for improved living conditions escalates and public propensity to demonstrably register discontent amplifies.[97] If these historical lessons are valid for China, general public support for transitioning away from pollutive electricity generation technologies will favor wind power over nuclear power as the dominant emergent utility-scale technology.

From a technological perspective, the scale of wind power expansion in China will likely be influenced to the greatest extent by technological inroads made in commercializing CCS technology and political inroads made in convincing the general public that nuclear power is a safe technology at high levels of penetration. As mentioned earlier, the Xian Thermal Power Research Institute has already developed a CCS system that is purportedly capable of recovering more than 85% of CO_2 emissions;[98] however, progressing from CO_2 capture capabilities to commercial-scale capture and storage represents technological progress that has yet to be proven in any nation. Regarding nuclear power, the disaster in Fukushima, Japan has heightened public sensitivities toward the use of nuclear power, both in China and around the Asian region. However, there is no indication that what transpired in Japan has altered the perspective of Chinese policymakers in regard to long-term plans to progressively ramp-up installed nuclear power capacity. Therefore, current indications suggest that nuclear power will continue to challenge wind power for the title of fastest growing alternative technology in China. Nevertheless, barring any currently unanticipated groundbreaking developments in CCS research, one can expect a continued amplification of support for wind power, provided that Chinese power grids continue to improve.

Power grid coverage and resilience represents a vexing bottleneck. The current grid fortification projects outlined in this chapter will enable enhanced wind power contributions to a certain extent; however, China's grid infrastructure is for the most part still rudimentary and will require significant technological upgrading if wind power in China is to have a chance of surpassing the 30% penetration level.

Although the outlook for continued wind power development in China appears rosy, there are two geopolitical concerns that could significantly suppress wind power diffusion in the short run—continued global economic stagnation and the collapse of the Kyoto Protocol. There are indications that the global economic slowdown that began in 2007 is beginning to have a corrosive effect on Chinese fiscal health.[99] If domestic expansion continues to outpace export growth, the ability of the CPC to finance the necessary power grid infrastructure fortification projects that are needed to support further wind power development will be weakened. Consequently, if the government is forced to rein in its domestic expenditures in order to maintain fiscal health, all domestic infrastructure projects—including wind power developments— face enhanced scrutiny and diminished growth prospects. If the Kyoto Protocol collapses and there is no other international agreement to replace it, the political pressure for China to expedite a transition away from coal-fired power could weaken. In conjunction with continued global economic stagnancy, it may be that China's ambitious wind power development plans will be implemented at a slower pace.

In summary, is China going on a carbon-free diet? Indications are that the answer is yes. Just as every successful nutritionist knows that extreme diets seldom yields lasting results, China's leaders appear intent on applying the gradualist approach that has served them so well in other sectors to electricity generation sector reform. Barring any unforeseen events which may alter the policy landscape in China, the CPC's gradualist approach to wind power development policy may wind up being a model for other nations that try to emulate China's accomplishments because indications are that by 2030, China's performance in wind power development will be worthy of emulation.

NOTES

1. Data taken from British Petroleum (BP). 2011. *Statistical Review of World Energy 2011*. London: British Petroleum and International Energy Agency (IEA). 2011. *World Energy Outlook 2010*. Paris: International Energy Agency. www.iea.org/Textbase/nppdf/free/2007/WEO_2007.pdf.
2. Global Wind Energy Council website: http://www.gwec.net/global-figures/graphs/.
3. Powell, Bill. 2009. "China: Serious About Climate Change?" *TIME Magazine*, September 24. www.time.com/time/world/article/0,8599,1925859,00.html.
4. Data from Global Wind Energy Council website: http://www.gwec.net/global-figures/graphs/
5. Zhang, Na, Noam Lior, and Hongguang Jin. 2011. "The Energy Situation and its Sustainable Development Strategy in China." *Energy* 36 (6): 3639–3649.
6. Liu, Wen, Henrik Lund, Brian Vad Mathiesen, and Xiliang Zhang. 2011. "Potential of Renewable Energy Systems in China." *Applied Energy* 88 (2): 518–525.
7. Wang, Yanjia, Alun Gu, and Aling Zhang. 2011. "Recent Development of Energy Supply and Demand in China, and Energy Sector Prospects through 2030." *Energy Policy* 39 (11): 6745–6759.
8. Energy Information Administration (EIA). 2011. *International Energy Outlook 2011*. Washington, DC: US Energy Information Administration.
9. Ma, L., Z. Li, F. Fu, X. Zhang, and W. Ni. 2009. "Alternative Energy Development Strategies for China Towards 2030." *Frontiers of Energy and Power Engineering in China* 3 (1): 2–10.
10. Rong, Fang, and David G. Victor. 2011. "Coal Liquefaction Policy in China: Explaining the Policy Reversal Since 2006." *Energy Policy* 39 (12): 8175–8184.
11. For more on this see Zhang, Na, Noam Lior, and Hongguang Jin. 2011. "The Energy Situation and its Sustainable Development Strategy in China." *Energy* 36 (6): 3639–3649; and for the impact on jobs see Cai, Wenjia, Can Wang, Jining Chen, and Siqiang Wang. 2011. "Green Economy and Green Jobs: Myth or Reality? The Case of China's Power Generation Sector." *Energy* 36 (10): 5994–6003.
12. Zhang, Na, Noam Lior, and Hongguang Jin. 2011. "The Energy Situation and its Sustainable Development Strategy in China." *Energy* 36 (6): 3639–3649.

13. International Energy Agency (IEA). 2012. *CO2 Emissions from Fuel Combustion: Highlights 2012*. Paris: International Energy Agency.

14. International Energy Agency (IEA). 2012. *CO2 Emissions from Fuel Combustion: Highlights 2012*. Paris: International Energy Agency.

15. Data combined from British Petroleum (BP). 2011. *Statistical Review of World Energy 2011*. London: British Petroleum (BP) and International Energy Agency (IEA). 2012. *CO2 Emissions from Fuel Combustion: Highlights 2012*. Paris: International Energy Agency.

16. International Energy Agency (IEA). 2012. *CO2 Emissions from Fuel Combustion: Highlights 2012*. Paris: International Energy Agency.

17. Yuan, Jiahai, Junjie Kang, Cong Yu, and Zhaoguang Hu. 2011. "Energy Conservation and Emissions Reduction in China—Progress and Prospective." *Renewable and Sustainable Energy Reviews* 15 (9): 4334–4347.

18. This was described in Ma, Linwei, Pei Liu, Feng Fu, Zheng Li, and Weidou Ni. 2011. "Integrated Energy Strategy for the Sustainable Development of China." *Energy* 36 (2): 1143–1154. An English version of this law is available at www.renewableenergyworld.com/assets/download/China_RE_Law_05.doc.

19. Ma, Linwei, Pei Liu, Feng Fu, Zheng Li, and Weidou Ni. 2011. "Integrated Energy Strategy for the Sustainable Development of China." *Energy* 36 (2): 1143–1154.

20. Ibid.

21. Seligsohn, Deborah, and Angel Hsu. 2011. "How Does China's Twelfth Five-Year Plan Address Energy and the Environment?" World Resources Institute, March 17. www.wri.org/stories/2011/03/how-does-chinas-12th-five-year-plan-address-energy-and-environment.

22. A translation of this plan is available at http://cbi.typepad.com/china_direct/2011/05/chinas-twelfth-five-new-plan-the-full-english-version.html.

23. Martinot, Eric, and Junfeng Li. 2010. "Renewable Energy Policy Update for China." Renewable Energy World Online, July 21. www.renewableenergyworld.com/rea/news/article/2010/07/renewable-energy-policy-update-for-china.

24. Valentine, Scott Victor. 2011. "Towards the Sino-American Trade Organization for the Prevention of Climate Change (STOP-CC)." *Chinese Journal of International Politics* 4 (4): 447–474.

25. Chen, Yongjun, Yong He, Yidan Bao, and Jiehui Shen. 2008. "Present Situation and Future Development of Wind Power in China." Paper read at 3rd Institute of Electrical and Electronics Engineers (IEEE) Conference on Industrial Electronics and Applications, 2008. International Conference on Industrial Electronics and Applications2008, June 3–5.

26. Xia, Changliang, and Zhanfeng Song. 2009. "Wind Energy in China: Current Scenario and Future Perspectives." *Renewable and Sustainable Energy Reviews* 13 (8): 1966–1974.

27. Ma, Jinlong. 2011. "On-Grid Electricity Tariffs in China: Development, Reform and Prospects." *Energy Policy* 39 (5): 2633–2645.

28. Ibid.

29. Xia, Changliang, and Zhanfeng Song. 2009. "Wind Energy in China: Current Scenario and Future Perspectives." *Renewable and Sustainable Energy Reviews* 13 (8): 1966–1974.

30. Ma, Jinlong. 2011. "On-Grid Electricity Tariffs in China: Development, Reform and Prospects." *Energy Policy* 39 (5): 2633–2645; Zhao, Zhen-yu, Wen-jun Ling, George Zillante, and Jian Zuo. 2012. "Comparative Assessment of Performance

of Foreign and Local Wind Turbine Manufacturers in China." *Renewable Energy* 39 (1): 424–432.

31. Ma, Jinlong. 2011. "On-Grid Electricity Tariffs in China: Development, Reform and Prospects." *Energy Policy* 39 (5): 2633–2645.

32. As reported in: Ma, Linwei, Pei Liu, Feng Fu, Zheng Li, and Weidou Ni. 2011. "Integrated Energy Strategy for the Sustainable Development of China." *Energy* 36 (2): 1143–1154. An English version of this law is available at www.renewableenergyworld.com/assets/download/China_RE_Law_05.doc.

33. Ma, Guonan, and Wang Yi. 2010. "China's High Saving Rate: Myth and Reality." In *BIS Working Papers*. Basel: Bank for International Settlement.

34. Zhang, Na, Noam Lior, and Hongguang Jin. 2011. "The Energy Situation and its Sustainable Development Strategy in China." *Energy* 36 (6): 3639–3649.

35. Lasserre, Philippe, and Hellmut Schutte. 2006. *Strategies for Asia Pacific: Meeting New Challenges. 3rd ed.* New York: Palgrave Macmillan.

36. Evans, Richard. 1995. *Deng Xiaoping and the Making of Modern China.* London: Penguin Books.

37. For more on this story see McGivering, Jill. 2006. "Three Gorges Dam's Social Impact." BBC News, May 20. http://news.bbc.co.uk/2/hi/asia-pacific/5000198.stm.

38. Sovacool, Benjamin K., and Scott Victor Valentine. 2010. "The Socio-Political Economy of Nuclear Energy in China and India." *Energy* 35 (9): 3803–3813.

39. Valentine, Scott Victor. 2011. "Sheltering Wind Power Projects from Tempestuous Community Concerns." *Energy for Sustainable Development* 15 (1): 109–114.

40. Han, Suyin. 1995. *Eldest Son: Zhou Enlai and the Making of Modern China, 1898–1976.* New York: Kodansha USA.

41. Sovacool, Benjamin K., and Scott Victor Valentine. 2010. "The Socio-Political Economy of Nuclear Energy in China and India." *Energy* 35 (9): 3803–3813.

42. Liu, Cuihang. 2011. "Higher Education in China: A Brief Introduction. Report posted on the website of the Education Office of the Embassy of the People's Republic of China. Retrieved October 15, 2012, from http://washcouncil.org/documents/pdf/WIEC2011_Higher-Education-in-China.pdf.

43. Zhao, Zhen-yu, Wen-jun Ling, George Zillante, and Jian Zuo. 2012. "Comparative Assessment of Performance of Foreign and Local Wind Turbine Manufacturers in China." *Renewable Energy* 39 (1): 424–432.

44. Sovacool, Benjamin K., and Scott Victor Valentine. 2010. "The Socio-Political Economy of Nuclear Energy in China and India." *Energy* 35 (9): 3803–3813.

45. Cai, Wenjia, Can Wang, Jining Chen, and Siqiang Wang. 2011. "Green Economy and Green Jobs: Myth or Reality? The Case of China's Power Generation Sector." *Energy* 36 (10): 5994–6003.

46. Delman, Jørgen. 2010. "China's 'Radicalism at the Center': Regime Legitimation through Climate Politics and Climate Governance." *Journal of Chinese Political Science* 16 (2): 183–205.

47. For more on this topic see McGivering, Jill. 2006. "Three Gorges Dam's Social Impact." BBC News, May 20. http://news.bbc.co.uk/2/hi/asia-pacific/5000198.stm and Rong, F. 2010. "Understanding Developing Country Stances on Post-2012 Climate Change Negotiations: Comparative Analysis of Brazil, China, India, Mexico, and South Africa." *Energy Policy* 38 (8): 4582–4591.

48. This was certainly the case in neighboring Japan. See Barrett, Brendan F.D. (ed.). 2005. *Ecological Modernization in Japan.* Edited by U. N. University. New York: Routledge.

49. Bambawale, Malavika Jain, and Benjamin K. Sovacool. 2011. "China's Energy Security: The Perspective of Energy Users." *Applied Energy* 88 (5): 1949–1956.

50. Valentine, Scott Victor. 2010. "The Fuzzy Nature of Energy Security." In *The Routledge Handbook of Energy Security*, edited by B. K. Sovacool, pp. 56–73. New York: Routledge.

51. Partridge, Ian, and Shama Gamkhar. 2012. "A Methodology for Estimating Health Benefits of Electricity Generation using Renewable Technologies." *Environment International* 39 (1): 103–110.

52. Sovacool, Benjamin K., and Scott Victor Valentine. 2010. "The Socio-Political Economy of Nuclear Energy in China and India." *Energy* 35 (9): 3803–3813.

53. Wen, Bo. 2011. "Japan's Nuclear Crisis Sparks Concerns over Nuclear Power in China." Nautilus Institute. http://nautilus.org/napsnet/napsnet-policy-forum/japans-nuclear-crisis-sparks-concerns-over-nuclear-power-in-china/#ii-article-by-wen.

54. Ma, Linwei, Pei Liu, Feng Fu, Zheng Li, and Weidou Ni. 2011. "Integrated Energy Strategy for the Sustainable Development of China." *Energy* 36 (2): 1143–1154.

55. Yang, Mian, Dalia Patiño-Echeverri, and Fuxia Yang. 2012. "Wind Power Generation in China: Understanding the Mismatch between Capacity and Generation." *Renewable Energy* 41: 145–151.

56. Wang, Yanjia, Alun Gu, and Aling Zhang. 2011. "Recent Development of Energy Supply and Demand in China, and Energy Sector Prospects through 2030." *Energy Policy* 39 (11): 6745–6759.

57. Hvistendahl, Mara. 2008. "China's Three Gorges Dam: An Environmental Catastrophe?" *Scientific American*, March 25. www.scientificamerican.com/article/chinas-three-gorges-dam-disaster/

58. Taken from the United Nations Framework Convention on Climate Change (UNFCCC) website: http://cdm.unfccc.int/Statistics/Issuance/CERsIssuedByHostPartyPieChart.html.

59. Lin, Boqiang, and Aijun Li. 2011. "Impacts of Carbon Motivated Border Tax Adjustments on Competitiveness Across Regions in China." *Energy* 36 (8): 5111–5118.

60. Xu, Ming, Ran Li, John C. Crittenden, and Yongsheng Chen. 2011. "CO2 Emissions Embodied in China's Exports from 2002 To 2008: A Structural Decomposition Analysis." *Energy Policy* 39 (11): 7381–7388.

61. Energy Information Administration (EIA). 2011. *International Energy Outlook 2011*. Washington, DC: US Energy Information Administration.

62. China National Development and Reform Commission (NDRC). 2007. *China Medium and Long Term Development Plan for Renewable Energy*. Beijing: National Development and Reform Commission.

63. Kahrl, Fredrich, Jim Williams, Ding Jianhua, and Hu Junfeng. 2011. "Challenges to China's Transition to a Low Carbon Electricity System." *Energy Policy* 39 (7): 4032–4041.

64. Li, X., K. Hubacek, and Y. L. Siu. 2012. "Wind Power in China—Dream or Reality?" *Energy* 37 (1): 51–60.

65. Yang, Mian, Dalia Patiño-Echeverri, and Fuxia Yang. 2012. "Wind Power Generation in China: Understanding the Mismatch Between Capacity and Generation." *Renewable Energy* 41: 145–151.

66. Kahrl, Fredrich, Jim Williams, Ding Jianhua, and Hu Junfeng. 2011. "Challenges to China's Transition to a Low Carbon Electricity System." *Energy Policy* 39 (7): 4032–4041.

67. Li, X., K. Hubacek, and Y. L. Siu. 2012. "Wind Power in China—Dream or Reality?" *Energy* 37 (1): 51–60.
68. Liu, Wen, Henrik Lund, Brian Vad Mathiesen, and Xiliang Zhang. 2011. "Potential of Renewable Energy Systems in China." *Applied Energy* 88 (2): 518–525.
69. Wang, Yanjia, Alun Gu, and Aling Zhang. 2011. "Recent Development of Energy Supply and Demand in China, and Energy Sector Prospects through 2030." *Energy Policy* 39 (11): 6745–6759.
70. Liu, Wen, Henrik Lund, Brian Vad Mathiesen, and Xiliang Zhang. 2011. "Potential of Renewable Energy Systems in China." *Applied Energy* 88 (2): 518–525.
71. Zhang, Na, Noam Lior, and Hongguang Jin. 2011. "The Energy Situation and its Sustainable Development Strategy in China." *Energy* 36 (6): 3639–3649.
72. Ibid.
73. More on this topic see: Zhang, Na, Noam Lior, and Hongguang Jin. 2011. "The Energy Situation and its Sustainable Development Strategy in China." *Energy* 36 (6): 3639–3649 and Yuan, Jiahai, Junjie Kang, Cong Yu, and Zhaoguang Hu. 2011. "Energy Conservation and Emissions Reduction in China—Progress and Prospective." *Renewable and Sustainable Energy Reviews* 15 (9): 4334–4347.
74. Zhou, Yun. 2010. "Why is China Going Nuclear?" *Energy Policy* 38 (7): 3755–3762.
75. Source: World Nuclear Association: www.world-nuclear.org/info/inf63.html.
76. Zhou, Sheng, and Xiliang Zhang. 2010. "Nuclear Energy Development in China: A Study of Opportunities and Challenges." *Energy* 35 (11): 4282–4288.
77. Ma, Linwei, Pei Liu, Feng Fu, Zheng Li, and Weidou Ni. 2011. "Integrated Energy Strategy for the Sustainable Development of China." *Energy* 36 (2): 1143–1154.
78. Caprotti, Frederico. 2009. "China's Cleantech Landscape: The Renewable Energy Technology Paradox." *Sustainable Development Law and Policy* 9(3): 6–10.
79. Shen, Lei, Tian-ming Gao, and Xin Cheng. 2012. "China's Coal Policy Since 1979: A Brief Overview." *Energy Policy* 40: 274–281.
80. For more on this see: Shen, Lei, Tian-ming Gao, and Xin Cheng. 2012. "China's Coal Policy Since 1979: A Brief Overview." *Energy Policy* 40: 274–281 and International Energy Agency (IEA). 2009. *Cleaner Coal in China*. Geneva: International Energy Agency.
81. Shen, Lei, Tian-ming Gao, and Xin Cheng. 2012. "China's Coal Policy Since 1979: A Brief Overview." *Energy Policy* 40: 274–281.
82. Zeng, Shaojun, Yuxin Lan, and Jing Huang. 2009. "Mitigation Paths for Chinese Iron and Steel Industry to Tackle Global Climate Change." *International Journal of Greenhouse Gas Control* 3 (6): 675–682.
83. Kahrl, Fredrich, Jim Williams, Ding Jianhua, and Hu Junfeng. 2011. "Challenges to China's Transition to a Low Carbon Electricity System." *Energy Policy* 39 (7): 4032–4041.
84. Ma, Guonan, and Wang Yi. 2010. "China's High Saving Rate: Myth and Reality." In *BIS Working Papers*. Basel: Bank for International Settlement.
85. Valentine, Scott Victor. 2011. "Towards the Sino-American Trade Organization for the Prevention of Climate Change (STOP-CC)." *Chinese Journal of International Politics* 4 (4): 447–474.
86. Yang, Dennis Tao. 2008. "China's Agricultural Crisis and Famine of 1959–1961: A Survey and Comparison to Soviet Famines." *Comparative Economic Studies* 50 (1): 1–29.
87. Caprotti, Frederico. 2009. "China's Cleantech Landscape: The Renewable Energy Technology Paradox." *Sustainable Development Law and Policy* 9 (3): 6–10.
88. Zhou, Yun, Christhian Rengifo, Peipei Chen, and Jonathan Hinze. 2011. "Is China Ready for its Nuclear Expansion?" *Energy Policy* 39 (2): 771–781.

89. Xia, Changliang, and Zhanfeng Song. 2009. "Wind Energy in China: Current Scenario and Future Perspectives." *Renewable and Sustainable Energy Reviews* 13 (8): 1966–1974.

90. Chien, John Chung-Ling, and Noam Lior. 2011. "Concentrating Solar Thermal Power as a Viable Alternative in China's Electricity Supply." *Energy Policy* 39 (12): 7622–7636.

91. Yang, Mian, Dalia Patiño-Echeverri, and Fuxia Yang. 2012. "Wind Power Generation in China: Understanding the Mismatch Between Capacity and Generation." *Renewable Energy* 41: 145–151.

92. Caprotti, Frederico. 2009. "China's Cleantech Landscape: The Renewable Energy Technology Paradox." *Sustainable Development Law and Policy* 9 (3): 6–10.

93. International Energy Agency (IEA). 2010. *World Energy Outlook 2009*. Paris: International Energy Agency 2010.

94. For more on this see DeCarolis, Joseph F., and David W. Keith. 2006. "The Economics of Large-Scale Wind Power in a Carbon Constrained World." *Energy Policy* 34 (4): 395–410; and Morthorst, Poul-Erik, and Shimon Awerbuch. 2009. *The Economics of Wind Energy*. Edited by S. Krohn. Belgium: European Wind Energy Association.

95. Zhao, Zhen-yu, Wen-jun Ling, George Zillante, and Jian Zuo. 2012. "Comparative Assessment of Performance of Foreign and Local Wind Turbine Manufacturers in China." *Renewable Energy* 39 (1): 424–432.

96. Sovacool, Benjamin K., and Scott Victor Valentine. 2010. "The Socio-Political Economy of Nuclear Energy in China and India." *Energy* 35 (9): 3803–3813.

97. For more on this topic see Barrett, Brendan F.D. (ed.). 2005. *Ecological Modernization in Japan*. Edited by U. N. University. New York: Routledge; and Carter, Neil. 2004. *The Politics of the Environment: Ideas, Activism and Policy*. Cambridge: Cambridge University Press.

98. Zeng, Shaojun, Yuxin Lan, and Jing Huang. 2009. "Mitigation Paths for Chinese Iron and Steel Industry to Tackle Global Climate Change." *International Journal of Greenhouse Gas Control* 3 (6): 675–682.

99. Xie, Ye, and Christine Harvey. 2012. "China Economy Heading for 'Hard Landing' as Exports Falter, Shilling Says." *Bloomberg News*, February 2. www.bloomberg.com/news/2012-02-02/china-economy-heading-for-hard-landing-as-exports-decline-shilling-says.html.

CHAPTER 7

Wind Power in the United States

7.1. INTRODUCTION

The Obama administration has poured billions into subsidizing its favored green energy sources. . . twenty years of subsidizing wind is more than enough.[1]
　　—letter from 47 house Republicans to John Boehner (September 21, 2012)

Governor Romney even explained his energy policy this way: I'm quoting here: "You can't drive a car with a windmill on it," that's what he said about wind power. . . Now I don't know if he's actually tried that. I know he's had other things on his car.[2]
　　—President Barack Obama (August 15, 2012)

There is a lot of money on the line in America's energy sector and where there is money, there is politics. In 2011, Exxon reported revenues of US\$486 billion and after-tax profits of US\$41 billion. Only 27 nations generated more GDP than Exxon generated in revenues. As of 2011, Exxon reported over US\$214 billion invested into property, plant, and equipment.[3] In short, there are a lot of sunk costs to defend. In the coal sector, America's Peabody Energy, which is the world's largest private sector coal company, posted US\$8.077 billion in revenue in 2012.[4] Understandably, America's energy sector is one of the most hotly contested marketplaces in the world and in this marketplace, fossil fuel interests rule the roost.

On the other hand, 9/11 and the ensuing military response have engendered a change in the ideological underpinnings of American energy security efforts. Even conservative factions that have typically supported a free trade energy policy have now begun to talk about the importance of ensuring control over domestic energy security. One study by Oak Ridge National Laboratory estimated that between 1970 and 2004, American dependence on foreign oil has cost the country \$5.6–\$14.6 trillion.[5] This reflects both the cost of the oil and the direct economic consequences of macroeconomic shocks and transfers of wealth. Another more recent study estimated that oil dependence in the United States exceeded US$500 billion for 2008 alone.[6] These claims are supported by trade data. The United States purchases more than 60% of its oil from foreign sources each year and the cost of petroleum products is the single largest contributor—48%—to the country's US$700 billion trade deficit.[7] Supply costs aside, one study recently concluded that the military costs in the Persian Gulf needed to protect oil assets and infrastructure range from US$50 billion to $100 billion per year;[8] a second, independent study put the figure at between US$29 billion and $80 billion per year.[9] The United States is spending billions each year to protect a supply chain that is in part responsible for financing terrorist activities such as the 2001 attack on New York's World Trade Center buildings. In fact, the wealth accrued by the family of Osama Bin Laden came largely from the provision of infrastructure for oil operations in the Persian Gulf area. Understandably, many energy policymakers now see the folly of preserving the status quo.

This chapter attempts to document the evolution of American energy policy, particularly in regard to its influence on wind power development. Evidence suggests that a change is indeed afoot, but it will be a contested affair. On the one hand, after two decades of hibernation, wind power development in the United States is experiencing a reawakening. On the other hand, there is vociferous political support in the United States to exploit domestic shale gas reserves. In fact, the US Energy Information Administration (USEIA) projects that shale gas production will increase from 5.0 trillion cubic feet in 2010 to 13.6 cubic feet in 2035.[10] Moreover, there are emerging indications that political support for nuclear power is on the rise in the United States.[11] As this chapter will describe, the resurgence of wind power is predominantly due to economic trends that are beginning to cast wind power in a favorable light and altered perspectives regarding what constitutes energy security.

This case study illustrates how changes to two or three STEP variables can radically alter development trajectory. Historically, decisions about electricity generation have been made at the state or even municipal level with federal incentives playing a linchpin role in elevating the pace of development. As this chapter will document, the pace of wind power development has historically been linked to the federal wind power production tax credit. However, the recent decade long escalation of fossil fuel prices along with progressive advances in wind power technology have improved the competitiveness of wind power, allowing wind power developers to compete directly with fossil fuel electricity generation providers in many states. Wind power is now seen as a viable option for enhancing domestic energy security and abating greenhouse gas (GHG) emissions. This economic change has largely negated the strongest defense of fossil fuel advocates and shifted the main policy challenge from "Should we invest in wind power?" to "How much wind power is optimal?" As the two quotes at the beginning of this chapter imply, there is still a great deal of ideological difference of opinion in regard to how this new question is answered; however, there is no denying that wind power is enjoying a phase of unprecedented development.

In order to understand how wind power fits into the nation's electricity generation plans, it is useful to gain a clearer perspective on the scope and crux of challenges the United States faces in the sector.

7.2 AN OVERVIEW OF ELECTRICITY GENERATION IN THE UNITED STATES

Mammoth. This is an appropriate adjective for describing America's electricity generation infrastructure. In 2011, the nation boasted 982.8 GW of installed electricity generation capacity,[12] designed to meet annual electricity consumption of over 4.1 trillion kilowatt hours, a level of consumption that is approximately 20% more than all 25 European OECD nations put together. Moreover, demand for electricity is expected to increase by 16% between 2011 and 2040.[13]

Although one would be correct in assuming that new electricity generation infrastructure will be needed to keep pace with demand, the challenges of adding new power infrastructure does not adequately convey the enormity of America's power provision dilemma. The nation is currently saddled with an aging power grid and a fleet of electricity generation plants that are nearing obsolescence. Investment in transmission and distribution (T&D) infrastructure has significantly lagged behind generation capacity expansion over the past two decades. One study has estimated a deterioration of transmission capacity in the United States of 11% between 2002 and 2012. In addition to lagging T&D investment, the infrastructure that currently exists is aging. For example, it purportedly costs over US$1 billion each year just to replace corroded cabling in New York's T&D network. In terms of existing electricity generation capacity, there are plans to build approximately 120 coal fired power stations between 2008 and 2018, largely to replace dilapidated units. Moreover, operating permits for 40% of America's nuclear power plants are set to expire by 2020. Altogether, refurbishments and upgrades to existing generation facilities, transmission infrastructure and distribution networks to keep up with demand will likely cost trillions of dollars.[14]

Although this poses a threat to national fiscal health, it also represents an opportunity in that many of the nation's existing power plants that are nearing obsolescence could be replaced by clean energy technologies, resulting in rapid decarbonization of the nation's electricity mix. Decisions on what type of technology is chosen to replace these aging units will shape the electricity profile of the nation for decades to come and could radically alter the cost profiles for some of the renewable energy technologies that stand to benefit from this.

The electricity generation profile of the United States as of 2011 is depicted in Table 7.1. As the table illustrates, electricity generation is still dominated by carbon-intensive technologies, with coal-fired power still amounting to 44.92% of all electricity generation. The second most prominent electricity generation technology utilizes natural gas (20.04%). Adding the minor contribution from oil-fired power plants (0.72%) brings the total contribution from fossil fuel power plants to 66%.

The continued dominance of fossil fuel makes electricity provision a key contributor to the nation's prolific GHG emissions. In 2010, the nation emitted 5,415 million tonnes of CO_2 equivalent GHG, constituting 17.4% of global GHG emissions.[15] Of this total, electricity and heat generation accounted for 42.7% of all US GHG emissions. Therefore, it should come as no surprise that the Obama Administration is keen to support a transition

Table 7.1 ELECTRICITY GENERATION BY SOURCE IN THE UNITED STATES, 2011

Source	Contribution (%)
Coal	44.92
Nuclear	20.56
Natural Gas	20.04
Hydro	7.90
Wind	2.91
Wood	0.83
Petroleum	0.72
Waste	0.72
Geothermal	0.41
Other	0.40
Electricity imports	0.32
Other gases	0.23
Solar PV	0.04
TOTAL	**100.00**

Source: Energy Information Administration website - www.eia.gov, 2013.

to cleaner electricity generation in order to support its December 2009 pledge made at the Copenhagen COP15 (Conference of the Parties to the UNFCCC) conference to reduce US GHG emissions to 17% below 2005 levels by 2020. As of 2010, US GHG emissions were 92.6% of 2005 levels, suggesting that a lot of work is still yet to be done to reach this watered-down target. In order to meet the commitment of 6% below 1990 levels that it agreed to in the lead up to the Kyoto Protocol, the United States would have to reduce GHG emissions by 15.6% from 2010 levels.

As Figure 7.1 (next page) suggests, the transition toward cleaner electricity production has been a slow, counterproductive process. In 1990, coal-fired power supplied 51.84% of the nation's energy. By 2011 this percentage had dropped to 44.92%, but over the same 21-year period, aggregate energy use increased by 26%. The net result was an aggregate 10% increase in the amount of energy provided by coal-fired production from 16.5 million British Thermal Units (BTUs) to 18 million BTUs.[16]

These aggregate trends mask more commendable developments at the state level. Table 7.2 breaks down the power generation profile in states with the highest levels of electricity generation in 2010. Indiana, Ohio, and Florida all generated more than 86% of electricity from fossil fuel sources. At the other extreme, Washington State generated only 19% of its electricity from fossil fuel sources thanks to a rich endowment in hydropower. Illinois generated nearly half of its electricity by nuclear power while Indiana did not use any nuclear power. Washington, California, and New York utilized sizable contributions from renewable energy, while Ohio had negligible renewable energy capacity. When the US electricity generation sector is broken down into state-level statistics, it becomes apparent that the state electricity mixes differ significantly. In the case of a state like Washington, a high level of renewable energy capacity stems from its abundant water resources which enable exploitation of hydropower. However, as this chapter will illustrate, for most states, the difference in electricity profile is explained more by political decisions made at the state level than by geographic attributes. As a testament to this, as of March 2013, the Texas Renewable Energy Industries Association reports that the total contribution from renewable energy to the state electricity grid is now at 8.9%, an expansion of approximately 75% since 2010. This is largely attributed to wind power capacity expansion, which has been driven by aggressive state-level wind power development policy.

Nevertheless, despite pressure to diversify the electricity mix to reduce GHG emissions, there has been profuse and effective resistance from fossil fuel advocates. Consequently, in addition to large-scale investment in natural gas projects, US oil production has also been on the rise, with production

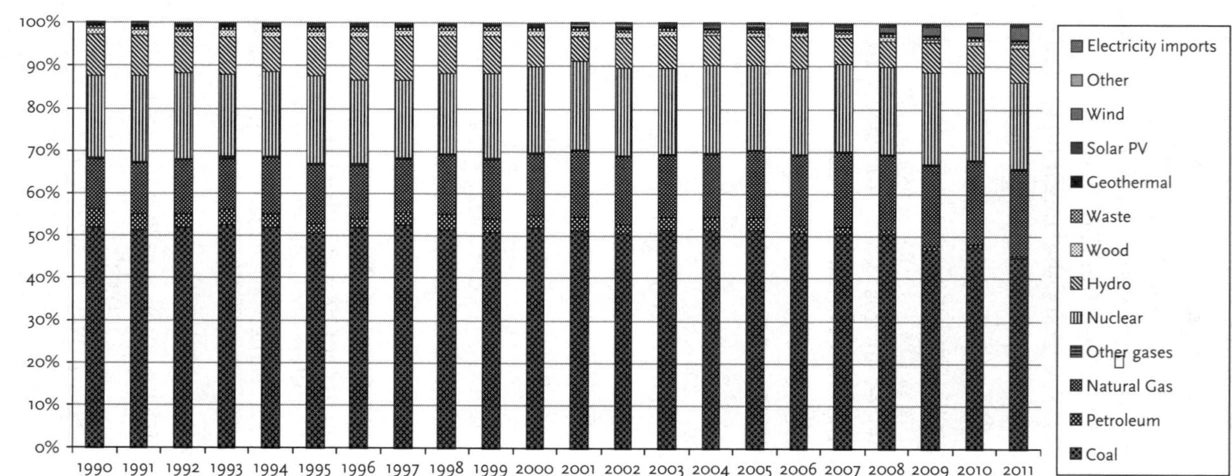

Figure 7.1. Percentage of Contribution to Generation by Energy Source, 1990–2011
Source: Energy Information Administration (2013).

Table 7.2 POWER GENERATION PROFILES OF LEADING ENERGY CONSUMING STATES, 2010

	Net Generation (Gigawatt Hours)	%Generated by Fossil Fuel	%Generated by Nuclear
Texas	411,695	82.84	10.04
Pennsylvania	229,752	63.20	33.87
Florida	229,096	86.28	10.45
California	204,126	55.05	15.78
Illinois	201,352	49.47	47.77
Alabama	152,151	67.54	24.94
Ohio	143,598	88.20	11.01
New York	136,962	47.10	30.57
North Carolina	128,678	62.71	31.66
Indiana	125,181	96.74	0.00
Michegan	111,551	70.40	26.56
Washington	103,473	18.57	8.93

Source: Energy Information Administration, www.eia.gov.

increasing from 5.1 million barrels per day in 2007 to 5.5 million barrels per day in 2010. By 2020, it is expected that domestic oil production will top 6.7 million barrels per day, a production level not seen since 1994.[17] Research has documented fossil fuel special interest groups pouring sizable amounts of money into strategies designed to perpetuate the status quo. Some of these organizations have been known to disseminate biased analyses that highlight the disadvantages of renewable technologies while obfuscating the benefits. Climate change expert James Hansen asserts that the damage caused by these misinformation campaigns has been so obstructive to climate change mitigation efforts that the perpetrators of these tactics should be tried for "high crimes against humanity and nature."[18]

The change in 2009 from a Republican regime led by George W. Bush to a Democratic regime led by Barack Obama has catalyzed a noticeable shift in federal support for a transition to clean energy. The American Recovery and Reinvestment Act of 2009 included commitment to the Advanced Energy Manufacturing Tax Credits program that provided over US$300 billion in tax breaks and benefits for manufacturing clean energy products.[19] In February 2011, the government announced the Sun Shot Initiative, which is a federally funded effort to restore American leadership in solar PV manufacturing by supporting initiatives to reduce the cost of solar PV energy by 75% by 2020.[20] If successful, it is envisaged that the contribution of solar power to the electricity grid will be 14% by 2020. Wind power has also

Table 7.3 EVOLUTION OF THE US ELECTRICITY GENERATION MIX, 2010–2035		
	2010	2035
Coal	45%	39%
Natural Gas	24%	27%
Renewable Energy	10%	16%
Nuclear Power	20%	18%
Oil and other liquids	1%	1%

benefitted from the favorable clean energy business climate that is springing up in various US states, as the next section will detail.

The uptake of all these developments in regard to the electricity sector is that both natural gas and renewable technologies are projected to play an increasing role in electricity provision, while contributions from coal-fired technologies are expected to decrease (see Table 7.3).[21]

7.3. HISTORY OF WIND POWER DEVELOPMENT IN THE UNITED STATES

As is the case with wind power development in most industrialized nations, the oil price shocks of the 1970s served to fuel political fervor in diversifying the nation's electricity mix. In 1975, with oil prices double the level of just two years previous, the US Energy Policy and Conservation Act was passed, with the primary intention being to improve energy efficiency. However, embedded within the act were a series of policy initiatives designed to encourage R&D in alternative fuels. In 1976, about US$22 million was earmarked for wind R&D; and by 1978, with oil prices still hovering at historical highs, funding had increased to US$60 million.[22]

In 1978, the government passed the National Energy Act (USNEA), which contained two statutes that proved to be instrumental in supporting the diffusion of renewable energy. The first statute was the Public Utility Regulatory Policies Act (PURPA), which mandated electric utilities to facilitate grid access for renewable power providers. PURPA also compelled utilities to purchase electricity generated by renewable technologies at avoided cost. Therefore, not only did this act facilitate grid access, it also forced utilities to calculate avoided cost, and in doing so forced utilities to consider the economics of including electricity from technologies that had previously not been part of the portfolio. The second statue was an Energy Tax Act (ETA), which provided tax credits for private development of alternative energy technologies.[23]

The USNEA also precipitated an amplified R&D budget for wind power development. By 1981, with a barrel of oil at approximately US$90—4.5 times the level of a decade earlier—expenditures on wind R&D rose to approximately US$125 million; by 1982, the wind R&D budget topped US$150 million. By comparison, Germany, the nation with the second highest wind R&D expenditures during that era, spent about US$50 million on wind R&D in 1982.[24] This amplification of financial support proved to be instrumental in kick-starting the development of an American wind turbine manufacturing sector and bringing down the cost of wind power.

Unfortunately, the structure of the USNEA suffered from two weaknesses. First, the tax credits under the ETA were granted for capital investment rather than for power produced. As result, the ETA became a tax shelter for wealthy individuals and less attention was given to the reliability of these early turbines. Second, PURPA's stipulation that power from renewable technologies be purchased by utilities at avoided cost made it difficult for private turbine owners to generate project profits. The larger utilities had the ability to construct larger wind power projects, reaping the benefits of economies of scale. This then became the avoided cost that owners of smaller projects would receive. The result of these two weaknesses was that the USNEA failed to catalyze much in the way of wind power development on a national scale. By 1980, there was still only 8 MW of installed wind power capacity in the country (see Figure 7.2).

There was, however, one state where wind power development took hold. In 1978, the California legislature, which had passed a 1976 state act to provide private investors with a 10% tax credit to cover the expenses of developing solar energy projects, expanded the tax credit to include wind power projects. Unfortunately, early wind power projects were fraught with planning and execution problems that deterred investment.[25] As result, the state

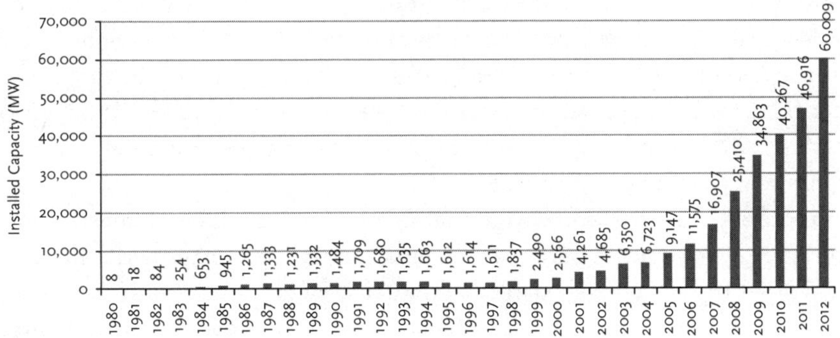

Figure 7.2. Wind Power Development in the United States (1980–2012)
Source: Combined data from GWEC, Earth Policy Institute, Worldwatch, AWEA.

tax credit did little to influence installed capacity until 1980–1981, when California's two largest utilities—Pacific Gas and Electric Company (PG&E) and Southern California Edison Co. (SCE)—responded to US$100 per barrel oil prices by voluntarily adopting plans to encourage wind power development. PG&E announced a target of 120 MW of wind power by 1990, and SCE signed a letter of intent to develop a 320 MW wind farm. Due in large part to the competition between these two firms for leadership in green energy, a "wind rush" developed in California and by 1982, California was host to over 1000 wind turbines. By 1985, half of the world's wind power production came from the Altamont Pass Wind Farm alone and by the end of 1986, about 6,700 turbines were in operation in Altamont.[26]

In the early 1980s, there was another notable state that began to lay the foundation for wind power development—Iowa. In 1983, the state introduced a renewable energy standard that required investor-owned utilities in the state to purchase 105 MW of wind power. This renewable portfolio standard (RPS) was the first in a successive line of development policies that would lead Iowa to become the state with the third most installed wind power capacity in the nation.

What transpired in California and Iowa in the early 1980s epitomizes the historical importance of state policy in catalyzing wind power development in the United States. States like Texas, California, Iowa, Illinois, Oregon, South Dakota, North Dakota, Minnesota, and Wyoming all have achieved success from proactive wind power development policies that leveraged federal policy to achieve more optimal results.

By late 1985, oil and natural gas prices were once again declining and a political debate had emerged within the US Congress in regard to extending the federal tax credit that was due to expire at the end of the year. In the end, renewable energy proponents were unable to convince Congress to grant an extension and were only successful in extracting a transitional concession which allowed tax credits for projects started beyond 1985, provided that a power purchase contract and a wind farm site were procured by the end of 1985.[27] Consequently, 1986 marked the end of the first boom period for wind power in the United States. Between 1980 and 1986, 1257 MW of wind power capacity was installed in the nation; it would take 14 more years to replicate this feat.

In 1989, with the knowledge that costly and unreliable domestic wind turbines were contributing to phlegmatic wind power growth, the federal government published the Renewable Energy and Energy Efficiency Competitiveness Code, which included goals to improve design standards to create more reliable and efficient wind turbines. Specific goals for the Wind Energy Research Program stemming from this code included targets to reduce wind energy costs to US$0.03-$0.05 per kWh by 1995, reduce capital costs of wind systems to $500-$750 per kW of installed capacity by 1995, reduce operation and maintenance costs for wind systems to under US$0.01 per kWh by 1995, and increase capacity factors for wind systems to 25 to 35% by 1995.[28]

With wind power research goals firmly in place and the price of oil still hovering at around US$40 per barrel, the government then turned its focus to stimulating market development. In this regard, two bills introduced in the early 1990s bear mention. First, in 1990–1991, the U.S. Congress passed the Solar, Wind, Waste, and Geothermal Power Production Incentives Act.[29] This act essentially amended PURPA to allow renewable energy facilities of any size to receive PURPA benefits. This amendment set the stage for a new subsidy that would be announced the following year to replace the USNEA tax credits, which had expired back in 1985. Second, in 1992, an Energy Policy Act was introduced. This included a renewable electricity production tax credit (PTC) which provided wind power providers with a US$0.015 per kWh tax credit (in inflation adjusted 1993 dollars) for power produced over the first 10 years of operation.[30] Eligibility for applying for the PTC would extend until July 1999. In addition to the PTC, the Energy Policy Act also established a federal renewable energy production incentive (REPI) amounting to US$0.015 per kWh (in inflation adjusted 1993 dollars) for the first 10 years of operation to state or nonprofit electric cooperatives for wind power production. This added incentive was slated to run until October 1, 2016, with annual appropriation limits established to limit the fiscal burden.

With oil prices dropping to below US$30 per barrel (and the price of natural gas following suit), these incentives were not large enough to catalyze a wind power renaissance; however, they were sufficient to fuel enough wind power development to replace older turbines, which were nearing obsolescence. Consequently, although installed wind power capacity did not grow between 1992 and 1997 (see Figure 7.2), the wind power industry emerged from this era in far better condition. The new turbines being erected around the country were of far superior quality and the reliability concerns experienced in the early 1980s were largely attenuated.

It bears noting that in the seven-year period between 1993 and 1999, national wind power capacity expanded by only 855 MW. The implication of such phlegmatic progress was that the financial impact of the PTC was not enough to attract staunch political opposition. Therefore, when the PTC expired in July of 1999, advocates of extending the incentive managed to garner enough support for the subsidy to be resurrected in December 1999 and extended until December 31, 2001 (under the auspices of the Ticket to Work and Work Incentives Improvement Act). This strategy of subsuming PTC extensions under other acts would become standard practice and is indicative of the pork barrel politics under which the PTC (and other government programs) is renegotiated.

In 1999, even though oil prices had returned to the historic trading range of US$20–30 per barrel, Texas, which had 184 MW of wind power capacity, announced a new strategy to help abate greenhouse gas emissions. It announced the Renewable Portfolio Standard (Senate Bill 7), which required utilities (based on respective market share) to jointly incorporate 2000 MW of renewable energy by 2009. Overnight a wind industry was born in the state. These laudable state-level initiatives notwithstanding, by the end of 1999, there was still only 2490 MW of installed wind power capacity in the nation and 94% of these facilities were in California (1616 MW), Minnesota (273 MW), Iowa (242 MW), and Texas (184 MW).[31]

In 2001, more state level initiatives were announced. Minnesota announced the Renewable Energy Objective, which legally mandated utilities to make a good faith effort to ensure 10% of retail sales came from renewable sources by 2015. This initial policy would eventually catapult the state of Minnesota to the fourth spot in national wind power capacity.[32]

In the same year, the first US offshore wind farm proposal—the Cape Wind farm—in Massachusetts was tabled. From its inception the project ran into opposition from Indian tribes, business interests and homeowners. It would take nearly a decade before the opposition could be thwarted. Completion of the project is now slated for 2015–2016, making it potentially the first offshore wind farm in the United States.[33]

On September 11, 2001 (9/11), terrorist attacks on New York's World Trade Center complex and the Pentagon thrust the national security dimension of energy to popular attention. Suddenly, backroom discussions on the importance of enhancing domestic energy security became a main stream media topic. Amidst this backdrop, debate once again resumed in congress over whether or not to extend the PTC, which was set to expire on December 31 of that year. Spurred on by oil prices averaging US$37 a barrel and the declining cost profile of wind power, 2001 was already proving to be a banner year despite the turmoil of 9/11. By year end, 1695 MW had been added, increasing national installed wind power capacity by 66% to 4261 MW. Opponents to the extension argued that the PTC represented an excessive subsidy to wind power providers. However, in March 2002 arguments for enhancing domestic energy security prevailed, and the PTC was extended to December 31, 2003 by the Job Creation and Worker Assistance Act of 2002. The extension was made retroactive to include projects started after December 31, 2001.[34] However, extension delay and a general economic downturn had forced many developers to postpone plans—only 424 MW were added in 2002 (Figure 7.2).

In hindsight, 9/11 proved to have a catalytic affect on wind power development in many states. Between 2001 and 2004, eight states initiated wind power programs (Illinois, Montana, North Dakota, South Dakota, Ohio, Oklahoma, Washington, and West Virginia) and eleven more states significantly ramped up wind power development.[35] An example of a policy driving the latter was the 2003 policy of Iowa Governor Tom Vilsack to deploy 1,000 MW of renewable energy—a doubling of state capacity, which would subsequently be achieved only three years later.[36]

By 2003, the wind power market was growing in leaps and bounds with 1665 MW of additional capacity being added, an annual increase of 36% (see Figure 7.2). However, once again congressional debate ensued over the necessity of the PTC and it was allowed to lapse on December 31, 2003. It was eventually renewed (retroactive to cover projects in the lapsed period) for two years in October 2004, under the American Jobs Creation Act.[37] As a result of the extension delay, developments dropped off once again in 2004, with only 373 MW being added. After its renewal, the market once again responded favorably and 2424 MW of installed capacity was added in 2005—the largest single year increase in the history of the US wind power program.

In 2005, the boom and bust cycle of wind power development caused by the PTC renewal delays was interrupted with the passage of the Energy Policy Act (EPACT). The act sent a strong market message by establishing renewable energy goals for federal government energy consumption. The target for the 2007–2009 period was set at 3% or higher, the target for 2010–2012 was set at 5% or higher, and a 2013 and beyond target of 7.5% or higher was announced (Sec. 203).[38] The EPACT also extended the PTC through to December 31, 2007 (Sec.1301). The market message sent by the EPACT fostered an unprecedented growth phase in wind power capacity. In 2006 and 2007 respectively, 2428 MW and 5332 MW of installed capacity were added.

The EPACT also served to mobilize state-level initiatives. In 2005, Texas passed a new senate bill (Bill 20) that increased the renewable portfolio target from 2000 MW to 5880 MW by 2015 and 10,000 MW by 2025. As a result, by the end of 2006 Texas had surpassed California as the state with the most wind power capacity. In 2006, Washington State announced an ambitious renewable portfolio standard (RPS) that required all utilities that serve over 25,000 customers to acquire 15% of their electricity from qualifying renewable resources by 2020 and to meet biennial energy efficiency targets. A fine of $50 for each MWh below the target would be levied on utilities which failed to meet progressively increasing targets of 3% by January 1, 2012, 9% by January 1, 2016, and 15% of its load by January 1, 2020.[39]

In December 2006, one year ahead of the December 31, 2007 PTC expiry date, George W. Bush's administration further stabilized the wind power market by pushing through a further extension of the PTC to December 31, 2008—incorporating the extension into the 2006 Tax Relief and Healthcare Act.[40] Accordingly, by 2008, market confidence was such that a record 8503 MW of new capacity was added. In sum, in the three year period since the EPACT was announced, the US wind power market had grown by 178% to 25,410 MW in total installed capacity (see Figure 7.2).

By 2007, even more states were elevating support for wind power. In Minnesota, a renewable energy standard was announced that required utilities to provide 25% of generated power from renewable energy (including wind) by 2025.[41] In Oregon, an RPS was announced that required 25% of retail electricity revenues to come from renewable energy by 2025. In Illinois, a renewable energy portfolio was announced. Under Public Act 095-0481, Illinois utilities Ameren and ComEd were required to purchase 25% of their electricity from qualified renewable sources by 2025. The legislation stipulated that of the 25% portfolio, 75% was to come from wind energy.[42] In July 2008, Texas approved funds for a $4.93 billion project to expand the state's electricity grid to transmit wind energy to its major cities.[43]

In July 2008, the US Department of Energy released an important study titled "20% Wind Energy by 2030: Increasing Wind Energy's Contribution to US Electricity Supply." The study described a scenario whereby the United States could achieve 20% wind electricity by 2030, and concluded it was technologically and economically viable.[44] This study has since served as an important reference for political proponents of wind power development and underpins the perceived role of wind power in helping the nation achieve its GHG emission reduction target of 17% below 2005 levels by 2020, as announced at the end of the UNFCCC COP15 conference in Copenhagen.

In 2009, another landmark act was passed: the American Reinvestment and Recovery Act (ARRA).[45] The act was significant in three respects. First, it earmarked a significant amount of money for supporting renewable energy diffusion. US$2 billion was set aside for advanced battery development and US$4.5 billion was apportioned for grid enhancement investment. Second, ARRA extended the PTC to December 31, 2012 (Sec. 1101)—a staggeringly long extension compared to previous cases. Third, ARRA introduced a provision for a 30% investment tax credit to be claimed in lieu of the production tax credit (Sec. 1102). Although this provision contradicted a global policy trend which favored production incentives over investment subsidies, in hindsight this option helped stimulate investment in an economic climate where investment funds were diminished. By the end of 2010, there was 40,267 MW of installed wind power capacity in the nation.

By 2010, there was clear evidence that energy policymakers and developers were beginning to focus efforts on exploiting offshore wind power potential. In October 2010, the US Department of the Interior finally signed off on the Cape Wind project, approving the nation's first lease for wind energy development on the outer continental shelf. In the same month, a consortium led by Google announced a plan to finance a project known as the Atlantic Wind Connection—a 350-mile-long electricity transmission cable with a capacity of 7,000 MW that would be set 15–20 miles offshore from New Jersey to Virginia, to facilitate offshore wind power development.[46] This project would be instrumental in providing the infrastructure necessary to further develop offshore projects in the Atlantic region. In February 2011, the US Department of Energy published the "National Offshore Wind Strategy," which declared an intent (OSWInD initiative) to facilitate "54 gigawatts (GW) of offshore wind generating capacity by 2030, at a cost of energy of $0.07 per kWh, with an interim scenario of 10 GW of capacity deployed by 2020, at a cost of energy of $0.10 per kWh."[47] The strategy also advocates funding for technology development, market barrier alleviation and constructing a next generation wind turbine drive train.

In 2011 and 2012, the onshore wind program continued to gather steam. Nationally, 19,742 MW of new capacity was added, bringing the total to 60,009 MW, an increase of 49% over the two-year period. Unfortunately, once again political wrangling delayed extension of the PTC, which expired on December 31, 2012. This time there was more to the story than political bickering and criticisms that the PTC was unnecessary and excessive. The ruling Democrats and the Republicans (who enjoyed a house majority) had been locked in a contest of brinkmanship over how the 2013 federal budget should be apportioned, with the end result being a delay in approving the fiscal budget for 2013. This political battle made it difficult for the Obama administration to authorize a timely extension of the PTC. Fortunately for the wind power community, the PTC was once again extended for one year on January 2, 2013 under the auspices of the American Taxpayer Relief Act; however, the delay in signing an extension has adversely affected the pace of wind power development in 2013 only 1,098 MW were added.

7.4 UNDERSTANDING THE GENERAL FORCES FOR CHANGE

7.4.1 Sociocultural Landscape

There's a great deal of general support for alternative sources of energy. A 2005 Harris Interactive poll found that 91% of American respondents supported the enhanced financing of R&D to develop alternative sources of energy.[48] In 2009, a poll taken by Rasmussen Reports revealed that 60% of American respondents identified finding new sources of energy as being more important than improving energy efficiency.[49] Part of the rationale stems from the fallout from 9/11 and the prevailing political desire to enhance domestic energy security, while other justifications stem from anti-nuclear sentiments and climate change mitigation concerns.

Prior to 9/11, climate change was a back page issue. Although the Clinton administration played a key role in shaping the final text of the Kyoto Protocol, it was widely understood even at that time that it was unlikely the United States would ratify the document because there was not enough political support in Congress. The terrorist attacks in New York and Washington in 2001 elevated the topic of energy security to a front page issue. Critics were questioning the wisdom of transferring billions of dollars of wealth from the United States to political regions that enable terrorism.[50] So from 2001 onward, one can say that there was enhanced support for improving energy security through domestically available resources.

Given that the United States has the most installed capacity of nuclear power in the world, it makes sense that nuclear power would stake a place on the agenda of domestic energy security. Interestingly, despite the fact that no new nuclear plants have been built in the United States since 1979, there is a high degree of general support for expanding nuclear power capacity. In a 2008 poll by Rasmussen Reports, 55% of US respondents asserted that more nuclear power plants should be built in the United States, with only 27% opposed.[51] Even after the Fukushima disaster, a 2012 Angus Reid poll indicated that 47% of Americans surveyed supported building more nuclear power stations in the United States, with 38% opposed.[52] However, general support for nuclear power does not filter down to community-level support when it comes to siting reactors and this explains why there has been no nuclear power plants built in the United States for decades.

Nevertheless, there is still general affinity in the United States for enhancing domestic energy security. There are proponents in favor of expanding nuclear power, proponents in favor of increased oil and gas exploration within US borders and proponents in favor of renewable energy technologies.

Climate change concerns, which have been so influential in catalyzing wind power development in other nations, have had less sway over energy policy in the United States. This is because perhaps more than anywhere else in the world, in the United States the science of climate change has been aggressively challenged by fossil fuel special interest groups.

Since the Obama administration has been in power, climate change policy in the United States has somewhat gelled, culminating in the 2009 Copenhagen Accord where the United States committed to a 2020 GHG emission reduction target of 17% below 2005 levels. However, public perception has not necessarily followed suit. In 2009, an Angus Reid poll determined that 80.1% of American respondents were either completely or mostly convinced that global warming is a reality.[53] Yet in a separate poll the same year, only 33% of Americans polled saw global warming as *very serious*, and only 34% attributed global warming to human activity, with 48% attributing global warming to long-term planetary trends.[54] Paradoxically, yet another Angus Reid poll determined that 59% of Americans surveyed felt that the United States should take action on global warming even if other nations do less.[55] If these surveys are accurate (the margin of error is estimated at 3.5%), despite the fact that only 34% of Americans believe that global warming is caused by human activity, a majority (59%) feel that action should still be taken to mitigate the problem.

In sum, public support for enhanced domestic security, NIMBY opposition to nuclear power, and a prevalent undercurrent of support for mitigating climate change has engendered support for wind power development. Even during a political era when a high degree of polarization exists between Democratic and Republican voters, there is a high degree of general support for clean, domestic energy development, provided that a transition can be facilitated in an economically effective manner. However, this does not mean that there is uniform support for wind power. Indeed, states vary significantly in terms of endorsing wind power development. As of 2011, there were still 14 states that had less than 2 MW of cumulative installed wind power. On the other hand, Texas currently boasts over 12,000 MW of wind power capacity, followed by California and Iowa with over 5000 MW each and Illinois and Oregon with over 3,000 MW each. These five states together constitute about 50% of America's total wind power capacity.

Although geographic conditions and settlement patterns play roles in framing wind power development prospects, there are other sociocultural influences that also impact political support for wind power in a given state. For example, wind power in Texas, Iowa, and Minnesota is seen as an economically advantageous investment for farm owners and has therefore received broad public support.[56] In California, there is strong environmental support for renewable energy; as a result, electricity consumers in the state exhibit a willingness to pay more for renewable energy. Conversely, in a state like Massachusetts, despite the existence of strong environmental lobbyists, there is also a degree of NIMBY opposition to projects which may adversely impact community aesthetics, particularly in regard to offshore wind power development.[57]

7.4.2 Economic Landscape

Despite national affluence, the United States is still a nation that prioritizes economic growth. In a June 2009 poll by Angus Reid Strategies, 53% of American respondents indicated that the economy was the most important issue facing the United States. By contrast, the second and third issues of most concern were healthcare (9%) and the federal budget deficit (7%).[58] In fact, when George W. Bush explained the rationale behind the US decision not to ratify the Kyoto Protocol, the main justification was that the United States would not enter into an agreement that would undermine the competitiveness of US firms.[59]

Since 9/11, the economic environment underpinning the energy sector has changed significantly. As Figure 7.3 illustrates, since 2011 the price of oil has increased from a trading range of approximately US$30 per barrel to a range that capriciously fluctuates around US$80–90 per barrel. The price of oil is significant, because it tends to serve as a benchmark for other fossil fuel commodity prices. For example, the cost of North Appalachia coal swelled from a trading range of US$40–45 per short ton between December 2005 and December 2007, to US$150 per short ton in September 2008. Although the

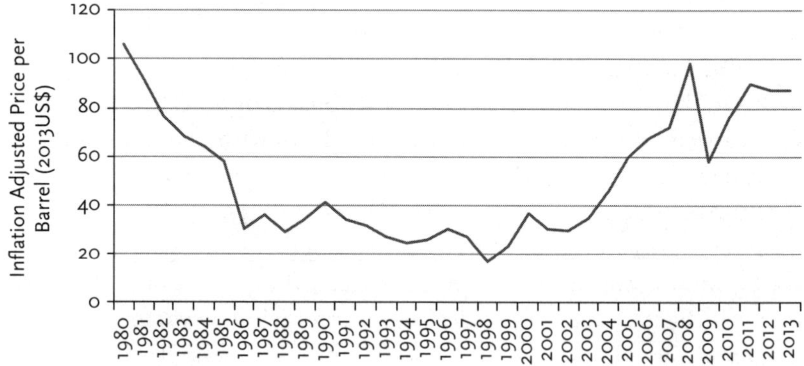

Figure 7.3. Price of Oil, 1980–2013
Source: www.inflationdata.com.

cost retreated to approximately US$60 per ton in response to the autumn 2008 global economic slowdown which quashed demand for coal, the cost is still 30–50% higher than historic levels (US$61 as of February 8, 2013). What this suggests is that it has become considerably more costly to generate a kWh of electricity from fossil fuel technologies than it did 10 to 15 years ago.

Conversely, the cost of wind power has declined significantly over the past two decades. According to one recent study, modern wind energy systems in the United States can now generate power for about US$0.05–0.08 per kWh, making these systems "competitive with the cost of fossil fuel generation in many markets."[60] The Department of Energy concurs. According to a 2008 report, the cost of wind power ranged between US$0.03–0.065 per kWh in 2006.[61] Therefore, even if the GHG emission reduction imperative is ignored, the combination of escalating fossil fuel costs and declining wind power generation costs is chipping away at the foundation of the greatest barrier to political support for expanding wind power capacity.

The wind power sector is also beginning to receive attention as a source of new jobs. Already over 75,000 people are employed in the US wind industry and the US Department of Energy estimates that the industry will support 500,000 American jobs by 2030.[62] By comparison, the US Energy Information Administration (EIA) reports that the US coal industry employed 91,611 people in 2011, despite accounting for 15 times more electricity generated. As further testament to the job creation benefits of wind power over coal-fired power, the World Resource Institute estimates that wind power generates 50% more employment per kWh.[63] As an aside, it merits noting that 7 of the 12 states that do not currently have any installed wind power are Appalachian states—a region which accounts for over 60% of national coal production.

Given these positive economic attributes of wind power, the strategic hold that fossil fuel technologies have over electricity generation is weakening. When combined with carbon tax regimes, which are emerging throughout the United States, there appears to be a positive economic climate forming in support of wind power.

7.4.3 Technological Landscape

Massive wind power potential in the United States is a technological enabler. The National Renewable Energy Laboratory (NREL) has estimated that 10,956 GW of economically viable onshore wind power potential exists in the United States (at 80 m).[64] Assuming a capacity factor of 30%, if fully realized, this amount of wind power could produce 38.5 petawatt hours annually, almost 10 times the total amount of electricity consumed in 2011.[65] The NREL also estimates that there is 4,150 GW of economically viable offshore wind power potential (at 80 m.).[66] Moreover, thanks to superior wind quality, the United States generates more electricity than either Germany or China for the same installed capacity. Such potential gives rise to both commercial interest and political support for the technology.

Consequently, the market potential in the United States has fostered a burgeoning wind power industry. According to the American Wind Energy Association, there are over 500 facilities in the United States that manufacture wind system components and this number is expanding each year. The industry already employs 75,000 individuals and some estimates indicate that it could employ as many as 500,000 by 2030.[67] Over the past five years, average annual private investment has topped US$15 billion. With over 80 universities nationwide offering courses in aeronautics and so much manufacturing activity, the United States is a hotbed for wind power technological innovation.[68]

In diametric opposition to the rosy prospects for wind power system innovation, there is one technological concern that could severely impair the pace and scale of industry development. The US electricity grid is unstable, inefficient, and inadequate. Investment in transmission infrastructure has failed to keep pace with increased energy demand. The result is that the ratio of transmission capacity to electricity demand has alarmingly decreased, particularly in the 1990s where the ratio declined by 16%.[69] This poses a problem for expanding wind power capacity due to the stochastic nature of wind power flows. Grids that do not have sufficient slack capacity, as the US grid exemplifies, become less stable at heightened levels of wind power contribution. The US electricity grid is also inefficient due to aging and structural inadequacies. It has been estimated that transmission and distribution (T&D) losses in the American grid can exceed 25% in extreme cases and US electricity consumers pay over US$200 billion annually to subsidize electricity that they never receive because it dissipates en route. Energy policy expert Benjamin Sovacool explains the inefficiencies in this way: "imagine purchasing a dozen eggs at your local grocery store, but having between one and three eggs break every time you transported them to your home, year after year."[70] Finally, the US electricity grid is inadequate in terms of coverage, particularly in remote areas where wind farms would most likely be developed.[71] Although one purpose of the American Reinvestment and Recovery Act was to earmark funds for electricity grid reinforcement and expansion, these activities take time and there is no guarantee that enhancing grid coverage will necessarily occur in areas where wind power is likely to develop.

There is one other structural aspect of the US electricity grid that is hindering wind power development. In many cases, wind power is most economically harnessed in rural areas where low property values reduce siting costs. However, such rural sites are frequently geographically separated from demand centers and in many cases, transmission of electricity from wind farms to demand centers must traverse sections of the patchwork collective grid that is owned by different entities. In the absence of laws which provide for electricity grid easements, utilities can hinder wind power from traversing grids to reach demand centers. To illustrate, Kansas purportedly has the potential for 950 GW of wind power capacity. Assuming wind system heights of 100 m and 35% or better capacity factors, this is potentially enough to satisfy all current electricity demand in the United States. However, as of the end of 2011, Kansas possessed only 1.3 GW of wind generation capacity. The trouble is that there's not enough electricity demand in Kansas to support wind power supply expansion, and for wind power generated in Kansas to reach distant demand centers, it would have to traverse grids owned by other T&D entities, which would not be keen to accommodate transient transmissions of stochastic energy flows. To put the scale of the problem into perspective, in 2006 the North American Electric Reliability Corporation reported 1,901 transmission request denials due to congestion.[72]

7.5 INFLUENCES ON GOVERNMENT POLICY

7.5.1 Sociocultural → Political

Broadly speaking, high levels of public support for wind power development make it very difficult for politicians to oppose wind power, even at the behest of lobbyists from fossil fuel special interests. According to the American Wind Energy Association, nearly 9 of 10 voters (Republicans, Democrats, and Independents) support wind power expansion. Wind power appeals to advocates of enhanced domestic energy security and opponents of nuclear energy and aligns with the majority view in the United States that the United States should do more to abate greenhouse gas emissions.

Although there are still pockets of political resistance, as epitomized by the letter of opposition from 47 House Republicans to John Boehner on September 21, 2012, the rationale for opposition now stems generally from concerns that wind power is being overly subsidized.

It is telling to note that renewable energy technologies, in general, and wind power, in specific, enjoy such broad public and political support that the 2013 extension of the PTC was passed in a relatively noncontentious manner, despite the heated political conflict over the government's fiscal cliff that was transpiring at the same time. Even in times of fiscal austerity, the government has demonstrated a propensity to continue to support wind power development.

The only specks of trouble on an otherwise cloudless landscape relate to concerns expressed by avian special interest groups over bird mortality and the existence of NIMBY opposition in some locales, particularly offshore.[73] As wind power capacity continues to expand, wind farms will become more aesthetically invasive and government officials can expect amplified public resistance—a challenge that needs to be effectively managed. In a litigious society such as that of the United States, mismanagement of opposition to wind power could result in the same legal blockades that nuclear power providers have faced for decades.

7.5.2 Economic → Political

The emergent commercial viability of wind power has had a profound effect on political behavior. Although there will always be a degree of political resistance toward a transition away from coal-fired power in states that are endowed with coal resources, wind power is now seen as a technology that is capable of both enhancing domestic energy security and mitigating greenhouse gas emissions, in an economically effective manner. In the 20% wind scenario put forth by the US Department of Energy, the cost of achieving 20% contribution from wind power, when passed on to retail consumers, has been estimated at about US$0.06/kWh or about $0.50 per month per household. The rationale for supporting an expansion of wind power is politically irresistible.

This is especially true given the realization that wind power has become a major industry, employing over 75,000 workers. For evidence that the government views the wind industry as strategically important, one needs only to look to the actions taken by the US Commerce Department in July 2012. Responding to evidence that China had been dumping steel towers into the US market, the Commerce Department imposed punitive tariffs on steel towers from China, claiming that these activities were undermining the economic health of domestic firms.

On the other hand, there is also evidence that political resistance is growing in regard to subsidizing wind power to the extent that it has been subsidized in the past. Critics can point to an EIA report indicating, in 2010, that the wind power industry in the United States was on the receiving end of nearly US$5 billion in federal support for electricity generation. This compares to US$1.36 billion in support for coal, US$2.8 billion in support for natural gas and other fossil fuel liquids, and US$2.5 billion in support for nuclear.[74]

However, such a critique fails to recognize that this scale of support for wind power is a recent phenomenon. Just three years previous, in 2007, federal support for wind power was only US$476 million, compared to US$3.9 billion in support for coal, US$2.0 billion in support for natural gas and other fossil fuel liquids, and US$1.7 billion in support for nuclear power. Indeed, many of the fossil fuel technologies have enjoyed decades of federal financial support far in excess of any support wind power has received. For example, between 1943 and 1999, federal subsidization of nuclear power R&D and operations has been estimated at US$144.5 billion. Moreover, annually, US$4.9 billion are channeled from the federal budget into a fund to support the legal risk for nuclear plants and US$3.6 billion go to the maintenance of the strategic petroleum reserve. According to a study from the OECD, in 2001 US coal subsidies accounted for 70% of all subsidies for coal worldwide.[75]

It also merits noting that federal support for wind power is largely aimed at supporting capacity expansion not R&D. In 2010, only US$166 million of the US$5 billion in federal funding for wind power went to R&D activities. In contrast, US$663 million of the US federal energy support budget was spent on supporting R&D in advanced coal technologies, US$1.17 billion was channeled to nuclear power R&D, US$537 was channeled to biomass R&D and US$348 million was channeled to solar R&D. In short, it appears that the commercial viability of wind power is such that the government views capacity expansion as the best approach for simultaneously addressing the nation's energy supply needs and reducing wind power costs (through improved economies of scale).

7.5.3 Technological → Political

As described earlier, the technological (and economic) potential for harnessing massive amounts of wind power has already been recognized by the US Department of Energy.[76] This has provided the foundation upon which to craft policies in support of wind power development. According to the Department of Energy there are already 29 states (plus Washington, DC and two territories) that have adopted renewable portfolio standards aimed at increasing renewable energy capacity.[77]

However, as also outlined earlier, significant investment will be needed to reinforce the US electricity grid in order to accommodate amplified contributions from wind power. The US$4.5 billion that was earmarked under the American Reinvestment and Recovery Act for electricity grid enhancement represents an important first step in regard to overcoming the transmission and distribution inadequacies. However, this represents a drop in the bucket compared to what is needed. According to the Department of Energy, the capital costs to accommodate a 20% contribution from wind power are estimated at US$197 billion. However, the department further argues that the true net present value of achieving a 20% contribution from wind power is US$43 billion, because wind power will result in approximately US$154 billion in fuel expenditure savings.[78] Yet unless a system is established to cover the capital costs out of electricity revenues, the capital costs will somehow have to be fronted by some combination of the 240 investor-owned utilities, 3187 other private utilities, 2012 public utilities, 2168 nonutility generating entities, and nine federal utilities that own and manage individual portions of the US electricity grid.[79] Given that the US electricity grid has arrived at its current suboptimal state under the current disparate regime of grid operators, there is little evidence to suggest that appropriate grid enhancements can be effectively made without a federal mandate and universally agreed and enforced standards.

7.6 POLITICAL INFLUENCES ON POLICY

7.6.1 National Political Structure

The United States is a federal constitutional republic where the federal government and state governments share the responsibility for governance. Federal governance is carried out through three separate branches—the legislative, the executive, and the judicial.

The legislative branch is bicameral, consisting of the House of Representatives and the Senate. The House of Representatives consists of 435 voting members, each of whom represents a congressional district. Together the House of Representatives has a constitutional right to levy and collect taxes, establish post offices and roads, issue and manage patents, create the courts, and undertake decisions related to national defense. The House of Representatives is also where bills are put on the path to becoming a law. The other part of the legislative branch is the Senate, which is made up of two senators from each state, who serve six-year terms. The Senate is tasked with the provision of legislative oversight and is responsible for providing advice to cabinet officers, federal judges, department secretaries, senior military officers, and ambassadors. It also reviews or rejects bills originating from the House of Representatives. Upon Senate approval of a bill, the bill then goes on to become law when signed by the president.

In the case of a presidential veto, a bill can become law if it is passed by a two-thirds majority of each chamber.

The executive branch of the federal government is headed up by the president of the United States, who presides over more than 2.7 million civil servants and 1.5 million military personnel who serve in federal government departments, agencies, and affiliated organizations. The president has the power to sign legislation into law or veto it, sending the bill back to both houses of Congress to vote to override the veto. In order to carry out day-to-day government activities, the president appoints a group of cabinet members who head up the 15 departments under the executive branch. This includes the US Department of Energy, which is currently headed up by Energy Secretary Steven Chu. The executive branch also oversees the activities of independent agencies such as US Post, NASA, the CIA, the Environmental Protection Agency, and US AID.

The third branch of the federal government is the judicial branch, which is responsible for interpreting and enforcing laws. The highest court in the nation is the US Supreme Court, which handles matters pertaining to the federal government, settles disputes between states, and provides final interpretation of the United States Constitution. It is the court of final recourse in the United States.

State-level governance is also carried out through three branches—the legislative, the executive, and the judicial—that are structured in a similar fashion to the federal branches. In all states except Nebraska, the legislative branch is comprised of two chambers—an upper house (typically called the Senate) and a lower house (typically called the House of Representatives)—and it is in the state legislature where state bills are introduced. The executive branch of the state is led by a governor, who presides over day to day affairs of the state. Finally, there is a judicial branch that hears civil and criminal cases not under federal jurisdiction. The state Supreme Court is typically the court of final recourse for legal matters which fall under state jurisdiction.

Generally speaking, the state is tasked with providing electricity services to its residents. As a result, it is responsible for establishing the structure of the electricity sector in the state, it decides who will be responsible for electricity provision, and it ensures that public interests are protected in the course of providing such services. Once electricity is transmitted from one state to another, it becomes a federal matter. A commerce clause found in the US Constitution (Article I, Section 8, Clause 3) bars any state from interfering with interstate commerce, which has been interpreted as including interstate transmission and distribution of electricity. This interpretation has also led to federal authority over oil, natural gas and gasoline pipelines that cross state lines.[80] In short, although the states enjoy a high degree of sovereignty over electricity generation and distribution, in practice federal influence is rarely absent.

7.6.2 Governing Party Ideology

Two political parties—the Republican Party and the Democratic Party—have dominated American politics. Ideologically, the Republican Party can be considered as embracing right of center ideals while the Democratic Party embraces ideals considered more left of center. Among the electorate, support for the two parties is very close to being evenly split. This has been illustrated by the last four federal elections (in 2000, 2004, 2008, and 2012), where the popular vote separating winning and losing parties was 0%, 3%, 7%, and 4% respectively. In fact, it is only through the unique electoral vote system in the United States that either party has managed to establish clear legislative majorities. As a consequence of a divided electorate, the platforms of both parties tend to exhibit centrist leanings in the hope of enticing swing voters—moderate supporters within opposing parties who might switch alliances.

Although both parties have centrist tendencies, there are clear policy preferences which tend to stem from party ideology. Republican ideology tends to favor (in theory at least) smaller government and less government interference in the economy, preferring to leave as much as possible to free market forces. Democratic ideology accepts the theoretical efficiencies attributed to free-market economies but also harbors a tenet that government should be more actively involved in the economy to ensure that negative externalities associated with free market activity do not harm less privileged segments of society.

In regard to energy policy, there tends to be a perception that Republicans favor conventional energy technologies, while Democrats are more supportive of a transition to renewable technologies. This is an inaccurate generalization that tends to be supported by anecdotes such as Republican Ronald Reagan's removal of solar panels (that were installed under the Democratic Carter Administration) from the roof of the White House and Democrat Barack Obama's decision to reinstall solar panels on the roof when he took office.[81] Although there is evidence that Republicans have historically favored conventional energy interests—and indeed, the majority of lobbyist funds from fossil fuel interests go to Republican politicians—the reality is that Republicans have historically supported conventional energy technologies because conventional energy has been the least expensive. When one harbors a free-market ideology, anything that enhances domestic competitiveness is desirable, including policies that reduce the cost of energy—a key factor of production. Therefore, with the cost of wind power decreasing and the cost of fossil fuel resources ratcheting up in a capricious manner, there is growing support for wind power even in the Republican camp. While it is true that most hard-core Republicans may not support renewable energy development for environmental reasons (unless it is economically advantageous to do so), it is not accurate to say that all Republicans, in this day and age, oppose wind power. As a testament to this, a special interest group called the Red State Renewable Alliance has recently been established to engender Republican support for wind power.[82]

Currently the executive branch of the US government is led by Democratic President Barack Obama. The Obama administration has supported renewable energy development in part for the same reason that Republicans support wind power, it is a commercially viable source of domestic energy that allows American industry to keep a cap on operating costs, while at the same time, providing the nation with a higher degree of national energy security. While there is likely more support for wind power on environmental/climate change grounds within the Democratic Party, it would be inaccurate to say that environmentalism is the key driver of Democratic support. If environmentalism was the key factor, the Obama administration would not be supporting an expansion of nuclear power capacity, would not be boasting about opening "millions of new acres for oil and gas exploration in the last three years, and we'll open more" and would not have an ambition to "develop a hundred year supply of natural gas that's right beneath our feet."[83] If climate change mitigation was the key driver, the Obama administration would be embracing a far greater role in leading global GHG emission reductions.

As it stands, the Obama administration has committed to a GHG reduction target aiming to reduce GHG emissions by 17% of 2005 levels by 2020. A key initiative for achieving such a goal is to facilitate the expansion of wind power capacity, thereby preserving industrial competitiveness, while at the same time addressing important environmental concerns. It bears noting that the 20% wind scenario that is advocated by the Department of Energy could contribute to reducing CO_2 emissions in 2030 by 25% compared to business-as-usual levels.[84]

7.6.3 Fiscal Health

In spite of the much-ballyhooed ideological differences, successive administrations in the United States have demonstrated a common propensity for failing to balance the annual fiscal budget. As Table 7.4 illustrates, in absolute terms the federal public debt has increased significantly under every

Administration	Size of Gross Debt	Size of Gross Debt to GDP (%)
Before Reagan	US$1 trillion	33
Ronald Reagan	US$2.9 trillion	53
George H. W. Bush	US$4 trillion	64
Bill Clinton	US$5.6 trillion	57
George W. Bush	US10.6 trillion	77
Barack Obama	US$16.4 trillion	105

Table 7.4 THE ASCENT OF PUBLIC DEBT

Source: Glen Kessler, January 4, 2013, Washington Post blog article based on White House data. www.washingtonpost.com/blogs/fact-checker/post/why-is-the-national-debt-16-trillion/2013/01/03/e2a85386-55fc-11e2-8b9e-dd8773594efc_blog.html.

president from Ronald Reagan onward. Of greater concern is the insight conveyed in the far right column of Table 7.4, which highlights how the size of the gross public debt has grown considerably in relation to GDP. The reason this is a concern is because higher levels of gross debt to GDP means that the road to debt reduction becomes a longer, more arduous proposition. As an analogy, a person who is earning US$50,000 per year stands a far greater chance of expediently paying off a debt of US$1,000 compared to the challenge of paying off a debt of US$50,000.

This escalation of public debt is troubling for two reasons. First, it is suggestive of a pervasive lack of fiscal fiduciary responsibility inherent to both political parties. Other than the three-year period following 9/11 where government revenues fell by 17% and the three-year period between 2006 and 2009 where government revenues fell by 21%, total government revenues have increased annually from US$1.642 trillion in 1992 to US$2.902 trillion sequestered for 2013. Yet despite this 77% increase in government revenues, presidents since 1970 have successively posted annual federal budget deficits (the exception being 1998 to 2001 under the Clinton administration when the global economy was booming).[85] Governing administrations have a marked propensity to spend beyond their financial means, accumulating debt (and debt service obligations) that is beginning to adversely affect fiscal health. This has been cited as a key justification for Standard & Poor's downgrade of the nation's debt rating in 2011.[86]

Although there have been some extenuating circumstances (predominantly a costly military campaign in the Middle East and a global economic slowdown), the Obama administration has been operating deep into the red since taking office in 2008. In the first four years (2009–2012), the Obama administration racked up US$5.8 trillion in fiscal deficit, more than any other administration in US history including the massive deficit incurred in eight years of Republican rule under George W. Bush. Both parties recognize such large deficits are damaging to the US economy. Consequently, to rein in spending, a Budget Control Act was passed in 2011 that places limits on discretionary spending. Since then virtually any issue that is subject to discretionary funding has come under intense political scrutiny, and this includes expenditures like energy R&D and subsidies for renewable energy. The debate over what to fund has become so politicized that the parties could not reach agreement on a 2013 budget, resulting in a provisional budget (through the end of March 2013) being approved at the end of September 2012. The budget for the remainder of the year was not signed by President Obama until March 26, 2013, five days before the government was slated to lose its authority to collect and disburse funds.

7.6.4 Policy Regime

Energy policy is implemented at both the federal and state level. The US Department of Energy is responsible for coordinating federal energy policy. A key function and indeed a key reason for the formation of the department back in 1977 has been nuclear power oversight. There is an undersecretary of energy for nuclear security who oversees the National Nuclear Security Administration. Annually, this branch of the department receives about one-third of the total Department of Energy annual budget. The department also includes the Office of Energy Efficiency and Renewable Energy, the Office of Fossil Energy, the Office of Science, the Energy Information Administration (EIA), the Federal Energy Regulatory Commission (FERC), and numerous laboratories and research centers scattered across the nation. It employs about 16,000 people, and manages an elaborate network of 24 national laboratories that employ a further 60,000.[87] It is responsible for over 65% of all energy related federal expenditures and had a 2012 budget of US$30.6 billion.

In regard to federal governance of the electricity sector, FERC is the key agency for regulatory oversight. It has jurisdiction over interstate electricity sales, wholesale electricity rates, and hydroelectric project licensing, to name but a few important areas. The Energy Policy Act of 2005 expanded the authority of FERC to allow it to draft and impose regulations to enhance national grid resilience. This included granting additional authority over the transmission and wholesale pricing of electricity across state borders. FERC also has the authority to prosecute entities that violate FERC directives.

At the state level, decisions regarding electricity system management tend to rest with public utility commissions that make strategic decisions and set retail prices subject to state legislature or state executive approval. Since the electricity grid in most states is owned and managed by an array of public (regional or municipal) and private utilities, much of the work of state regulatory bodies relates to standard setting and enforcement, the issuance of plant permits, pricing regulation, and overall system coordination.[88]

The interplay between federal and state energy authorities is complex and complicated. As an example, FERC plays a key role in governing wholesale electricity prices but states or substate agencies are responsible for retail electricity price setting. The commission also oversees the transport of gas and oil across state borders and grants permission for the siting of liquid natural gas (LNG) terminals, but the states are responsible for setting retail gas rates and for licensing in-state power plants. Under this system of divided authority, both federal and state governments share authority in a form of interactive federalism. It has been estimated that the electricity sector is regulated by 53 federal, state, and substate entities and subject to more than 44,000 state or substate codes.[89]

Climate change mitigation policy also influences what happens in the electricity sector, and in a similar manner to energy policy, both federal and state policies interact to weave together what eventually becomes national climate change policy. At the federal level, the overarching climate change mitigation target stems from the commitment made by the Obama Administration in Copenhagen in 2009. As outlined earlier, as a party to the Copenhagen Accord, the United States departed from its previous aversion to concrete GHG emission reduction commitments by agreeing to work toward reducing emissions by 17% of 2005 levels by 2020.

Prior to the US commitment in Copenhagen, a number of frustrated state and substate policymakers stepped up, in the absence of a federal commitment, to lead subnational initiatives to reduce GHG emissions. For example, in 2006, the California legislature passed the Global Warming Solutions Act, which aims to reduce GHG emissions to 1990 levels by 2020—a 25% reduction from 2006 levels.[90] In some cases states have even banded together to form regional coalitions. An example is the Regional GHG Initiative of 2003, which developed a cap and trade program for GHG emissions from power plants in states in the Northeastern United States and Eastern Canada. Another example is the recent coalition of Western state governors that is aiming to coordinate renewable energy targets across state boundaries. There are even examples of substate leadership such as the US Conference of Mayors Climate Protection Agreement which aims to comply with the US commitment made during the Kyoto Protocol negotiations. As of May

2013, 1,060 mayors of US cities that represent over 88 million citizens have signed the agreement.[91]

7.7 THE CULMINATION OF INFLUENCES

During George W. Bush's two terms in power, the policy priority was to expand domestic fossil fuel reserves. Upon leaving office this torch was passed on to John McCain, who was nominated as the Republican presidential candidate. Many people who witnessed the Republican National Convention in 2008 likely still have etched in their minds the spectacle of grown adults chanting in unison "drill baby drill" in support of the Republican platform to drill for oil in the Arctic National Wildlife Refuge.

On the other side of the political spectrum, Barack Obama's presidential campaign in 2007–2008 included numerous references to the importance of expanding renewable energy capacity. Upon Obama's ascent to the presidency, statements highlighting the importance of addressing climate change and support for renewable energy have been commonplace in public addresses given by the president, including in his State of the Union addresses.

Up until recently, these two divergent perspectives on how to improve energy security played an influential role in shaping public attitudes and perceptions regarding energy policy. Generally, Republicans tend to view the expansion of domestic fossil fuel resources as a top priority, while Democrats tend to view the development of clean energy technologies to be of higher imperative. As a sweeping generalization, Republicans that are concerned about climate change tend to harbor more support for technologies such as carbon capture and sequestration, while Democrats tend to oppose technologies that extend the dominance of fossil fuel energy in the nation's electricity mix.

These ideological divides tend to become entrenched due to the pervasive influence of mass media on public perception. In too many cases the objective nature of news reporting has been usurped by political bias, with the public becoming unwitting recipients of a distorted analysis. On the one hand, television networks such as CBS and newspapers such as the *New York Times* have been said to possess a liberal bias that favors Democratic positions.[92] On the other hand, a conservative bias has been attributed to television stations such as Fox and newspapers such as the *Wall Street Journal* that tends to favor Republican positions.

In 1999, in an attempt to directly influence public awareness of wind power, the US Department of Energy launched the Wind Powering America (WPA) initiative that is a program designed to "educate, engage, and enable critical stakeholders to make informed decisions about how wind energy contributes to the U.S. electricity supply."[93]

At the state level, there is a wide variety of compulsory and voluntary green purchase programs that are promoted to electricity consumers, thereby enhancing awareness of the pros and cons of the various energy technologies. For example, in Washington State, there are seventeen utilities that offer consumers the chance to designate that their electricity comes from wind power. Similarly, in Texas, six utilities offer plans which allow consumers to purchase electricity from wind power.[94]

As has been inferred throughout this chapter, the PTC has had an enormous impact on wind power development, particularly over the past five years or so. If the estimates by the US Department of Energy are accurate and wind power generation prices in the United States range between US$0.03–0.065 per kWh, then a production tax credit of US$0.02/kWh represents a catalytic subsidy. Indeed, many analysts have commented on the trend that when the PTC expires, the pace of wind power development tends to drop off precipitously, and when the PTC is renewed, a rigorous pace of wind power development resumes.[95] Moreover, as was suggested in the last section, because of the expansionist effect that the PTC has had on the US wind power market, many firms have relocated to new production facilities in the United States to conduct R&D. This has created a thriving wind power sector and has fueled wind power technology innovation, which in turn has helped reduce the cost of wind energy generation.

Yet the federal PTC has not been sufficient in itself to catalyze wind power development everywhere. As outlined earlier, there are a dozen US states that do not have wind power programs. Conversely, there are also 15 states that host over 1000 MW of installed wind power capacity. In short, it is apparent that state policy has played a major role in influencing market development. For most states, the central policy for catalyzing development has been the establishment of renewable portfolio standards (RPS). According to the AWEA, 29 states (plus Washington, DC and territories) have announced RPS schemes. The establishment of such policies is usually accompanied by regulatory oversight measures to ensure that the standards are met. As such, the introduction of an RPS tends to break down any non-economic barriers that may exist.

The influence of the state is also felt in national political circles. For example in August 2011, a coalition of 24 governors entreated the Obama administration to do more to support the development of wind power by extending the PTC, fortifying the grid, improving collaboration between state and federal authorities, improving siting standards, and advocating more offshore wind power development.[96] As another example, the Western

Governors Association has recently announced an intention to coordinate energy policy to facilitate development of western renewable energy zones. The aim is to provide leadership in clean energy development in the absence of a cohesive federal policy.[97] These types of initiatives serve both a commercial and a political purpose. Commercially, they serve as a beacon to attract new wind power projects. Politically, they tend to inspire similar efforts, engendering a level of political competition that can create conditions for universal change. On the other hand, the development of state initiatives also takes the pressure off of federal policymakers. If the state is financing wind power development, the federal government can save on fiscal expenditure by ceding such responsibility to the states. Therefore, both state and federal leadership is essential for expediting the pace of diffusion.

The technological development of wind systems has been directly influenced by financial support for R&D both at the federal and state level. As stated earlier in this chapter, federal support for wind power technology development was instrumental in enabling the wind power boom in the early 1980s. Since then, the Department of Energy has been influential in helping supplement private wind power research and its numerous laboratories around the country have directly contributed to technological innovation.

Recently, the US Department of Energy has entered into a partnership with six leading wind turbine manufacturers to facilitating exchange of information related to R&D to enhance system reliability and efficiency, develop effective siting strategies, encourage advances in manufacturing and design, and improve process automation and fabrication techniques.[98] All of this is in support of the initiative intended to achieve a 20% contribution from wind power by 2030. One concrete contribution that the US Department of Energy's National Renewable Energy Laboratory will make to this collaborative effort will come from its state of the art wind turbine blade test facility in Ingleside, Texas. The department will also be tasked with disbursing US$2 billion earmarked under ARRA that will go largely to supporting research in advanced battery development.

It merits pointing out that political influence on technological innovation in wind systems is far greater than just the funding that is disbursed under federal and state programs. For example, if PURPA did not mandate access for wind power developers to the electricity grid and if the PTC did not provide the added financial incentive to encourage wind power development, it is very possible that sufficient market demand would not have existed to support the R&D that was done by private wind power firms. Similarly, the RPS that are currently in place in many states and the standards, codes, and regulations administered by local utility commissions also play an instrumental role in generating sufficient market demand to encourage R&D.

In terms of technological support, FERC has served an enabling role for wind power technology because standards, codes and regulations administered by FERC attenuate many of the technological barriers to wind power connection. Moreover, government investment in grid enhancement, such as the US$4.5 billion sequestered under ARRA in 2009, provides necessary infrastructure to accommodate higher levels of wind power. However with that said there is still a lot of work that needs to be done. According to the American Wind Energy Association (AWEA), there is an estimated 200,000 MW of proposed projects that are currently in a developmental cue because there is not enough transmission capacity to carry the electricity that these installations would produce.[99] Some of this additional capacity will be realized through new private sector ventures as the economics of supporting wind power continue to progress. For example, a project known as Green Power Express has recently been proposed to transmit up to 12,000 MW of wind power from the Dakotas, Minnesota, and Iowa to the Chicago area. This project would add 3000 miles of extra high-voltage transmission lines to the transmission system.

Turning to the challenges facilitating offshore wind power development, section 388 of the Energy Policy Act of 2005 has been a key enabler of offshore wind power development in that it authorizes the Secretary of the Interior to grant leases, easements and rights of passage on the outer continental shelf. This one policy will be instrumental for facilitating offshore wind power development because through policy comes certainty, and in the presence of certainty, commercial commitment thrives. Although this is not a political initiative that directly supports technological development, like many of the other initiatives outlined in this section, it is a critical enabler for technological innovation to take place.

7.8 WHAT TO EXPECT GOING FORWARD

The commercial viability of wind power has attenuated many of the barriers inhibiting wind power diffusion in the United States. Wind power is now seen as a way to solve three problems through one technology. First, it is a domestic source of energy; therefore, it enhances national energy security. Second, the cost of wind is competitive with fossil fuels; therefore, it has engendered support from both industry and the general public. Third, CO_2 emissions associated with wind systems are among the lowest of all energy technologies; therefore, wind systems have a key role to play in reducing GHG emissions. In short, looking to the near future, the question is not really whether or not wind power will continue to expand in the United States; but rather, at what pace will it continue to expand?

In addressing this question, a key technological bottleneck must be addressed—grid enhancement. Achieving even the 20% wind power scenario proposed by the US DoE will require upgrades to the nation's T&D system, the creation of larger electric load balancing regions, improved regional planning and integration, enhancements to the manufacturing supply chain, and strategies for alleviating community concerns in relation to site selection, wildlife habitat preservation, and other environmental issues.[100] Although the government is responding to the need for grid reinforcement through investment, this will be an expensive and time-consuming endeavor. Infrastructure investment needs aside, FERC will be called on to provide a pivotal role in ensuring that cooperation exists between grid management entities to allow wind power to be distributed from supply centers to demand centers.

Intriguingly, the ongoing initiatives designed to expand domestic natural gas production will likely yield a net benefit in terms of wind power diffusion. Although competition from cheap natural gas will dampen the pace of wind power diffusion in the short term, the added peak-load capacity that natural gas represents will also enhance the capacity of America's grids to support higher levels of wind power.

Moving forward, state leadership can only take the US wind power development program so far. If the United States hopes to achieve the commitment made in the North American Leaders' Declaration of Climate Change and Clean Energy to reduce CO_2 emissions by 80% by 2050, wind power will likely have to contribute at least 40% or more to the electricity grid. This cannot be achieved through the historical practice of cobbling together a patchwork of state electricity grids that vary in terms of system resilience and carrying capacity. To allow the United States to exploit the wind power potential in the Great Plains, the Midwest, the Pacific Northwest, and the Northeast, a higher level of strategic grid integration will be necessary.

In regard to offshore development, what will transpire is still up in the air. There is enormous potential. The NREL has estimated that within 50 nautical miles, at a height of 90 meters, there is 4,150 GW of offshore wind potential with wind speeds greater than 7 m/s. The national strategy, which was released in February 2011, announced intentions to encourage 10 GW of offshore wind power by 2020 and 54 GW by 2030. According to the US DoE, if this is achieved, it will lead to the development of 43,000 permanent jobs.[101] A regional study estimated that the installation of 3.2 GW of offshore wind power in Virginia alone would create 9,700 to 11,600 jobs and would attract US$403 million in investment to the local economy, regardless of where the wind systems were manufactured. These types of figures suggest that ample incentive exists for more proactive federal support. Although history has shown that the central government tends to cede the strategy for local development to state and private interests, the recent involvement of the federal government in enabling the Cape Wind project to proceed indicates a more proactive federal approach may be evolving in regard to offshore wind power diffusion.

Another issue of concern is the extent to which the federal government will continue to proactively support the PTC. In 2013, extending the PTC was delayed, to the detriment of the wind power market[102] and was eventually extended only to the end of 2013, when it once again expired. As of June 2014, efforts to revive the PTC have stalled in the senate.[103] So the question is whether the Obama administration has fallen back into a reactive strategy for managing the PTC, or whether it will foster a bull market in the future by reintroducing the PTC. With so much opposition to extending the PTC every time it comes up for renewal, one also wonders what will transpire in the lead up to the expiry of the renewable energy production incentive (REPI) on October 1, 2016.

If the recent past provides any indication of how all of this will play out in the US wind power sector, it would appear that economic, industrial, and public interest all point to a rosier outlook regarding the future pace of development. To a large extent, what transpires in regard to fossil fuel prices and how the perils attributed to climate change unfold will likely play a major role in influencing federal government energy strategy. There is little to suggest that fossil fuel prices are going to return to the low levels seen in the 1990s. Meanwhile, there is mounting evidence that the perils attributed to climate change will continue to intensify, fueling public alarm. In short, with or without the PTC, market momentum is on the side of wind power in the United States.

NOTES

1. As reported on the Huffington Post website, September 24, 2012. "John Boehner Pressured By Conservatives To End Wind Production Tax Credit." www.huffington post.com/2012/09/24/john-boehner-wind-production-tax-credit_n_1910687.html.

2. Will Oremus. 2012. "The Latest Wedge Issue in the Presidential Campaign: Wind Power." *The Slate*, August 15. www.slate.com/blogs/future_tense/2012/08/15/wind_power_growing_fast_in_u_s_but_romney_would_end_production_tax_credit.html

3. Exxon. *2011 Summary Annual Report*. http://corporate.exxonmobil.com/~/media/Reports/Summary%20Annual%20Report/2011/news_pub_sar2011.pdf.

4. *Peabody Energy Annual Report 2012*. http://www.peabodyenergy.com/mm/files/Investors/Annual-Reports/PE-AR2012.pdf

5. Greene, David and Sanjana Ahmad. 2005. *Costs of U.S. Oil Dependence: 2005 Update (January, 2005)*. Report to the US DOE, ORNL/TM-2005/45. Washington, DC: Department of Energy.

6. Greene, David L. 2010. "Measuring Energy Security: Can the United States Achieve Oil Independence." *Energy Policy* 38 (4): 1614–1621.

7. Vaclav Smil. 2010. *Energy Myths and Realities: Bringing Science to the Energy Policy Debate*, p. 3 Washington, DC: Rowman and Littlefield.

8. O'Hanlon, Michael. 2010. "How Much Does the United States Spend Protecting Persian Gulf Oil?" In *Energy Security: Economics, Politics, Strategies and Implications*, edited by Carlos Pascual and Jonathan Elkind, pp. 59–72. Washington, DC: Brookings Institution Press.

9. Delucchi M. A., Murphy J. J. 2008. "US Military Expenditures to Protect the Use of Persian Gulf Oil for Motor Vehicles." *Energy Policy* 36 (6): 2253–2264.

10. US Energy Information Administration (EIA). 2012. *AEO2012 Early Release Overview*. Washington, DC: US Energy Information Administration.

11. Krancer, Michael. 2014. "Obama Energy Official: Nuclear Plants Essential to Our Carbon Reduction Goals." *Forbes Online*, February 12. www.forbes.com/sites/michaelkrancer/2014/02/12/obama-energy-official-nuclear-plants-essential-to-our-carbon-reduction-goals/.

12. US Energy Information Administration (EIA). 2013. *Annual Energy Outlook 2013*. Washington, DC: US Energy Information Administration.

13. Ibid.

14. Sovacool, Benjamin K. 2008. *The Dirty Energy Dilemma: What's Blocking Clean Power in the United States*. New York: Praeger Publishers.

15. International Energy Agency (IEA). 2012. *CO_2 Emissions from Fuel Combustion: Highlights 2012*. Paris: International Energy Agency.

16. Data extracted from the EIA website: www.eia.gov.

17. US Energy Information Administration (EIA). 2012. *AEO2012 Early Release Overview*. Washington, DC: US Energy Information Administration.

18. Hansen, James. 2008. "Global Warming Twenty Years Later: Tipping Points Near." Paper read at the National Press Club, June 23, in Washington.

19. More on this program at www.recovery.gov/News/featured/Pages/Recovery-Tax-Credits-for-Clean-Energy.aspx.

20. More on this program at www1.eere.energy.gov/solar/sunshot/about.html.

21. US Energy Information Administration (EIA). 2012. *AEO2012 Early Release Overview*. Washington, DC: US Energy Information Administration.

22. Harborne, Paul, and Chris Hendry. 2009. "Pathways to Commercial Wind Power in the US, Europe and Japan: The Role of Demonstration Projects and Field Trials in the Innovation Process." *Energy Policy* 37 (9): 3580–3595.

23. Van Est, Rinie. 1999. *Winds of Change: A Comparative Study of the Politics of Wind Energy Innovation in California and Denmark*. Utrecht, Netherlands: International Books.

24. Ibid.

25. Rivkin, David, and Laurel Silk. 2013. *Wind Energy: The Art and Science of Wind Power*. Burlington, VT: Jones and Bartlett Learning.

26. Smith, D R. 1987. "The Wind Farms of the Altamont Pass Area." *Annual Review of Energy* 12 (1): 145–183.

27. Van Est, Rinie 1999. *Winds of Change: A Comparative Study of the Politics of Wind Energy Innovation in California and Denmark*. Utrecht, Netherlands: International Books.

28. The code can be accessed at http://uscode.house.gov/download/pls/42C125.txt.

29. This act can be found at http://thomas.loc.gov/cgi-bin/bdquery/z?d101:H.R.4808:.

30. A copy of this act can be obtained at http://thomas.loc.gov/cgi-bin/query/z?c102:H.R.776.ENR.

31. For more on US wind statistics visit www.windpoweringamerica.gov/wind_installed_capacity.asp.

32. More on efforts in Minnesota can be found at www.renewable.state.mn.us/.

33. More on this project at: www.capewind.org/index.php.

34. A summary of this act can be obtained at www.jct.gov/x-22-02.pdf.

35. For more on US wind statistics visit www.windpoweringamerica.gov/wind_installed_capacity.asp.

36. More on this policy at http://apps1.eere.energy.gov/states/pdfs/57712.pdf.

37. A copy of this act can be obtained at http://thomas.loc.gov/cgi-bin/query/D?c108:6:./temp/~c108lZZfDm.

38. A copy of this act can be obtained at www.gpo.gov/fdsys/pkg/BILLS-109hr6enr/pdf/BILLS-109hr6enr.pdf.

39. More on this initiative is found at www.dsireusa.org/incentives/incentive.cfm?Incentive_Code=WA15R&re=1&ee=1.

40. A copy of this act can be obtained at http://otexa.ita.doc.gov/PDFs/Africa_Investment_Incentive_Act_of_2006_Title_VI.pdf.

41. More on this at www.renewable.state.mn.us/.

42. More on this at http://heartland.org/policy-documents/policy-tip-sheet-no-11-illinois-renewable-energy-mandate.

43. Galbraith, Kate. 2008. "Texas Approves a $4.93 Billion Wind-Power Project." *New York Times*, July 19. www.nytimes.com/2008/07/19/business/19wind.html?_r=0

44. A copy of this report is available at www.nrel.gov/docs/fy08osti/41869.pdf.

45. A copy of this act can be obtained at www.gpo.gov/fdsys/pkg/BILLS-111hr1enr/pdf/BILLS-111hr1enr.pdf.

46. More on this project can be found at http://atlanticwindconnection.com/.

47. US Department of Energy. 2011. *A National Offshore Wind Strategy: Creating an Offshore Wind Energy Industry in the United States*. Washington, DC: US Department of Energy.

48. Angus Reid Global Scan. 2005. "More Americans Favor Energy Conservation." February 3. www.angus-reid.com.

49. Angus Reid Global Scan. 2009. "Americans Keen on New Sources of Energy." March 1. www.angus-reid.com.

50. For popular examples of such critiques, see Campbell, Kurt M., and Jonathon Price, eds. 2008. *The Global Politics of Energy*. Washington: The Aspen Institute; and Friedman, Thomas L. 2008. *Hot, Flat and Crowded*. New York: Farrar, Strauss and Giroux.

51. Angus Reid Global Scan. 2008. "Majority in US Wants More Nuclear Plants." November 27. www.angus-reid.com.

52. Angus Reid Poll. 2012. "Nuclear Power Continues to Split Views in the United States." February 16. www.angus-reid.com.

53. Angus Reid Poll. 2009. "Global Warming Happening, Say Americans." June 28. www.angus-reid.com.

54. Angus Reid Poll. 2009. "Fewer in US Blame Humans for Global Warming." April 22. www.angus-reid.com.

55. Angus Reid Poll. 2009. "Americans Assess How to Fight Climate Change." June 30. www.angus-reid.com.

56. Jones, Tim. 2007. "Dusty Farms Reap Wind as Growing Cash Crop." *Chicago Tribune*, December 21. http://articles.orlandosentinel.com/2007-12-21/news/windfarm21_1_wind-turbines-wind-power-wind-farm

57. Firestone, Jeremy, and Willett Kempton. 2007. "Public Opinion About Large Offshore Wind Power: Underlying Factors." *Energy Policy* 35 (2007) 35 (3): 1584–1598.

58. Angus Reid Global Monitor. 2009. "Economy Remains Top Worry for Americans." June 26. www.angus-reid.com.

59. NBC new story on this position at www.nbcnews.com/id/8422343/ns/politics/t/bush-kyoto-treaty-would-have-hurt-economy/.

60. Pernick, Ron, Clint Wilder, and Trevor Winnie. 2012. *Clean Energy Trends 2012.* San Francisco: Clean Edge.

61. US Department of Energy. 2008. *20% Wind Energy by 2030: Increasing Wind Energy's Contribution to US Electricity Supply.* Washington, DC: US Department of Energy.

62. According to the American Wind Energy Association website: www.awea.org.

63. World Resources Institute (WRI). 2010. *Fact Sheet: Policy Design for Maximising US Wind Energy Jobs.* Washington, DC: World Resources Institute.

64. A chart of this is available at www.windpoweringamerica.gov/pdfs/wind_maps/us_contiguous_wind_potential_chart.pdf.

65. More on this study at www.windpoweringamerica.gov/filter_detail.asp?itemid=2542.

66. This study is available on the NREL website: www.nrel.gov/news/press/2010/885.html.

67. All data taken from the American Wind Energy Association website: www.awea.org.

68. American Wind Energy Association website: www.awea.org.

69. Sovacool, Benjamin K. 2008. *The Dirty Energy Dilemma: What's Blocking Clean Power in the United States.* New York: Praeger Publishers.

70. Ibid.

71. US Department of Energy. 2008. *20% Wind Energy by 2030: Increasing Wind Energy's Contribution to US Electricity Supply.* Washington, DC: US Department of Energy.

72. Abel, Amy. 2008. *Electric Transmission: Approaches for Energizing a Sagging Industry.* Washington, DC: Congressional Research Service.

73. Firestone, Jeremy, and Willett Kempton. 2007. "Public Opinion About Large Offshore Wind Power: Underlying Factors." *Energy Policy* 35 (3): 1584–1598.

74. Both tables are available at www.eia.gov/analysis/requests/subsidy/.

75. Sovacool, Benjamin K. 2008. *The Dirty Energy Dilemma: What's Blocking Clean Power in the United States.* New York: Praeger Publishers.

76. US Department of Energy. 2008. *20% Wind Energy by 2030: Increasing Wind Energy's Contribution to US Electricity Supply.* Washington, DC: US Department of Energy.

77. Information on state policies can be obtained from the Database of State Incentives for Renewables and Efficiency at www.dsireusa.org/.

78. US Department of Energy. 2008. *20% Wind Energy by 2030: Increasing Wind Energy's Contribution to US Electricity Supply.* Washington, DC: US Department of Energy.

79. Sovacool, Benjamin K. 2008. *The Dirty Energy Dilemma: What's Blocking Clean Power in the United States.* New York: Praeger Publishers.

80. Sovacool, Benjamin K. 2011. "National Energy Governance in the United States." *Journal of World Energy Law & Business* 4 (2): 97–123.

81. The timeline can be found at http://usgovinfo.about.com/od/thepresidentandcabinet/tp/History-of-White-House-Solar-Panels.htm.

82. Their website: http://redstaterenewables.org/.

83. More on his comments at http://nawindpower.com/e107_plugins/content/content.php?content.10368.

84. US Departement of Energy. 2008. *20% Wind Energy by 2030: Increasing Wind Energy's Contribution to US Electricity Supply.* Washington, DC: US Department of Energy.

85. Budget data is available from the Whitehouse website at www.whitehouse.gov/omb/budget/Historicals.

86. More on this analysis at www.washingtonpost.com/blogs/fact-checker/post/why-is-the-national-debt-16-trillion/2013/01/03/e2a85386-55fc-11e2-8b9e-dd8773594efc_blog.html.

87. Sovacool, Benjamin K. 2011. "National Energy Governance in the United States." *Journal of World Energy Law & Business* 4 (2): 97–123.

88. Ibid.

89. Ibid.

90. More on this initiative available at www.arb.ca.gov/cc/cleanenergy/cleanenergy.htm.

91. More on this program can be found at www.usmayors.org/climateprotection/list.asp.

92. An interesting article on this topic can be found at http://archive.mrc.org/biasbasics/biasbasics2admissions.asp.

93. More on this initiative at www.windpoweringamerica.gov/.

94. A mapping and list of these programs can be accessed through the EPA website: www.epa.gov/greenpower/pubs/gplocator.htm.

95. As reported by both Barradale, Merrill Jones. 2010. "Impact of Public Policy Uncertainty on Renewable Energy Investment: Wind Power and the Production Tax Credit." *Energy Policy* 38 (12): 7698–7709; and Lu, Xi, Jeremy Tchou, Michael B. McElroy, and Chris P. Nielsen. 2011. "The Impact of Production Tax Credits on the Profitable Production of Electricity from Wind in the U.S." *Energy Policy* 39 (7): 4207–4214.

96. More on this proposal is available at http://cleantechnica.com/2011/08/26/24-governors-ask-obama-to-focus-on-wind-energy-deployment/.

97. More on this at www.westgov.org/wga/publicat/WREZ09.pdf.

98. For more on this story see "DOE Announces Effort to Advance U.S. Wind Power Manufacturing Capacity," June 2, 2008. http://energy.gov/articles/doe-announces-effort-advance-us-wind-power-manufacturing-capacity.

99. The American Wind Energy Association provide useful updates on the state of play for wind power in the US on its website: www.awea.org.

100. US Department of Energy. 2008. *20% Wind Energy by 2030: Increasing Wind Energy's Contribution to US Electricity Supply.* Washington, DC: US Department of Energy.

101. US Department of Energy. 2011. *A National Offshore Wind Strategy: Creating an Offshore Wind Energy Industry in the United States.* Washington, DC: US Department of Energy.

102. Crooks (ed.). 2013. "US Wind Power Investment Is Set to Fall." *The Financial Times,* January 2.

103. More on this story is available at the 9News website: http://www.9news.com/story/news/politics/2014/05/15/wind-energy-tax-credit-stalls-in-us-senate/9151947/.

CHAPTER 8

Wind Power in Canada

8.1 INTRODUCTION

Canada regards herself as responsible to all mankind for the peculiar ecological balance that now exists so precariously in the water, ice and land areas of the Arctic archipelago. We do not doubt for a moment that the rest of the world would find us at fault, and hold us liable, should we fail to ensure adequate protection of that environment from pollution or artificial deterioration.

—Canadian Prime Minister Pierre Elliott Trudeau, House of Commons Debate (October 24, 1969)

In August 2007, the ice sheets choking off Canada's Northwest Passage receded, permitting passage without the aid of an icebreaker for the first time in Canada's 150-year history.[1] Although this development presents economic opportunities, it also exposes enormous ecological threats that, 50 years ago, former Prime Minister Pierre Trudeau professed Canada should strive to avoid. Lamentably, Canada has played a role in this environmentally invidious development due to the greenhouse gas (GHG) emissions it has produced in prolific quantities over the course of its comparatively short history.

This chapter highlights the barriers to developing a cohesive national energy strategy in a federal system where the states—or in Canada's case, the provinces—enjoy constitutional sovereignty over electricity generation. More than any other case study covered in this book, this study on Canada demonstrates how political institutions can produce conditions that make it difficult to fully exploit wind power potential, despite public support for such an outcome.

As of the end of 2012, Canada boasts the ninth highest amount of installed wind power capacity in the world.[2] Based on this statistic alone, it is tempting to conclude that Canada's wind power development policies merit recognition for being comparatively successful. However, in order to equitably assess performance in stimulating wind power development, one must also take into consideration the contextual factors which influence wind power development potential. When one does so, it becomes apparent that when it comes to wind power, Canada is a Ferrari in a world dominated by Fords. Three factors, in particular, bestow Canada with an astonishing high degree of realizable wind power potential.

First, although geographically Canada is the world's second-largest nation, it enjoys one of the lowest population density ratios in the world demand. The strategic benefit of Canada's sheer size is that wind farms could be geographically dispersed to significantly attenuate the threats posed by wind intermittency.[3] Wind conditions are impacted by disparate atmospheric conditions as one traverses the nearly 6,000 km from Canada's east coast to west coast. The strategic benefit of Canada's low population density is that Canada possesses enormous tracts of undeveloped land and open farmland that could accommodate tens of thousands of wind turbines. It has been estimated that harnessing the wind power potential of just 0.25% of Canada's landmass would be sufficient to provide enough electricity to fully satisfy Canada's aggregate electricity demand.[4] The untapped potential that exists in Canada is best put in context by observing that Canada is 28 times larger than Germany; yet, by the end of 2012, Germany had five times more installed wind power capacity than Canada.[5]

Second, hydropower currently dominates Canada's electricity mix. In fact, hydropower capacity in Canada is bested only by hydropower capacity in China and Brazil. Approximately 60% of all electricity generated in Canada comes from hydropower—and hydropower, of course, is a perfect complement to wind power because it can most expediently compensate for power fluctuations arising from wind intermittency.[6]

Third, Canada's only land-connected neighbor also happens to be the world's largest consumer of electricity. Not only is US demand for electricity expected to increase by 28% between 2011 and 2040,[7] as outlined in the previous chapter, the United States is currently confronting the challenge of financing refurbishments and upgrades to existing generation facilities, transmission infrastructure, and distribution networks that will be measured in the trillions of US dollars.[8] Already Canada is the predominant supplier of electricity to the United States, with exports of electricity via established cross-border transmission conduits amounting to C$3.2 billion in 2007.[9] Given the substandard state of electricity infrastructure in the United States and increasing international pressure for facilitating a transition away from CO_2 intensive electricity generation, the likelihood that the United States will desire heightened imports of renewable energy in order to meet future demands is high. For Canadian utilities, the potential for selling off surplus power to the United States presents a useful conduit for attenuating the stochastic nature of wind power. For Canadian wind power generators, the US electricity market represents a lucrative emergent market opportunity.

These three factors alone suggest prodigious opportunities for wind power development. There have been jurisdictions in Australia, Denmark, and Germany where wind power contributions to the electricity mix have exceeded 40% without serious disruption to grid stability or excessive investment in back-up capacity.[10] However, none of these jurisdictions possess anywhere near the hydropower generation capacity of Canada. If Canada's provincial electricity grids were more effectively integrated and management of power generation were coordinated, a nationwide electricity generation system comprised only of hydropower (60%) and wind power (40%) is conceivable.

In order to satisfy demand in 2040, approximately 264,000 MW of installed wind power capacity would be needed to meet a 40% contribution from wind power (assuming 30% capacity factor)[11]. Assuming 3 MW rated capacity for the average wind system, meeting the 40% target would require the installation of approximately 88,000 turbines. Although this seems like a vast amount, it merits mention that at the turn of the twentieth century, at least 600,000 turbines were in use for farm irrigation in North America.[12] Clearly, a modern day, utility-scale wind turbine is larger and more aesthetically invasive than the wind powered irrigation systems that were employed in the 1900s; however, the amount of land required to accommodate a modern turbine is not significantly greater. The point is if there is a will to reach 40% wind power contribution, there would be a way.

There are half a dozen good reasons to cultivate a will to adopt much stronger wind power development policies in Canada. First, the combination of escalating demand in the United States for clean energy, difficulties in expanding US electricity generation infrastructure capacity to keep pace with demand growth,[13] a beneficial trade agreement (NAFTA), and 6416 km of shared border gives rise to conditions that allow Canada to benefit economically by exporting low-carbon electricity to the United States.

Second, whether Canada chooses to install wind power for domestic use, for export or for both purposes, employment prospects in the wind power sector far eclipse the employment prospects in the conventional power sector. In 2010, Canada's electrical utilities employed a little over 110,000 people, amounting to 1.15 jobs per megawatt of installed capacity. Conversely, based on global estimates of employment in wind power,

26,000 new jobs would be created for every 10% contribution by wind power to Canada's electricity grid.[14] If this is true, wind power generates 2.4 times more jobs per installed megawatt than Canada's current electricity system does.

Third, enhancing wind power capacity represents a more sustainable energy strategy—one that will attenuate the rate at which fossil fuel and uranium resources are depleted. It is estimated that Canada's oil sands contain 173 billion barrels of recoverable petroleum. This makes Canada second only to Saudi Arabia in terms of total oil reserves.[15] However, in addition to comparatively high levels of CO_2 emissions associated with the extraction and refinement process, one study estimated that at current rates of oil production, Canada's abundant reserves will be depleted in 158 years. The same study concluded that Canada's natural gas reserves, which amounted to 56.1 trillion ft^3 in 2005, will be depleted in less than a decade at current rates of consumption and Canada's coal reserves, estimated at 7.3 billion short tons of recoverable coal will only last 100 years at current rates of consumption.[16] In short, although Canada is currently seen as a fossil fuel superpower, at the current rate of extraction and consumption, conditions of scarcity may be only a couple of generations away.

Fourth, Canadian policymakers are just now coming to grips with the magnitude of hidden health and environmental costs associated with fossil fuel combustion. For example, one study by the Ontario Medical Association estimated that pollution-induced health problems in the late 1990s, stemming predominantly from coal-fired power generation, cost Ontario C$1 billion each year and contributed to over 1900 annual deaths.[17] A more recent study in 2005 estimated the health-related damages associated with coal-fired power to be in the neighborhood of C$3 billion each year.[18]

Fifth, a program which focuses on twinning wind power capacity with hydropower represents an economical way to temper a mounting nuclear waste storage dilemma. Currently, due to the absence of a long-term storage strategy, 2 million 24 kg bundles of highly reactive spent uranium fuel (enough to fit into six ice hockey rinks) has accumulated since the 1950s and are now stored on an interim basis at six nuclear facilities in Canada.[19] Although Canada clearly has enough land to store nuclear waste at a comparatively isolated facility, public opposition, technological challenges, and cost concerns have to date confounded an acceptable resolution.[20]

Last, but certainly not least, without a more robust commitment to decarbonizing electricity generation, Canada will be hard pressed to meet its overall GHG emission reduction targets. The government has announced an intention to reduce GHG emissions to 20% below 2006 levels by 2020 and 60–70% below 2006 levels by 2050.[21] However, between 2008 and 2035, electricity generation capacity in Canada is projected to increase by almost 50% (128GW to 180 GW).[22] Even if all 60 GW of the required capacity expansion came from renewable energy sources, many of Canada's existing coal-fired power plants would also have to be replaced by cleaner sources of electricity in order to render the 2050 emission reduction target achievable.

In summary, in concert with Canada's significant potential for large-scale wind power development, there are numerous international, political, economic, and environmental reasons to strive to exploit this potential. Yet installed wind power capacity in Canada at the end of 2012 was only 6,201 MW, which at an assumed capacity factor of 30% represented only about 1.5% of Canada's electricity generation capacity. Therefore, this chapter seeks to explain why wind power in Canada has fallen so far short of its potential.

As has been the case with the other nations studied, this chapter will outline the social, technological, economic and political forces which influence wind power development. Many of the barriers discussed in relation to Canada are barriers which are found in many of the other case study nations. However, one obstacle that is unique stems from Canada's federal political structure. As the reader will discover, sovereignty over electricity generation (and fossil fuel resources) is constitutionally assigned to the provinces. This poses significant challenges in regard to cobbling together a national energy strategy that could fully exploit Canada's wealth of hydropower resources and wind power potential.

8.2 AN OVERVIEW OF ELECTRICITY GENERATION IN CANADA

It should come as no surprise that Canada consistently ranks as one of the top electricity consuming nations in the world, given that it is the world's fourteenth largest economy and experiences severe winter conditions throughout much of the country. In fact, in 2010, Canada consumed the sixth highest amount of electricity in the world. A 2008 study indicated that on a per capita basis, Canadians consumed an average of 1910 watts per hour (w/h) of electricity. This was the fourth highest level of consumption in the world, trailing only Iceland (3150 w/h), Norway (2012 w/h), and Finland (1918 w/h).[23]

Canada's electricity grid is underpinned by six core technologies. As Table 8.1 demonstrates, hydroelectric power is far and away the dominant source of power, contributing over 60% to Canada's aggregate power supply. In 2010, only China and Brazil generated more hydropower than Canada.[24] However, despite contributions of 63% from hydro and 15% from nuclear power, Canada's electricity carbon footprint is still sizable due to the nation's

Table 8.1 CANADA'S ELECTRICITY MIX 2011		
(in Terawatt Hours)		
	2011	% of Total
Hydro	372.78	62.9
Nuclear	90.03	15.2
Conventional steam (coal)	95.65	16.1
Internal combustion	1.12	0.2
Combustion turbine (oil, NG)	25.10	4.2
Tidal	0.03	0.0
Wind	7.56	1.3
Solar	0.05	0.0
TOTALS	**592.32**	**100.0**

Source: Canadian Centre for Energy Information.

high aggregate consumption levels and comparatively heavy reliance on fossil fuel-fired power in provinces that are not well-endowed with hydropower resources. As of 2010, Canada had 269 fossil fuel thermal generation plants including 10 coal-fired power plants, 114 natural gas-fired plants, and 136 oil-fired plants.[25]

With over 20% contribution from fossil fuels, the electricity sector is a candidate for reform within Canada's climate change mitigation strategy. Under the Copenhagen Accord, Canada committed to reducing GHG emissions to 17% below 2005 levels by 2020 (3% above 1990 levels). The government has also announced a domestic goal of reducing GHG emissions to 60 to 70% below 2006 levels by 2050. As of 2009, GHG emissions had fallen to 690 million tonnes of CO_2 equivalent. This represents a 6% decrease from 2005 levels.[26] However, for Canada to meet its 2020 and 2050 targets, emissions will have to be reduced considerably amidst economic expansion, which will increase energy demand by an average of 0.7% per year between 2008 and 2035.[27] In 2009, electricity generation accounted for 14.2% of all GHG emissions. A further 9.3% of all GHG emissions for 2009 came from fossil fuel production activities.[28] Therefore, if Canada is to achieve its lofty 2050 GHG emission reduction, goal, oil, and coal-fired electricity generation represent obvious candidates for reduction. Promisingly, Ontario, which is responsible for 23.9% of Canadian GHG emissions, has recently declared an intention to phase-out all coal-fired power plants by 2014.[29]

In order to understand why Canada's uses so much fossil fuel for electricity generation, one must first understand the essence of Canada's constitution. Constitutionally, sovereignty over electricity generation and natural resources lies with the provinces. In other words, Canada does not really have one national electricity grid, but rather, 10 provincial grids that have been loosely, and in many cases inadequately, interconnected. This also suggests that Canada does not have a national electricity generation strategy;[30] rather, it muddles through with an amalgamation of strategic decisions made at provincial levels.

Provincial electricity systems are largely influenced by provincial resource profiles. Table 8.2 (next page) demonstrates just how disparate electricity generation profiles of each province really are. As the table indicates, there are four provinces and one territory—Newfoundland and Labrador, Québec, Manitoba, British Columbia, and the Yukon—where hydropower provides 89% or more of the province's electricity. For these regions, the carbon footprint associated with electricity generation is low; therefore, there is no imperative to embrace wind power to mitigate CO_2 emissions associated with electricity. Conversely, four other provinces—Nova Scotia (89%), New Brunswick (49%), Saskatchewan (69%), and Alberta (74%)—generate the majority of electricity through fossil fuel thermal generation systems (coal, oil, and natural gas). These provinces do not have sufficient hydropower capacity of satisfy electricity demand. Moreover, three of these provinces (Alberta, Saskatchewan, and Nova Scotia) possess a wealth of fossil fuel reserves. These three provinces view fossil fuels as cheap sources of domestic energy, which bolster the provincial economy and fortify provincial energy security. Last but not least, Canada's largest provincial economy, Ontario, cannot satisfy its robust demand through provincial energy resources and so has embarked on an ambitious nuclear power development program, which currently provides about 50% of the province's electricity and constitutes 95% of the nation's installed nuclear power capacity.

The disparate electricity generation profiles of the provinces hold promise for a transition to cleaner electricity generation technologies; however, the challenge lies in incentivizing hydropower rich provinces to share the bounty of this peak power source and bolster interprovincial connections to develop a more resilient power grid. As it stands now, provincial utilities dominate electricity generation in Canada. As Figure 8.1 outlines, electricity generation is dominated by private firms in only three provinces (Prince Edward Island, Nova Scotia, and Alberta). In the rest of the provinces and territories, public-owned utilities dominate. If it is true that it is easier to encourage collaboration between public organizations, there is scope for these provinces to conflate strategies to enhance national energy security, lower the electricity carbon footprint and better exploit wind power potential. Until collaboration improves, success or failure of wind power diffusion in Canada will depend predominantly on provincial initiatives.

Table 8.2 CANADA'S ELECTRICITY MIX BY PROVINCE

Sources of Electricity Generation by Canadian Utilities and Industry in 2007 and % of Provincial Electricity Mix (dominant sources in boxes)

	Total Megawatt Hours	Hydro (%)	Wind and Tidal (%)	Steam (%)	Nuclear (%)	Internal Combustion (%)	Combustion Turbine (%)
Newfoundland & Labrador	41,583,313	96	0	3	0	0	1
Prince Edward Island	44,732	0	89	12	0	0	0
Nova Scotia	12,574,042	7	1	89	0	0	2
New Brunswick	17,638,847	16	0	49	23	0	12
Quebec	191,962,098	94	0	1	2	0	2
Ontario	158,234,410	22	0	22	50	2	4
Manitoba	34,402,502	97	1	1	0	0	0
Saskatchewan	20,574,449	21	3	69	0	0	6
Alberta	67,432,359	3	1	74	0	1	21
British Columbia	71,833,012	89	0	7	0	0	3
Yukon	354,694	93	0	0	0	7	0
Northwest Territories	686,252	36	0	0	0	43	20
Nu navut	148,881	0	0	0	0	100	0

Source: Statistics Canada (2009).

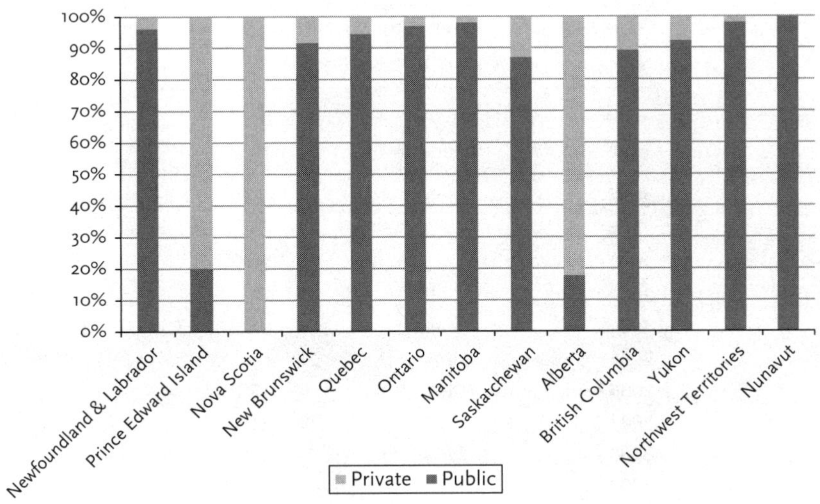

Figure 8.1.   Public and Private Electricity Generation by Canadian Province
Source: Statistics Canada. 2009. *Electric Power Generation, Transmission and Distribution*, edited by Ministry of Industry: Government of Canada.

8.3 HISTORY OF WIND POWER DEVELOPMENT IN CANADA

Wind power development in Canada is a comparatively recent phenomenon, lagging behind many of the advanced wind power nations by about a decade. In 2000, when installed wind capacity in Germany breached the 6000 MW plateau, installed wind power capacity in Canada amounted to only 137 MW. However, as Table 8.3 outlines, by the end of 2012 there was 6,201 MW of installed wind power capacity in Canada, putting Canada ninth in the world in total installed wind power capacity.[31] In the past five years alone, installed capacity has tripled.

Table 8.4 breaks down the total amount of installed wind power capacity by province as of the end of 2012, and illustrates how disparate the pace of development has been between provinces. The top three provinces for wind power are Ontario (2,043 MW), Québec (1,349 MW), and Alberta (1,116 MW). In aggregate, these three provinces host 73% of national installed wind power capacity.[32]

In 2002, the federal government launched the Wind Power Production Initiative (WPPI), which offered a comparatively small financial subsidy of C$0.012 per generated kWh for projects approved between April 1, 2002 and March 31, 2003; C$0.01 per kWh for projects approved between April 1, 2003 and March 31, 2006; and C$0.008 per kWh for projects approved between April 1, 2006 and March 31, 2007. As Table 8.3 suggests, this policy had minor impact on the pace of wind power development.

Table 8.3 WIND POWER DEVELOPMENT IN
CANADA (IN MW)

Year	Capacity	Annual Growth (%)
2000	137	
2001	198	45
2002	236	19
2003	322	36
2004	444	38
2005	684	54
2006	1460	113
2007	1770	21
2008	2369	34
2009	3319	40
2010	4008	21
2011	5265	31
2012	6201	18

Source: Canada Wind Energy Association (www.canwea.ca).

Table 8.4 INSTALLED WIND POWER CAPACITY IN CANADA BY
PROVINCE (IN MW AS OF DECEMBER 2012)

	Installed Capacity	% of Canada Total
Newfoundland and Labrador	54.70	0.9
Prince Edward Island	163.60	2.6
Nova Scotia	324.00	5.2
New Brunswick	294.00	4.7
Quebec	1,349.20	21.8
Ontario	2,043.20	32.9
Manitoba	258.50	4.2
Saskatchewan	197.60	3.2
Alberta	1,116.60	18.0
British Columbia	389.70	6.3
Yukon	0.81	0.0
Northwest Territories	9.20	0.1
Nunavut	0.00	0.0
TOTALS	**6,201**	**100.0**

Source: Canada Wind Energy Association (www.canwea.ca).

Subsequently, the WPPI was terminated prematurely in 2006 to give way to a C$1.4 billion ecoENERGY for Renewable Power program which guaranteed a similar productive incentive of C$0.010 per kWh over a 10-year period for all eligible renewable energy projects commissioned between April 1, 2007 and March 31, 2011.[33] Additionally, special tax regulations were established to permit wind power developers to accelerate write-offs of capital equipment.[34]

Overall, federal subsidies have been too trifling to make wind power a commercially attractive investment in most parts of Canada. To this point in time, the policy initiatives of the provincial governments have driven Canadian wind power development, with policies in Ontario, Quebec, and Alberta warranting special attention.

The roots of the wind power development boom in Ontario sprung from the province's Integrated Power System Plan, first released by the Ontario Power Authority (OPA) in 2007 to guide strategic development of provincial electricity infrastructure. The plan called for commissioning 4,600 MW of wind power by 2020. In support of this target, the OPA issued a series of requests for proposal (RFP), which in 2009 elicited about 1,400 MW of new wind power installation contracts. In March 2009, the Ontario government changed policy direction and decided to discontinue the RFP approach, instead opting for a feed-in tariff (FIT) program that was legally mandated under a new Green Energy Act. This FIT became North America's first guaranteed renewable energy development program. Currently, the FIT guarantees a purchase price of C$0.11 per kWh for wind power, plus incentives up to an additional C$0.02 per kWh, contingent upon community or aboriginal participation in any given project. As of December 2012, the FIT has induced another 600 MW of installed wind power capacity; however, the program has not been without controversy. In January 2013, the Ontario provincial government declared a moratorium on offshore wind power projects due to environmental concerns.

There has also been considerable opposition to further onshore wind power development led by groups such as Wind Concerns Ontario and Ontario Wind Resistance. Aside from the typical charges that wind power disrupts avian ecosystems, impairs landscape aesthetics and lowers land values (all of which have not been empirically substantiated), there has been a renewed focus in Ontario on the impact of wind system noise on human well-being. There have been additional calls for a moratorium on wind power development until a study into the matter by Health Canada is released in 2014. To date, the province has rejected such appeals.[35] However, opposition has undeniably dampened the fervor surrounding wind power development in the province.

In Québec, the electric utility is publicly owned; and so, wind power development has been fueled primarily by provincial political will. The government has established a target of 4,000 MW of installed wind power capacity by 2015.[36] Almost 1000 MW of capacity was added between 2009 and 2012 alone. Contracts for new wind power projects are determined through a RFP system. Given that Québec's electricity mix is dominated by 94% hydropower, there is potential for the province to create an electricity grid powered by 100% renewable energy. In 2013 there will be a fourth RFP tender period for wind power development in Québec, and depending on developments in Ontario, Québec may emerge as the province with the most installed wind power by 2015.

In Alberta, wind power inroads have been remarkable given the policy environment. Alberta is the fossil fuel center of Canada. The Western Canada Sedimentary Basin that runs down the Alberta side of the Canadian Rockies stores 39% of Canadian conventional oil reserves and is purported to house 90% of Canada's commercially viable coal reserves. Alberta enjoys a per capita GDP that is 70% higher than the Canadian average, thanks in large part to fossil fuel exports. In fact one study, by a group at the University of Calgary, estimated that without oil and gas revenues, Alberta's GDP would be halved. In other words, there is significant political support for fossil fuel in the province. As the provincial government summarized in its 2008 energy policy paper, "the development of clean hydrocarbons is essential to Alberta's energy future."[37]

Although there is a degree of political pressure to respond to public sentiment in favor of more proactive climate change mitigation policies, the only substantive policy that has encouraged Albertan wind power development has been a carbon tax of C$15 per metric tonne of CO_2 levied on major energy generators. Yet wind power has blossomed thanks to this tax, along with the small federal subsidy of C$0.01 per kWh, superb wind quality, and a provincial electricity market that guarantees unfettered access to the electricity grid for any technology that can successfully sell into the market. The success of wind power development in Alberta under relatively scant support is testament to the commercial viability of wind power.

In other provinces, there are some recent wind power initiatives that also merit mention. In British Columbia, 390 MW of wind power capacity has been installed since 2009. This represents the first wave of development in a government strategy to achieve energy self-sufficiency by 2016, with 50% of new generation coming from clean energy sources. In Manitoba, the government is aiming to add 1,000 MW of wind power capacity by 2016 and intends to achieve this through a series of RFPs. Similarly, the New Brunswick utility, NB Power, is planning to commission 400 MW of new wind power capacity by 2016. Although many of these initiatives are being driven by provincial climate change mitigation programs designed to support the national climate change mitigation strategy, some provinces are demonstrating more initiative than others in supporting expansion of decarbonized electricity systems.

8.4 UNDERSTANDING THE GENERAL FORCES FOR CHANGE

8.4.1 Sociocultural Landscape

Forces within Canada's sociocultural landscape tend to have an inconsonant influence on wind power development. On the positive side, there is a strong environmental ethic that underpins support for clean energy. For example, in a poll conducted by Angus Reid released on September 12, 2011, 55% of Canadians surveyed supported a view that it is important to protect the environment, even at the expense of economic growth. Conversely, only 22% deemed it permissible to foster economic growth that degrades the environment.[38]

Moreover, there is a high degree of support in Canada for initiatives to mitigate climate change. According to a December 17, 2009 survey, 85% of Canadians surveyed believed that global warming will impair the welfare of future generations and 40% agreed with the contention that global warming will impair the welfare of current generations.[39] Therefore, it should come as no surprise that a degree of voter dissatisfaction exists toward the Canadian government's handling of climate change negotiations. After the climate change summit in Copenhagen in 2009, 48% of Canadians surveyed expressed dissatisfaction with the performance of Prime Minister Stephen Harper at the meeting; while only 28% were satisfied with his performance. The same survey released on January 5, 2010, also revealed that the majority of Canadians surveyed (56%) would be in support of a legally binding climate change treaty and 46% supported the establishment of funds to help developing nations adapt to climate change.[40]

Despite these environmental sentiments, there are two other sociocultural factors that tend to foster support for fossil fuels. First, Canada's energy industry, which is dominated by fossil fuel enterprise, directly employs over 350,000 people (over 2% of Canada's labor force).[41] Many of these jobs are specialized and pay well because they are carried out in extreme conditions in remote locations. There is understandable concern within the ranks of those employed in the fossil fuel energy sector (and those who benefit from the resultant economic activity) that a transition to renewable energy may harm a great many employed in the conventional energy field. Second, although there is a high degree of public support for wind energy, there is also an increase in NIMBY-opposition in provinces where wind power is flourishing. Ontario in particular has faced well-organized resistance to larger wind power projects.[42]

It merits pointing out that many of the reasons for opposing wind energy are baseless. For example, fossil fuels are fungible and readily transportable, so a domestic shift to clean energy would merely allow more domestic fossil fuel reserves to be exported to overseas markets. Therefore, the perceived threat that alternative energy poses to employment in the conventional energy sector is without foundation. Similarly, global studies abound which demonstrate that wind power, once installed in a community, does not pose the envisaged level of threat to the environment, human health, or community aesthetics. In fact, research indicates the opposite is true; positive community perceptions on wind power tend to amplify once projects have been completed.[43]

8.4.2 Economic Landscape

Unsurprisingly, in addition to being a valuable source of jobs, the conventional energy sector is also an engine of national wealth. Since 2007 Canada's energy industry has accounted for over 5% of national GDP, representing about 20% of all export revenues. In 2009, energy exports amounted to C$80 billion and the importance of energy revenues to Canada's economy has actually increased since the 1990s.[44] Estimates of Canada's Albertan oil sands indicate that Canada now possesses the second most reserves of petroleum in the world (175 billion barrels), behind only Saudi Arabia.[45] In addition to oil, Canada possesses sizable natural gas reserves. In short, Canada is a fossil fuel super power, and this gives rise to political and business interests that have a vested financial interest in opposing any challenges to the status quo.

It should be noted that enhanced fossil fuel management is increasingly viewed as a new economic growth area. The Albertan government in particular has been unequivocal about the need to extract more value from the fossil fuel value chain.[46] This implies that fossil fuel extraction is considered to be a key foundation for supporting new business opportunities. Consequently, there is a degree of political sensitivity associated with any policy shifts that may adversely impact the fortunes of this golden goose.

One the other hand, Canada's abundance of natural resources has somewhat insulated the nation from the recent global economic downturn.

According to a poll released on December 28, 2012, 62% of Canadians surveyed rated current economic conditions in Canada as either *very good* or *good*.[47] This benefits wind power development in two ways. First, it has engendered a positive investment climate that improves financial access for wind power developers. Second, it potentially creates a degree of economic resilience, engendering a greater propensity for electricity consumers to accept higher prices in exchange for cleaner energy.

8.4.3 Technological Landscape

There are two distinctive geographic features that enhance technological potential of wind power in Canada. First, wind quality in Canada is comparatively favorable. Coastal areas in Canada's East and West tend to enjoy high quality wind thanks to the confluence of temperate and Arctic air masses and continental-oceanic temperature disparities. Canada's central prairies also exhibit high-quality wind profiles, thanks to large expanses of unbroken plains, which reduce wind turbulence. Second, Canada boasts immense tracts of undeveloped land and agricultural areas. This presents numerous options for wind farm siting and allows project developers to select sites where community opposition can be least expected. For example, the town of Halkirk, Alberta (population 122) was recently chosen as a site for an 83 turbine, 150 MW wind farm largely because of citizen support. In embracing the project, the town experienced a mini-economic boom during the construction phase, enhanced town revenues and has since seen 14 permanent jobs created in the community.[48]

On the downside, Canada's geographic size also poses some technical challenges for wind power development. Many of the most attractive wind power sites are geographically separated from the major demand centers, and in many cases the most attractive sites are a long way from existing power grids. Connecting remote sites to the electricity grid can undermine the financial viability of a given project. Moreover, transmitting electricity from remote wind farms to demand centers results in leakage (the dissipation of electricity as it courses through conduits), which can further inflate transmission and distribution (T&D) costs.

In terms of technological infrastructure, there is one factor that could significantly hinder widespread wind power development in many provinces. The grid connections between provinces have capacity limits. These grid connections have been designed to share relatively small amounts of surplus electricity or to top up supply during periods of high demand or system malfunction. However, in order to support wind power development in many provinces that do not enjoy high hydropower capacity, greater capacity limits will be needed to ensure that electricity can flow freely from provinces that are rich in hydropower to provinces which need peak load capacity support. In short, for wind power potential to be maximized, interprovincial grid connections would have to be expanded to closer resemble Quebec's profile in Table 8.5 (next page).

In terms of energy technology development, there are two significant trends worth highlighting. The first trend is that support for nuclear power R&D is waning, particularly post-Fukushima. Atomic Energy of Canada Ltd. (AECL) is the federal crown corporation that has historically been responsible for coordinating nuclear power R&D. In 2011, after posting losses in 2009 and 2010 totaling C$493 million, AECL's reactor design division was sold to a private buyer (SNC-Lavalin) for C$15 million. AECL still runs the Nuclear Laboratories division, which overseas research and production of isotopes for medical use at the Chalk River Laboratory. However, the sale of the reactor design division suggests that federal subsidies for nuclear power research, which purportedly amounted to over C$600 million in 2009, will be significantly reduced. The partial dismantling of AECL underscores years of unprofitable performance within Canada's nuclear power program and suggests a troubled road ahead for nuclear power proponents in Canada. The second technological trend worth noting is that there is growing support for carbon capture and sequestration (CCS) research across Canada in general, and in Alberta in specific. As the Albertan government outlined in its energy strategy, "Alternative and renewable energy sources will play a growing role in Alberta energy's future, but they cannot match the importance to Alberta of 'clean' fossil fuels."[49] The money flowing into CCS research from both public and private sources may allow fossil fuel-power firms to extend defense of their market positions.

8.5 INFLUENCES ON GOVERNMENT POLICY

8.5.1 Sociocultural → Political

Canada's sociocultural environment has had a net positive influence on political support for wind power development. The environmental ethic that pervades Canadian society tends to hold forces for unfettered economic development in check. In this regard, although there is considerable political support for exploiting Canada's stored fossil fuel wealth, there is also considerable public pressure for Canada's federal and provincial leaders to respond to climate change. As pointed out earlier, the majority of Canadians harbor

Table 8.5 PROVINCIAL GRID INTERCONNECTIVITY

Electricity Exchange between Provinces and between Canada and the United States in 2007 (in Megawatt Hours)

	Received from Other Provinces	Delievered to Other Provinces	Imported from US	Exported to US
Newfoundland and Labrador	16,947	30,096,817	0	0
Prince Edward Island	1,160,935	0	0	0
Nova Scotia	280,597	27,303	62,917	30,634
New Brunswick	1,466,014	1,556,758	641,755	1,780,259
Quebec	33,966,926	3,558,708	3,355,838	15,711,988
Ontario	3,711,520	4,501,487	7,070,359	11,089,756
Manitoba	173,568	1,782,187	534,285	11,092,806
Saskatchewan	1,031,828	840,178	203,343	391,579
Alberta	1,781,495	1,208,616	222,902	154,748
British Columbia	1,101,312	1,119,088	7,288,705	4,438,820
Yukon	0	0	0	0
Northwest Territories	0	0	0	0
Nunavut	0	0	0	0
TOTALS	**44,691,142**	**44,691,142**	**19,380,104**	**44,690,590**

Data source: Statistics Canada (2009).

concerns for both current and future generations over the perils attributed to climate change and feel that the administration of Prime Minister Stephen Harper has not done enough to respond to this threat.

On the other hand, public opinion has been far from homogeneous. There has historically been a degree of regional tension between Eastern and Western Canada, and energy policy is a source of animosity. At the heart of this issue is a widely held belief in Western Canada that the federal government uses taxes from the west to subsidize livelihoods in the east. Resentment came to a head in 1980 in response to the government's National Energy Policy, which introduced a Petroleum Gas Revenue Tax, in part to help reduce the federal budget deficit. This tax was seen as an encroachment upon provincial sovereignty over natural resource governance and engendered heated responses from western residents. A popular bumper sticker seen in the west during this period was "Let the Eastern bastards freeze in the dark."

This history is important for understanding modern sociocultural influences on energy policy for two reasons. First, the federal government has learned from experience and tends to avoid policies that might be construed as infringing on provincial sovereignty, particularly in regard to energy. Therefore, even though there is widespread support for GHG emission reduction and central coordination of provincial grids would both improve grid resiliency and reduce GHG emissions associated with electricity generation, the central government has shown a reluctance to challenge provincial sovereignty in this way. Second, most provincial leaders understandably view their primary responsibility as defending the interests of provincial voters, even if this conflicts with national interests. This creates provincial fiefdoms that are hardcoded to oppose federal involvement in provincial affairs. This combination of federal reticence to intrude on provincial sovereignty and provincial leadership which prioritizes provincial interests over national interests creates a political landscape where centralization of energy policy is unlikely.

8.5.2 Economic → Political

In a nation with the second highest amount of oil reserves and eleventh highest amount of coal reserves, it should come as no surprise that there's a degree of political support for fossil fuel production and utilization. Alberta and Saskatchewan, which together produce about 60% of the nation's coal, also have electricity networks that are dominated by coal-fired power (74% in Alberta and 69% in Saskatchewan).[50] This has engendered a high degree of support for CCS research, particularly in Alberta, which alone produced 46% of all coal mined in Canada in 2011. Alberta has recently announced an initiative to provide public support for three to five CCS projects, aiming to store approximately five million tonnes of CO_2 by 2015. Although this technology is intended primarily to help enhance oil recovery in the province, successful commercialization will also perpetuate coal-fired power production in the province.

With energy accounting for approximately 20% of Canada's exports, and further development of the Albertan oil sands expected to extend the commercial importance of fossil fuel production, there is a considerable amount of vested financial interest which is intent on preserving status quo. As noted earlier, given the fungible nature of oil and coal, it is not essential for these products to be consumed domestically for the industry to thrive; however, domestic reliance on fossil fuels tends to enhance industry profitability. Economic interests tend to exert a tremendous influence in the provinces where fossil fuels are extensively used in the electricity system—where change is most needed, industrial and political opposition are greatest.

Political support for fossil fuel resource development tends to be bolstered by recent trends in provincial development strategy, which are beginning to recognize the opportunities lost by the historical extract-and-sell approach to resource governance. Leaders in fossil fuel rich provinces such as British Columbia, Alberta, Saskatchewan, Manitoba, New Brunswick, and Prince Edward Island are beginning to actively seek strategies to add value to the fossil fuel resource chain. This economic upside garners political support and engenders political stakeholders who are favorably biased toward fossil fuel special interest appeals.

8.5.3 Technological → Political

Canada is unique in that technology-enabling environmental endowments support virtually any electricity generation technology. As pointed out earlier, Canada's oil reserves rank second in the world. Canada's coal reserves rank eleventh, and its natural gas reserves rank twenty-first in the world. Canada ranks third in the world in hydropower generation and seventh in the world in nuclear power generation, with proven uranium reserves that rank third in the world. Despite having no installed geothermal capacity, the Canadian Geothermal Energy Association estimates geothermal potential at 5,000 MW—half the current amount of installed geothermal capacity. In tidal power, the Pembina Institute in Canada estimates that the Bay of Fundy alone could produce 30,000 MW of power.

Despite enormous wind power potential, Canada could support virtually any technological platform for generating electricity. However, fossil fuel technologies have enjoyed political favor because the government is able to tax the resources as they are extracted and tax the electricity as it is provided. This double taxation characteristic is not a feature of renewable energy technologies.

Perhaps the greatest technological influence on energy policy in Canada is the prospect of equipping fossil fuel-fired power plants with CCS technology to preserve status quo. Although CCS is far from a commercially viable technology and a number of ecological risks associated with the concept have yet to be mitigated, the mere possibility that this technology may enable provinces to decarbonize the electricity generation mix, without actually transitioning away from technologies that make double taxation possible, tends to soften political support for alternative energy technologies. The Albertan government's energy strategy is indicative of this "have your cake and eat it too" outlook: "while (Alberta's oil sands) account for just four per cent of Canada's greenhouse gas emissions and less than one tenth of one per cent of all global greenhouse gases, (they) are a large fossil fuel resource and therefore provide a tremendous responsibility and opportunity for Alberta to lead."[51]

In Ontario, there is an additional technological phenomenon that influences energy policy—technological lock. Ontario is the vanguard province for nuclear power development and indeed, now hosts over 95% of the nation's nuclear generation capacity. In advocating for contentious technologies such as nuclear power, ideological positions are staked out, political battles are fought, and long-term investments are made.[52] Unless the supporting regime is replaced, it is difficult to enact change. Even if regime change does occur, sunken investment can preclude technological transition. For example, the nuclear power plants in Ontario, which are currently operational, represent long-term capital investments. They are progressively amortized annually. Until these plants are fully amortized, a switch to alternative forms of technology would require expensive capital write-offs.

Technological lock tends to be manifest in one other facet of Canadian electricity generation—rural service. Many communities are sparsely populated and remote. Historically, such communities have relied on highly responsive technologies such as hydropower when available, and in the absence of hydropower, combustion turbines operating on natural gas provided responsive, peak-load power flows. Electricity systems of remote communities such as these are less resilient than larger urban electricity grids because less surplus capacity exists in the grid. As such, further investment is required to incorporate intermittent energy flows into the grid—more backup capacity or additional generating capacity is required. Although strategies have emerged in electricity system engineering to attenuate the stochastic flows of wind power systems, support for fossil fuel generation systems still perpetuates.

8.6 POLITICAL INFLUENCES ON POLICY

8.6.1 National Political Structure

Canada is a federal parliamentary democracy that consists of 10 provinces and three territories. The motive for adopting a federal system arose from the challenge of trying to unify the culturally distinct provinces of Ontario (Anglophone-dominated) and Québec (Francophone-dominated). Without guaranteeing a degree of autonomous regional representation that is manifest in the federal system, it is likely that Canada would not have gelled as a nation.[53] In fact, many political experts would likely agree with the contention that Canada's separation of powers has been instrumental in preventing national breakup.[54]

The concept of separation of powers is the core of Canada's federal system. Constitutionally, political power is divided between a central federal government and provincial legislatures. In relation to energy, provincial legislatures possess exclusive authority over natural resource governance and electricity generation. This explains why Canada does not have a national electricity generation policy; the central government has no constitutional right to design such a policy.[55] In practice, this means that Canada has 10 provincial energy ministers, who get together periodically with the one federal energy Minister to discuss collaborative policy.

Although the provinces enjoy considerable autonomy in regard to electricity system development, the federal government has at least four strategies that it could adopt to influence provincial energy policy. First, the federal government has the right to raise funds through taxation and to exercise broad discretion in regard to how it uses these funds (Section 91(3) of the Constitutional Act).[56] These powers underpinned the Petroleum Gas Revenue Tax in the 1980s which, as described earlier, attracted so much ire from Alberta.

Second, the federal government is constitutionally authorized to enter into treaties with foreign powers and upon doing so, establish policies to compel provincial authorities to comply with the terms of the treaty. In short, it has the power to force the provinces to contribute in an equitable fashion to achieving Canada's emission reduction targets under the Kyoto

Protocol or any other climate change treaty if ratifies (Section 132 of the Constitutional Act).[57]

Third, the federal government enjoys residual powers, which include all rights that are not expressly allocated to the provinces under the Constitution (Section 91 of the Constitutional Act).[58] Among other residual powers, the federal government has authority over trans-provincial environmental governance. It has the right to pass legislation that regulates trans-boundary pollution, which suggests that the federal government possesses a constitutional right to set standards regarding CO_2 emissions from fossil fuel-fired power plants. Indeed, this environmental governance remit has manifested itself in the establishment of the Canadian Environmental Protection Act of 1999 (CEPA). This act authorizes the government to force industries (including utilities) to submit pollution prevention plans. However, to date the federal government has been reluctant to wield this coercive stick.

The fourth strategy that the federal government could adopt for influencing provincial electricity generation policy is to block electricity exports. In 2011, Canadian electricity exports to the United States exceeded US$2 billion.[59] The government has a constitutional right to regulate trade and commerce (Section 91(2) of the Constitutional Act) and must grant approval for exporting electricity (Section 92A(3) of the Constitutional Act). These two constitutional rights suggest that the federal government could make it very difficult for provinces to export energy and stabilize electricity grids through trade with the United States.

Despite all the constitutional arguments that the federal government could theoretically use to influence provincial energy policy, in practice the provinces enjoy other rights which could counter federal attempts to influence provincial energy strategy. Overall, the provinces could legally challenge any federal policies that appear to infringe on constitutionally guaranteed provincial authorities.[60] It could be argued that any federal initiatives to influence provincial energy policy contravene section 92A(1) of the Constitution, which grants authority to provinces to "exclusively make laws in relation to. . . development, conservation and management of sites and facilities in the province for the generation and production of electrical energy."[61] Whether or not the provincial governments would be successful in challenging federal attempts to apply taxes or regulations to fossil fuel emissions from power plants is somewhat of a moot point, because the legal challenge itself would be enough to block federal intrusion while the court considers the case. Such a legal delay could derail federal policy because the tides of politics are capricious.

In response to a federal attempt to invoke its environmental protection mandate in order to tax CO_2 emissions, there is a potential defense stemming from section 36 (1) of the Constitution, which mandates that the federal government must ensure equal opportunities and reduce disparity in the opportunities in regard to its policies.[62] Given the disparate nature of Canada's provincial electricity mix, the implication of this should be apparent. Any taxes or regulations applied to fossil fuel emissions would disproportionately disadvantage the five Canadian provinces that are dependent on fossil fuel electricity (Ontario, Alberta, Saskatchewan, Nova Scotia, and New Brunswick). A policy which disproportionately disadvantages some provinces over others would require some sort of transfer payments in order to equitably apportion the damage.

Another provincial defense to block federal attempts to initiate a carbon tax arises from Canada's representational political structure. Section 53 of the Constitution requires that federal policies which seek to apply taxes to goods or services must be approved by a simple majority within the House of Commons. The five Canadian provinces listed in the previous section as dependent on fossil fuel electricity generation hold the majority of the 308 seats in the House of Commons. Although Canada's political structure affords the prime minister considerable authority over day-to-day policy setting, when it comes to raising funds, this constitutional check suggests that members of Parliament from these provinces would be placed in the difficult position of having to choose between supporting the prime minister and his or her cabinet or appeasing the vested interests within his or her political riding.

In aggregate, the political tension that exists between the federal government and the provincial government has resulted in a national energy strategy which is more or less a conflation of provincial energy policies—policies which prioritize provincial interests. This is not to say that the status quo could not be changed; however, attempts to unify national energy policy would require considerable political will on the part of the ruling party to confront the challenges that the provinces could muster.

8.6.2 Governing Party Ideology

Since 2006, Canada has been governed by the Conservative Party of Canada, led by Prime Minister Stephen Harper. Canadian politics is characterized by centrist ideology with the Liberal Party of Canada claiming the left of center position and the Conservative Party claiming the right of center position. It has been said that the Conservative Party exhibits a stronger right wing ideology compared to previous decades due to the merger of the Progressive Conservative party and the further right-wing Canadian Alliance Party in 2003. In fact, Stephen Harper was leader of the Canadian Alliance party prior to the merger.

Generally speaking, the Conservative Party supports decentralization of government power and privatization of public services. This is evident in national energy planning. Since 2006, the government has demonstrated a propensity to leave energy strategy to the provinces. As outlined earlier, wind power development success in Canada has been largely driven by policy established at the provincial level. The tendency to support privatization is exemplified by the 2012 sale of Atomic Energy of Canada Limited's reactor division to a private Québec-based firm.

The Conservative Party is not supportive of the Kyoto Protocol. One of its election platforms in 2006 was that the Canadian Liberal Party erred in committing Canada to an ineffective treaty that has negative economic ramifications for Canada. At the conclusion of the Durban climate change talks in 2011, the Harper administration announced that it intended to repudiate the Kyoto Protocol. In justifying the decision, Environment Minister Peter Kent, explained that this "legacy of an incompetent Liberal government" would wind up costing Canadian taxpayers C$13.6 billion.[63] Since then, the Canadian government has embraced a policy of linking Canada's climate change mitigation targets to those of the United States.

8.6.3 Fiscal Health

Although Canada has one of the lowest debt to GDP ratios within the OECD, management of the public debt has become a political issue in the past few years, particularly in the lead up to the 2011 national election. Between 1996 and 2008, net public debt decreased from C$562 billion (63.8% of GDP) to C$457 billion (28.5% of GDP). Since 2008, net public debt has increased each year. As of February 2013, net public debt stood at C$605 billion.

One of the campaign promises made by Harper's Conservative Party in the lead up to the 2011 national election was to balance the budget by 2013–2014. For the fiscal period 2011–2012, fiscal debt amounted to C$31 billion. In 2012–2013, the expected deficit is C$21.1 billion. In other words, in the absence of windfall tax revenues, the government will have to make some strong cuts in its spending in order to live up to its campaign promise. On March 29, 2012, Finance Minister Jim Flaherty tabled a fiscal plan in a House of Commons speech that promised a reduction of the federal deficit

Table 8.6 PROVINCIAL BUDGET DEFICITS 2012–2013		
	Surplus (Deficit)— $Million	% of GDP
Newfoundland and Labrador	–258	25.0
Prince Edward Island	–75	35.6
Nova Scotia	–211	34.8
New Brunswick	–183	34.1
Quebec	–1500	35.2
Ontario	–15,300	37.2
Manitoba	–460	27.4
Saskatchewan	95	11.9
Alberta	–886	4.1
British Columbia	–968	26.2
Yukon	80	na
Northwest Territories	17	16.4
Nunavut	38	na

Source: www.cbc.ca/news/interactives/budgets-provinces/.

by an average of C$5.2 billion per year in order to return the government to the black by 2015.

Pressure to balance the fiscal budget will make it increasingly difficult for the federal government to finance any sort of transition away from fossil fuel-powered electricity generation. This leaves most of the hard work up to the provincial governments or private industry. As Table 8.6 illustrates, however, other than Saskatchewan and Canada's northern territories, all other provincial governments are also struggling to manage annual fiscal budget deficits. In short, on the heels of the recent global economic downturn, all of Canada's provinces are finding it difficult to balance budgets and will, therefore, be hard pressed to catalyze change in their respective electricity fiefdoms.

8.6.4 Policy Regime

Given the constitutionally guaranteed sovereignty that provinces enjoy over natural resource governance and electricity generation, it should come as no surprise that the federal government exhibits a proclivity toward "sermons and carrots over sticks"[64] in regard to selection of policy instruments for influencing developments in Canada's electricity sector.

In terms of sermons, the federal government, through the Ministry of Natural Resources, exerts soft power by coordinating the annual meeting of the Canadian Council of Energy and Mines Ministers. It also has set non-binding targets for national GHG emission reductions and through this program works with the provinces to coordinate achievement.

In terms of carrots, the predominant incentive offered by the federal government to spur on wind power development began with the wind power production initiative (WPPI), launched in 2002. Unfortunately, as detailed earlier, the incentives which ranged between C$0.008 and C$0.012 per generated kWh were too inconsequential to generate much activity. The C$1.5 billion ecoENERGY for Renewable Power program which was announced in 2006 as the replacement for the WPPI provided a similarly tame production incentive of C$0.010 per kWh over a ten-year period for all eligible renewable energy projects commissioned between April 1, 2007 and March 31, 2011.[65] Additionally, special tax regulations have been established to permit wind power developers to accelerate write-offs of capital equipment.[66] Overall, federal subsidies to date for wind power development can be considered to be top-ups to anything that is planned at the provincial level.

At the provincial level, RFPs (requests for proposal) have been the dominant tool for encouraging wind power development. This is because the electricity networks in the majority of Canada's provinces (with the exception of Alberta, Nova Scotia and Prince Edward Island) are dominated by public monopolies. RFPs historically meshed well with monopoly control of electricity networks. Simply put, RFPs have been an uncomplicated and fair way of expanding electricity networks. The only notable exceptions where a divergence from the RFP system has resulted in significant wind power development are in Alberta and Ontario. In Alberta, the wholesale electricity market has been liberalized and wind power developers compete with other sources of energy on an equal commercial basis. As mentioned earlier, thanks to carbon taxes applied to fossil fuel electricity plants and the high quality of wind in Alberta, wind power developers have been able to successfully compete on this basis. In Ontario in 2009, the Ontario government initiated the feed-in tariff system described earlier. To date, the tariff has been highly successful in encouraging wind power diffusion but rapid expansion of wind power capacity has also attracted a substantial amount of NIMBY opposition.

8.7 THE CULMINATION OF INFLUENCES

Canada is a moderate nation. The wealth of social programs found in Canada highlights an ideological perspective that supports efforts to mitigate any adverse impact associated with policy implementation, including policy affecting the energy sector. Put another way, when energy policy adversely influences a significant number of stakeholders, there tends to be public backlash, which compels attempts to mitigate any ensuing negative externality. For example, at the federal level, the government's opposition to the Kyoto Protocol has given rise to public pressure to play a more constructive role in international climate change mitigation efforts. As an example at the provincial level, Ontario's feed-in tariff has significantly increased wind power installations but has also engendered extensive NIMBY opposition to excessively invasive wind power build-up. In short, the centrist tendency on the part of Canadian voters tends to have a moderating influence on public policy.

It is perhaps for this reason that the federal government appears to be, at times, confused about the appropriate response to climate change mitigation. Whenever the government adopts a position in strong support of a particular initiative—for example, support for CCS at the federal level or support for nuclear power or wind power at the provincial level in Ontario—there tends to be critical public backlash that forces the government to moderate its position.

In addition to lackluster performance in supporting technological innovation in the energy sector, the federal government has had very little impact on renewable energy diffusion. As mentioned earlier, the WPPI, which has averaged less than C$0.01/ kWh, has had only a minor impact on wind power development. To the contraire, the absence of a federal carbon tax has provided coal-fired power plants with an artificial competitive advantage. In 2008, Liberal leader Stéphane Dion proposed the introduction of a national carbon tax as part of his party's national electoral platform. This proved to be unpopular with voters and played an influential role in his party's disastrous electoral defeat.

In contrast to federal efforts, provincial renewable energy initiatives have exhibited a modicum of success. The RFP programs put in place by the provinces of Québec, Nova Scotia, Manitoba, and British Columbia have clearly kick-started wind power development. Similarly, the Ontario government's feed-in tariff that was launched in 2009 has had a major impact on wind power development in the province.

In terms of carbon taxes, it is interesting to note that Alberta became the first state or province in North America to introduce a carbon tax in July 2007. Companies that emitted more than 100,000 tons of greenhouse gases were mandated to reduce emissions by 12%. Failure to do so would result in these entities having to either pay a penalty of C$15 per ton of CO_2 equivalent emissions or purchase offsets from another major emitter. This tax has positively influenced the fortunes of wind power development in the province.

The separation of powers between federal and provincial governments has also served to moderate federal involvement in the development of energy technologies. In fact, the Canadian government has not really been signifi-cantly involved in championing an energy technology since its involvement in the development of a nuclear power program at the tail end of World War II.[67]

Enumerating just a few of the federal government's energy technology support initiatives since 2004 highlights how phlegmatic federal support has been. The recent centerpiece of government support for research in energy technology is a clean energy fund that was established in 2006–2007, which earmarked C$800 million to be spent over five years on sup-porting clean energy research. The bulk of this funding has gone to support CCS pilot projects. In regard to renewable energy, the government currently manages a C$25 million Market Incentive Program (MIP), which encour-ages residential consumers and small businesses to adopt renewable energy technologies. It has also earmarked C$30 million for development and dem-onstration of decentralized energy systems and C$30 million for supporting adoption of renewable energy systems in aboriginal communities. The most recent investment has been a C$10 million program to support tidal energy research in Atlantic Canada. The only significant federal program designed for supporting wind power (aside from the WPPI) has been the develop-ment of a Canadian wind energy atlas.

On the surface, a C$800 million research program may not appear to be insignificant. However, given that 20% of all of Canada's export revenue comes from energy exports of some kind, C$800 million is a lackluster com-mitment to technological innovation for such a strategically important industry. This is less than what the government paid for four used diesel submarines purchased from the British government in 1998.

8.8 WHAT TO EXPECT GOING FORWARD

In terms of directly shaping market development, leadership is not likely to come from the central government in Canada, especially under the Harper administration, which tends to prefer leaving solutions to the free market. On the other hand, the Harper administration is on record as targeting 90% of electricity from renewable energy sources by 2020.[68] This is not as bold as it sounds given that the nation is starting from a foundation featuring a 62% contribution from hydroelectric power. Nevertheless, it is not unrea-sonable to expect the federal government to play an increasingly active role in trying to at least enhance collaboration between the provinces.

As climate change concerns become more apparent and public pressure mounts, the provinces can be expected to adopt increasingly robust renew-able energy programs. Given the commercial attractiveness of wind power, this means that wind power capacity in many of the provinces is bound to increase. The Canadian Wind Power Association forecasts that wind power will grow to 12,000 MW by 2016.[69] Already it is apparent that wind power development will continue to flourish in Québec and Alberta over the next few years. The fate of wind power in Ontario will largely depend on the outcome of a provincial investigation into the health effects of wind turbines. As mentioned earlier, this is a concern that has led to a moratorium on offshore development and significantly stymied further onshore development. In other provinces where wind power has shown potential—British Columbia, Saskatchewan, Manitoba, Nova Scotia, and New Brunswick—future development depends significantly on decisions made by the governing regime in each province. Currently, there appears to be at least a degree of support for expanding wind power in all these provinces.

Interestingly, the future of wind power development in Canada will likely be influenced most by what happens in the United States. As explained in the chapter on the United States, the electricity grid and electricity generation infrastructure in the United States is in dire need of technological enhancement. For northern US states, Canadian electricity exports represent a viable interim solution for bolstering grid resilience with clean energy. Policies developed in states like Washington, Minnesota, and Michigan could very well provide the added market inducement to enhance the pace of wind power development in Canada. Moreover, given that the Harper administration has linked Canada's climate change mitigation policy to US policy, developments in the United States could have a catalytic effect in Canada.

Former Prime Minister Pierre Elliot Trudeau is purported to have observed that being America's neighbor "is like sleeping with an elephant. . . no matter how friendly and even-tempered the beast, if one can call it that, one is affected by every twitch and grunt." In this case, a sizable twitch may significantly impact how the electricity sector in Canada evolves.

NOTES

1. This chapter draws from content from a previously published work: Valentine, S.V. 2010. "Canada's Constitutional Separation of (Wind) Power." *Energy Policy* 38: 1918–1930.
2. As reported by the World Wind Energy Association, www.wwindea.org.
3. Gil, Hugo A., Geza Joos, Jean-Claude DesLauriers, and Lisa Dignard-Bailey. 2006. *Integration of Wind Generation with Power Systems in Canada: Overview of Technical and Economic Impacts*. Ottawa: National Resources Canada.
4. Canadian Wind Energy Association (CanWEA. 2008). *Windvision 2025: Powering Canada's Future*. Ottawa: Canadian Wind Energy Association.

5. "Wind Energy Statistics 2012." 2013. Report. Global Wind Energy Council. February. www.gwec.net/wp-content/uploads/2013/02/GWEC-PRstats-2012_english.pdf.

6. Boyle, Godfrey, Bob Everett, and Janet Ramage, eds. 2004. *Energy Systems and Sustainability: Power for a Sustainable Future.* Oxford: Oxford University Press.

7. US Energy Information Administration (EIA). 2013. Annual Energy Outlook 2013. Washington, DC: US Energy Information Administration (EIA).

8. Sovacool, Benjamin K. 2008. *The Dirty Energy Dilemma: What's Blocking Clean Power in the United States.* New York: Praeger Publishers.

9. Statistics Canada. 2009. Report on Energy Supply and Demand in Canada: 2007. Ottawa, Canada: Manufacturing, Construction and Energy Division, Statistics Canada.

10. For more details see Gil, Hugo A., Geza Joos, Jean-Claude DesLauriers, and Lisa Dignard-Bailey. 2006. *Integration of Wind Generation with Power Systems in Canada: Overview of Technical and Economic Impacts.* Canada: National Resources Canada; and Valentine, Scott Victor. 2010. "A STEP Toward Understanding Wind Power Development Policy Barriers in Advanced Economies." *Renewable and Sustainable Energy Reviews* 14 (9): 2796–2807; and Valentine, Scott Victor. 2010. "Braking Wind in Australia: A Critical Evaluation of the Renewable Energy Target." *Energy Policy* 38 (7): 3668–3675.

11. The 2040 electricity demand for Canada was derived from the US Energy Information Agency web-site: http://www.eia.gov/oiaf/aeo/tablebrowser/#release=IEO2013&subject=0-IEO2013&table=16-IEO2013®ion=0-0&cases=Reference-d041117

12. Ackermann, Thomas, and Lennart Söder. 2002. "An Overview of Wind Energy Status 2002." *Renewable and Sustainable Energy Reviews* 6: 67–128.

13. Sovacool, Benjamin K. 2008. *The Dirty Energy Dilemma: What's Blocking Clean Power in the United States.* New York: Praeger Publishers.

14. Canadian Wind Energy Association (CanWEA). 2008. Windvision 2025: Powering Canada's Future. Ottawa: Canadian Wind Energy Association.

15. Alberta Provincial Government. 2008. *Alberta's Oil Sands: Opportunity. Balance.* Alberta: Alberta Provincial Government.

16. Huang, Liming, Emdad Haque, and Stephan Barg. 2008. "Public Policy Discourse, Planning and Measures toward Sustainable Energy Strategies in Canada." *Renewable and Sustainable Energy Reviews* 12 (1): 91–115.

17. Rowlands, Ian H. 2007. "The Development of Renewable Electricity Policy in the Province of Ontario: The Influence of Ideas and Timing." *Review of Policy Research* 24 (3): 185–207.

18. Government of Ontario. 2010. *Ontario's Long Term Energy Plan.* Toronto: Government of Ontario.

19. Nuclear Waste Management Organization (NWMO). 2008. *Moving Forward Together: Annual Report 2008.* Toronto: Nuclear Waste Management Organization.

20. Winfield, Mark, Alison Jamison, Rich Wong, and Paulina Czajkowski. 2006. *Nuclear Power in Canada: An Examination of Risks, Impacts and Sustainability.* Canada: The Pembina Institute.

21. Government of Canada. 2009. *A Climate Change Plan for the Purposes of the Kyoto Protocol Implementation Act.* Ottawa: Environment Canada, Government of Canada.

22. US Energy Information Administration (EIA). 2011. *International Energy Outlook 2011.* Washington, DC: US Energy Information Administration.

23. Valentine, Scott Victor. 2010. "Canada's Constitutional Separation of (Wind) Power." *Energy Policy* 38 (4): 1918–1930.

24. British Petroleum (BP). 2012. *Statistics Review of World Energy 2012*. London: British Petroleum (BP).
25. Data taken from the Canadian Centre for Energy Information, www.centreforen ergy.com/FastFacts-BySilo.asp?Template=About_Energy%2CElectricity&tid=12.
26. Environment Canada. 2011. *National Inventory Report: Greenhouse Gas Sources and Sinks in Canada*. Ottawa: Government of Canada.
27. US Energy Information Administration (EIA). 2011. *International Energy Outlook 2011*. Washington, DC: US Energy Information Administration.
28. Environment Canada. 2011. *National Inventory Report: Greenhouse Gas Sources and Sinks in Canada*. Ottawa: Government of Canada.
29. Government of Ontario. 2010. *Ontario's Long Term Energy Plan*. Toronto: Government of Ontario.
30. Huang, Liming, Emdad Haque, and Stephan Barg. 2008. "Public Policy Discourse, Planning and Measures toward Sustainable Energy Strategies in Canada." *Renewable and Sustainable Energy Reviews* 12 (1): 91–115.
31. Data taken from the World Wind Energy Association website, www.wwindea.org/.
32. Taken from the Canadian Wind Energy Association website, www.canwea.ca.
33. More on this program at www.nrcan.gc.ca/ecoaction/6444.
34. Huang, Liming, Emdad Haque, and Stephan Barg. 2008. "Public Policy Discourse, Planning and Measures toward Sustainable Energy Strategies in Canada." *Renewable and Sustainable Energy Reviews* 12 (1): 91–115.
35. Artuso, Antonella. 2012. "No More Wind Turbines Until Study Released, MPP Demands." *North Bay Nugget*, July 21.
36. Canadian Wind Energy Association (CanWEA). 2008. "Backgrounders on Wind Energy." In *Windvision 2025: Powering Canada's Future*. Ottawa: Canadian Wind Energy Association.
37. Government of Alberta. 2008. *Launching Alberta's Energy Future: Provincial Energy Strategy*. Edmonton: Government of Alberta.
38. Poll results available at www.angus-reid.com.
39. Poll results available at www.angus-reid.com.
40. Poll results available at www.angus-reid.com.
41. Government of Alberta. 2008. *Launching Alberta's Energy Future: Provincial Energy Strategy*. Edmonton: Government of Alberta.
42. Artuso, Antonella. 2012. "No More Wind Turbines Until Study Released, MPP Demands." *North Bay Nugget*, July 21.
43. More on this research can be found in Valentine, Scott Victor. 2011. "Sheltering Wind Power Projects from Tempestuous Community Concerns." *Energy for Sustainable Development* 15 (1): 109–114; and Wolsink, Maarten. 2000. "Wind Power and the NIMBY-Myth: Institutional Capacity and the Limited Significance of Public Support." *Renewable Energy* 21 (1): 49–64.
44. Government of Alberta. 2008. *Launching Alberta's Energy Future: Provincial Energy Strategy*. Edmonton: Government of Alberta.
45. Data taken from the Canadian Center for Energy Information's website: www.cen-treforenergy.com.
46. Government of Alberta. 2008. *Launching Alberta's Energy Future: Provincial Energy Strategy*. Edmonton: Government of Alberta.
47. Poll results available at www.angus-reid.com.
48. Mertz, Emily. 2012. "Alberta's Largest Wind Farm Nears Completion in Halkirk." *Global News*, September 20.

49. Government of Alberta. 2008. *Launching Alberta's Energy Future: Provincial Energy Strategy*, p. 22. Edmonton: Government of Alberta.

50. Data taken from the Energy Statistics Handbook (Statistics Canada): www.statcan. gc.ca/pub/57-601-x/57-601-x2012001-eng.pdf.

51. Government of Alberta. 2008. *Launching Alberta's Energy Future: Provincial Energy Strategy*. Edmonton: Government of Alberta.

52. Valentine, Scott Victor. 2010. "The Fuzzy Nature of Energy Security." In *The Routledge Handbook of Energy Security*, edited by B. K. Sovacool: Routledge.

53. Thorlakson, Lori. 2003. "Comparing Federal Institutions: Power and Representation in Six Federations." *West European Politics* 26 (2): 1–22.

54. Wimmer, A. 2007. "Institutions or Power Sharing: Making Sense of Canadian Peace." *Sociological Forum* 22 (4): 588–590.

55. Huang, Liming, Emdad Haque, and Stephan Barg. 2008. "Public Policy Discourse, Planning and Measures toward Sustainable Energy Strategies in Canada." *Renewable and Sustainable Energy Reviews* 12 (1): 91–115.

56. Government of Canada. 1867/1982. The Constitution Acts 1867 to 1982, Government of Canada.

57. Ibid.

58. Ibid.

59. Data taken from the Canadian Center for Energy Information's website: centre-forenergy.com.

60. Baier, Gerald. 2005. "The EU's Constitutional Treaty: Federalism and Intergovernmental Relations: Lessons from Canada." *Regional & Federal Studies* 15 (2): 205–223.

61. Government of Canada. 1867 / 1982. The Constitution Acts 1867 to 1982, Ottawa: Government of Canada.

62. Ibid.

63. BBC. 2011. "Canada to Withdraw from Kyoto Protocol." BBC, December 13.

64. This is an expression taken from Bemelmans-Videc, Marie-Louise, Ray Rist, and Evert Vedung, eds. 2003. *Carrots, Sticks, and Sermons: Policy Instruments and Their Evaluation*. City: Transaction Publishers. The authors group policy tools into three categories: carrots (incentives), sticks (regulations), and sermons (knowledge dissemination).

65. Data from the IEA Global Renewable Energy Database, www.iea.org/policiesandmeasures/.

66. Huang, Liming, Emdad Haque, and Stephan Barg. 2008. "Public Policy Discourse, Planning and Measures toward Sustainable Energy Strategies in Canada." *Renewable and Sustainable Energy Reviews* 12 (1): 91–115.

67. Sovacool, Benjamin K., and Scott Victor Valentine. 2012. *The National Politics of Nuclear Power: Economics, Security and Governance*. Milton Park, UK: Routledge.

68. For a critique on the likelihood of reaching this target, see Whittingham (ed.). "Budget 2012: Canada Won't Spare a Penny for Clean Energy, March 30, 2012, The Pembina Institute, www.pembina.org/blog/616.

69. Data taken form the Canadian Wind Power Association website, www.canwea.org.

CHAPTER 9

Wind Power in Japan

If we are going to reform the electricity market, then we should turn this extraordinarily terrible crisis into an opportunity. . . We will make our nuclear power generation increasingly safe while continuing to contribute to the nonproliferation regime. Japan has been at the forefront in both of these areas, and withdrawing from them is simply not an option for us.

—Shinzo Abe, June 2013[1]

9.1 INTRODUCTION

The story of wind power development in Japan is, at its essence, a subplot to a story of path dependency and the clout of a well-entrenched nuclear power regime. Path dependency refers to the tendency of an entrenched technology to evolve incrementally, primarily due to the existence of entrenched special interests that are committed financially and ideologically to a given technology. These special-interests spawn a regime that finances incremental technological evolution in order to keep pace with consumer demand and that is capable of mounting strong market defense of incumbent technology.[2] By achieving a high level of market penetration, an incumbent technology amasses both the market share necessary to undercut competitive offerings and the political support needed to create market entry barriers for competing technologies. In Japan, nuclear power has been such a technology. Prior to March 11, 2011, Japan laid claim to possessing the third-largest nuclear power program in the world. The nation's 54 nuclear power reactors were capable of providing almost 30% of the nation's electricity needs,[3] and the government was committed to a nuclear power expansion policy that would result in nuclear power capacity providing 40% of the nation's electricity supply by 2030.[4]

So how did a nation located in an extremely active seismic region that was on the receiving end of two atomic bombs (which killed between 150,000 and 250,000 people) wind up with such a well-entrenched nuclear power regime? The answer to this question helps explain why wind power developers in Japan have had such a difficult time penetrating the Japan market.

At the end of World War II, a defeated Japan found itself under the administrative oversight of the United States. On December 8, 1953, US President Dwight D. Eisenhower delivered a speech to an assembly of the United Nations, which came to be called the "atoms for peace" speech. In his address, Eisenhower announced a US initiative to "encourage worldwide investigation into the most effective peacetime uses of fissionable material."[5] As a first step toward implementing this vision, the United States deemed it to be of symbolic importance for the peaceful use of atomic energy to be first promoted in the nation that had been subject to the horrors of atomic energy used for military purposes. As US atomic energy Commissioner Thomas Murray callously summarized, "construction of such a power plant in a country like Japan would be a dramatic and Christian gesture which could lift all of us far above the recollection of the carnage of those cities."[6]

On the Japanese side, there was ardent political support for embracing a technology which, at the time, held the promise of eventually generating electricity that would be "too cheap to meter."[7] The nation already had a pool of experienced nuclear engineers who had been working on military applications of atomic energy during the war;[8] therefore, policymakers perceived the main challenge to implementing this vision to rest with the attenuation of public opposition.

In February 1955, the first concrete steps were taken to address this challenge. Matsutaro Shoriki, the colorful owner of the Yomiuri newspaper and founder of the Nippon Television Network, ran for a position in Japan's Lower House at the urging of political allies. He was elected, and nine months later appointed minister in charge of nuclear energy. Shoriki had no previous experience in energy governance but that is not what his supporters needed from him. He was a media expert and entrepreneur— a perfect person to lead a campaign to reverse adverse public opinion of atomic energy. This would be a formidable challenge given that, in 1956, 70% of Japanese perceived nuclear technology to be harmful.[9]

Between 1956 and 1958, a masterful PR campaign was executed by Shoriki and his team. This began with a ceremony at a Shinto shrine in Tokyo to "purify the atom," followed by a nationwide road show to promote the public benefits of atomic energy in eight cities around Japan, including Hiroshima—the epicenter of Japan's anti-nuclear movement. The social manipulation associated with many of the PR events is perhaps best exemplified by a campaign waged by the government that involved sending municipal workers dressed in white laboratory coats to speak authoritatively at schools and community centers about the beneficial peaceful applications of nuclear power.[10] Hundreds of newspaper articles heralding the benefits of nuclear power appeared throughout the 1950s.[11] These PR efforts culminated on April 1, 1958 with the opening of the Grand Exhibition of the Reconstruction of Hiroshima, which featured a pavilion dedicated to the peaceful use of atomic energy (ironically housed in the newly built A-bomb museum).[12] By 1958, a remarkable reversal of public opinion had taken place in regard to nuclear power—with only 30% perceiving nuclear technology as harmful.[13]

By the 1960s, Japan's fledgling nuclear power program had established irreversible momentum. The ground was already being prepared for Japan's first commercial project and numerous R&D activities were under way, including the construction of a demonstration reactor that would come online in 1963.[14] In 1966, Japan's first commercial nuclear power reactor became operational. By the end of the 1970s, the nation already hosted 19 commercial nuclear power reactors and was spending over US$2 billion per year on nuclear power R&D. The government would continue to channel at least US$2 billion per year into nuclear power research throughout the 1980s, and over US$2.5 billion per year into nuclear power research throughout the 1990s and beyond.[15]

Political support for nuclear power in Japan has extended well beyond R&D funding. The government has channeled billions into infrastructure support projects including new state-of-the-art waste reprocessing facilities in Rokkasho Village, Aomori Prefecture, which are scheduled to commence operations in October 2014.[16] In terms of financial support for siting nuclear plants, the government has strategically targeted villages experiencing economic decline and sweetened the appeal of hosting nuclear power plants by offering to bolster community infrastructure, build community centers, and provide other social benefits.[17] The utilities which own and operate these nuclear power plants provide similar incentives, including the annual provision of financial support to community members for hosting the plants.

This brief summary of the history of nuclear power development in Japan is necessary to understand wind power development because prior to the Fukushima disaster, the nuclear power regime had established itself as an indispensible and increasingly prominent component of the nation's electricity generation mix. It is a technology that has traditionally enjoyed full support of the ruling Liberal Democratic Party (LDP) and Japan's 10 regional utilities, which serve as gatekeepers to the nation's electricity grid. Moreover, the industry has enjoyed benefit-of-the-doubt oversight—lax regulatory supervision that depended extensively on self-regulation. After a series of mishaps, a tightening of regulatory supervision in 2002 resulted in the discovery that the Tokyo Electric Power Company (TEPCO), which managed 17 reactors, had falsified inspection reports and concealed damaged to reactor vessel shrouds in 13 of its 17 units.[18] Up until recently political opposition has been almost nonexistent, with even the anti-nuclear special interest groups seemingly acquiescent to the necessity of nuclear power, choosing to focus more on lobbying for improved transparency and governance of the technology.

Prior to the Fukushima disaster, such widespread support for nuclear power engendered a political climate that discouraged unbiased evaluation of alternative electricity generation technologies. This was particularly true for wind power, which was widely seen as a supplemental technology at best. The utilities opposed wind power because it was troublesome to integrate into grid operations. The government opposed wind power (or any other alternative energy technology) because billions of dollars had been committed to nuclear power development. Politically, the government had passed the point of no return in regard to nuclear power support. The general public was generally apathetic toward a transition away from nuclear power because the government had, for years, sold the public on the belief that nuclear power was the cheapest form of electricity and that a switch away from nuclear power would unnecessarily inflate electricity costs and damage industrial competitiveness. This of course all changed in March 2011.

On March 11, 2011 an earthquake measuring 9.0 on the Richter scale occurred approximately 70 km off the east coast of Japan's Tohoku region. The earthquake triggered a tsunami which in some places was over 40 meters high. The leading edge of the tsunami struck the Fukushima Daiichi nuclear power plant, causing serious (level 7) damage to three of the reactors. Boron-infused water was dumped onto the reactors to prevent meltdown of the cores, and this resulted in the leakage of radioactive materials into the ocean. Concerns over the extent of radioactive contamination induced widespread panic that extended 300 kms south to Tokyo, where there was considerable alarm over the prospect of tainted water supplies. Although officials are still struggling to contain the damage, in the first week alone this nuclear disaster had already become the second worst in human history.

In the weeks that followed the disaster, the government finally succumbed to public pressure and ordered that all but two of the nation's nuclear power reactors be shut down for safety checks. Removing so much electricity generation capacity from the nation's power network in one fell swoop created chaos in energy policy circles. TEPCO, the operator of the Fukushima nuclear power plant, was forced into financial insolvency by the event, eventually having to be financially propped-up and semi-nationalized by the Japanese government.[19] Throughout the country, universal appeals for energy conservation were made with particular consternation over the possibility that the increase in demand for electricity typically associated with summer months might result in blackouts in Tokyo. The expanded use of natural gas-fired power plants to fill the void caused by the nuclear plant shut down resulted in inflated energy costs, with TEPCO applying for permission to raise rates by nearly 20% to compensate for increased energy costs, which were estimated to be US$8.3 billion in the one-year period after the Fukushima disaster.[20] The switch from nuclear power to gas-fired power also scuttled the prospects of the nation meeting its Kyoto Protocol GHG emission reduction targets.

After the Fukushima disaster, the public apathy that enabled the ascendance of nuclear power and the political will that supported the industry and insulated it from scrutiny was supplanted by a climate of introspection. A June 5, 2012 survey by Pew Research indicated that 70% of the general public was in favor of at least a scaling down of the nuclear program.[21] The ruling Democratic Party of Japan (DPJ) initiated studies to explore the feasibility of reducing the nuclear power program to 0%, 15%, or 20 to 25% and declared an intention to phase-out nuclear power by 2050.[22] In July 2012, the DPJ announced an aggressive feed in tariff (FIT) that included wind power, and there were indications that the wind power industry was about to experience a golden age in Japan.

In December 2012, the prospects of a golden era for wind were upended by a return to power of the Liberal Democratic Party of Japan (LDP), the party that presided over the rise of nuclear power in Japan. Newly elected Prime Minister Shinzo Abe has since repudiated his campaign promise to phase-out nuclear power and declared his support for reviving the nuclear power program.[23] This has spawned two questions: how sizable will the nuclear power presence be, and what will the role of wind power be in the future energy scenario? This chapter will strive to shed light on both and in doing so tell the story of wind power development in Japan.

9.2 AN OVERVIEW OF ELECTRICITY GENERATION IN JAPAN

Figure 9.1 illustrates the degree to which Japan depends on foreign energy imports. Since the end of World War II, Japan has imported virtually its entire energy stock from overseas. Currently, almost 96% of Japan's primary energy supply comes from imported stocks of coal, oil, natural gas, and uranium. Although there are other countries—South Korea, France, and Italy—that have similar high rates of import dependence, Japan's reliance is more precarious because it is an island nation, which does not have

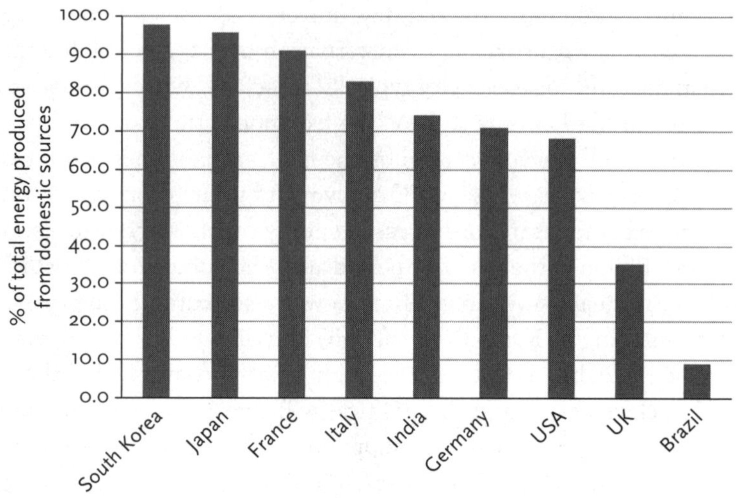

Figure 9.1. Japanese Dependence on Foreign Energy Supplies

Source: FEPC. 2013. *Electricity Review Japan: 2013*. Tokyo, Japan: The Federation of Electric Power Companies of Japan.

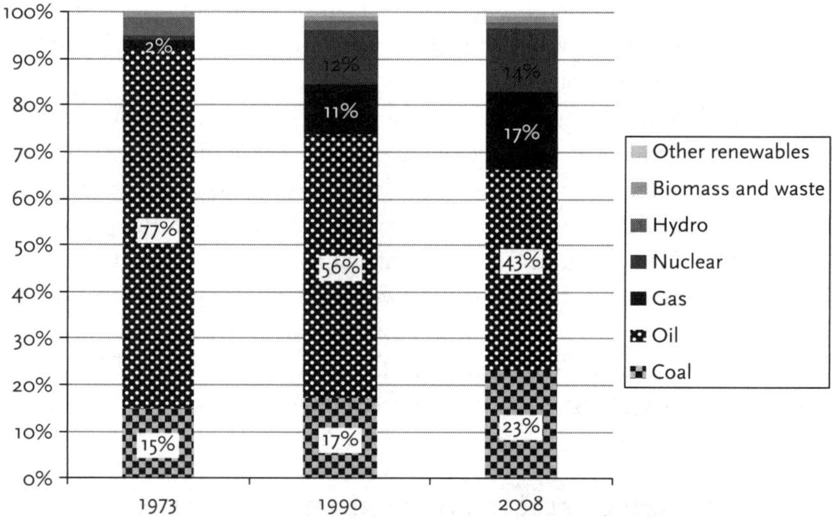

Figure 9.2. Japan's Transition Away from Oil for Primary Energy Supply

Source: World Energy Outlook 2007 and 2010, International Energy Agency.

the ability to link the national electricity grid to neighboring countries in order to improve stability and resilience.

Prior to the two oil shocks in the 1970s, oil was the dominant energy resource in Japan. As Figure 9.2 illustrates, in 1973 oil accounted for 77% of the nation's total primary energy consumption. Although all nations suffered from high oil prices during this period, Japanese energy consumers were particularly hard hit because of this high dependence. Like in other nations, this crisis precipitated a strategic shift away from dependence on oil. Unlike in other nations, Japan did not have much in the way of indigenous energy resources to exploit. Consequently, a common strategic thread in Japanese energy planning since the 1970s has been to facilitate a shift away from oil via a shift into natural gas, nuclear power, and coal electricity generation technologies.

Japan's commitment to greenhouse gas (GHG) emission reductions under the Kyoto Protocol has also influenced the evolution of the nation's energy mix. For Japan, the Kyoto Protocol, which was drafted at the Third Conference of the Parties (COP3) to the United Nations Framework Convention on Climate Change (UNFCCC) in 1997, represented a chance to demonstrate leadership in an area of international significance. Therefore, there was a considerable amount of political pressure attached to achieving Japan's 2008–2012 Kyoto Protocol round one (2008–2012) GHG emission reduction target of 6% below 1990 levels. To achieve this goal, the electricity sector was targeted as a priority area for facilitating GHG emission reductions. In the short term, the strategic intent was to replace oil-fired and coal-fired power plants with gas-fired power plants. In the medium term, the government announced a 2006 plan to expand the nation's nuclear power capacity to provide 40% of the nation's electricity needs by 2030.[24] Figure 9.3 depicts Japan's electricity mix as it stood in the lead up to the Fukushima disaster.

According to government documents, nuclear power was viewed as the darling technology of a new low-carbon Japanese economy because it also enhanced energy security and held economic appeal.[25] The government has

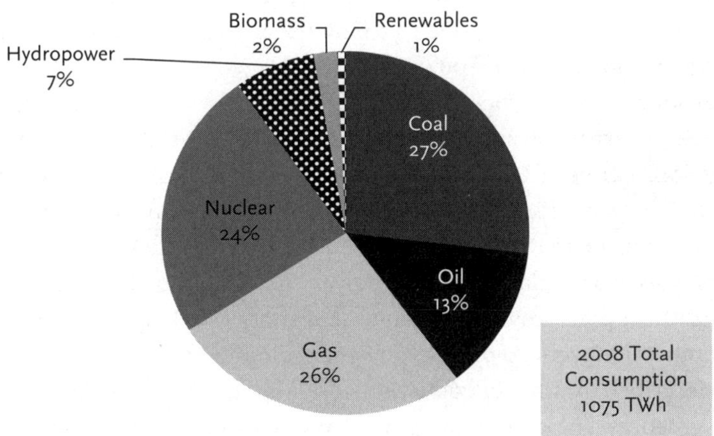

Figure 9.3. Japan's Electricity Mix, 2008 (% Contribution by Energy Source)
Source: World Energy Outlook 2009, International Energy Agency.

gone so far as to declare nuclear power to be a domestic energy resource,[26] despite Japan having no uranium deposits to speak of. This domestic perspective is derived from the fact that the nation has embraced a closed-cycle strategy to nuclear power development which, if achieved, would allow the nation to reprocess existing stockpiles of spent nuclear fuel.

Given the evident desire of Japan's energy policymakers to enhance domestic energy security and reduce carbon emissions, it begs the question of why the nation has neglected the development of renewable power technologies for electricity generation. Part of the explanation rests with the historical cost legacy of electricity generated by renewable technologies. Simply put, Japan's industrial lobby, which has a very strong influence on policymaking, has been opposed to high contributions from renewable energy out of concern that the added cost will undermine industrial competitiveness.[27]

This is not to say that the Japanese government has neglected the commercial promise of renewable energy. Immediately after the first oil crisis in 1973, the Japanese government ramped up its funding for renewable energy R&D. By the end of the 1970s, with oil still at inflated prices, the Japanese government amplified its support for renewable energy R&D (see Figure 9.4). Geothermal technology received priority funding because even in the early years, the cost profile of geothermal power and the untapped potential in the nation made it an obvious alternative energy technology to pursue commercially. By the 1990s, concerns over the adverse impacts of wide-scale expansion of geothermal plans on Japan's onsen industry had stymied utility-scale adoption.

Although the economics of solar technologies did not justify support for domestic use, solar power was embraced in the early 1980s as an enabling technology for existing (i.e., solar cells for electronics) and new (i.e., solar photovoltaic technology) industrial sectors. Meanwhile, support for wind power R&D was negligible, due to a belief that there was limited realizable potential in Japan for wind power.[28]

In absolute terms, the data from Figure 9.4, which shows that the Japanese government spent between US$100–200 million on renewable energy R&D between 1980 and 2000, appears to indicate a sizable commitment until it is compared to the investment made in nuclear power R&D. Figure 9.5 contrasts R&D investment in renewable energy to other energy technologies, particularly nuclear power. As the data demonstrates, the amount of support provided to renewable energy R&D was a mere sliver of the amount that went to nuclear power development.

As described in the introduction, on March 11, 2011, the relatively uncontested ascent of nuclear power in Japan ended. As a result of the Fukushima nuclear disaster, Japan has now shut down all of its nuclear reactors (although two units have been allowed to operate for a period of time).

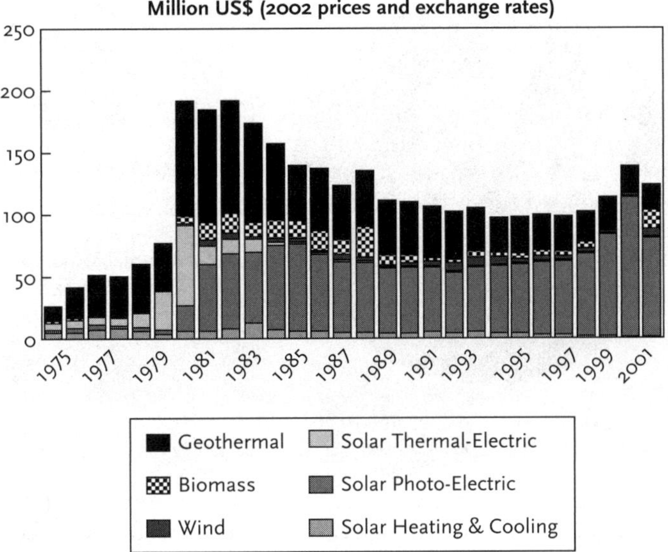

Figure 9.4. Japanese Investment in Renewable Technologies, 1975–2001

Source: IEA. 2004. Renewable Energy: Market and Policy Trends in IEA Countries, edited by I. E. Agency: International Energy Agency.

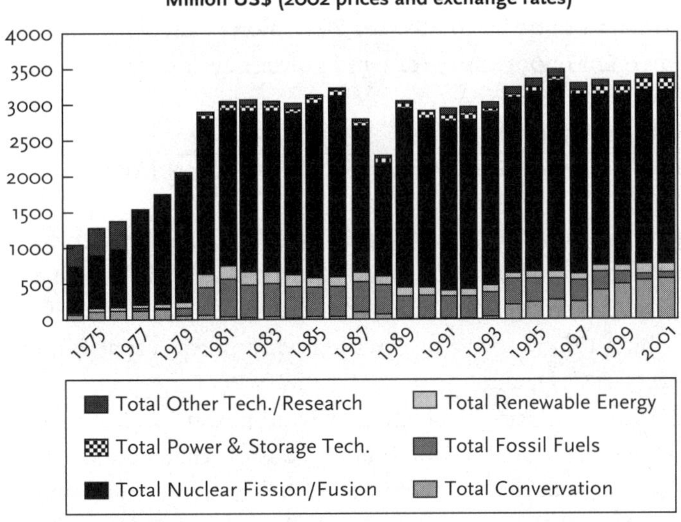

Figure 9.5. Total Government R&D Expenditure in Energy Technologies, 1973–2001

Source: IEA. 2004. *Renewable Energy: Market and Policy Trends in IEA Countries*, edited by I. E. Agency: International Energy Agency.

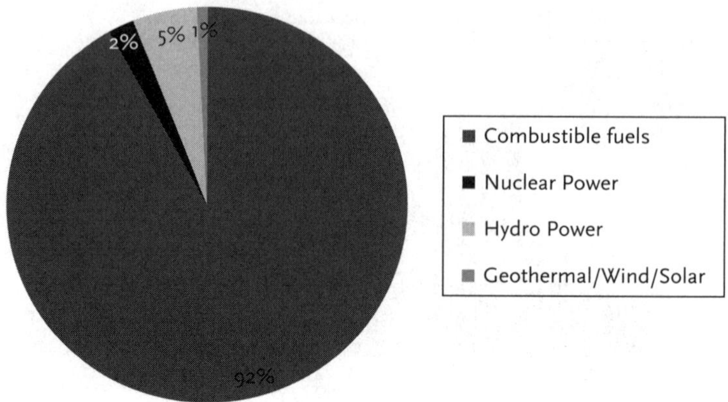

Figure 9.6. Japan's Electricity Mix January 2013
Source: International Energy Agency, Monthly Electricity Statistics, January 2013.

They will remain idled pending safety assessments and how effectively the radioactive waste leaks at Fukushima can be contained. As of January 13, 2013, Japan's electricity mix was as depicted in Figure 9.6.

Clearly, this implies that Japan's aspiration to play a leading role in facilitating global GHG emission reductions has suddenly become unhinged because the switch from nuclear power has predominantly been accomplished by expanding gas-fired electricity capacity. In addition to inflating GHG emissions, this shift has resulted in a substantial increase in energy costs, estimated to be 15 to 20% for 2012 alone. Amidst all of this turmoil lies an emergent opportunity for wind power developers.

9.3 HISTORY OF WIND POWER DEVELOPMENT IN JAPAN

When it comes to wind power development, Japan is a comparative underachiever. As of the end of 2012, Japan possessed 2,649 MW of installed wind power capacity. This is miniscule in comparison to other major economies such as China (75,564 MW of installed wind power capacity), the United States (60,007 MW), and Germany (31,332 MW).[29]

Some telling statistics illustrate how wind power has been something of an afterthought in the nation. This is a nation that uses a lot of energy. Japan has a population of 127 million (1.9% of the global population), yet it accounts for a little over 5% of global energy consumption on an annual basis.[30] Despite aggregate energy consumption that exceeds the global average, installed wind power capacity represents only 0.9% of the global total. In 2012, electricity consumption in Japan amounted to 859,700 GWh.[31] Assuming that Japan's wind power turbines operate at a capacity factor of

25%, the contribution that wind power makes to national electricity production amounts to about 5724 GWh, a mere 0.7% of domestic electricity output.

As Figure 9.7 depicts, Japan was a comparatively late adopter of commercial wind power. In 1992, the nation possessed only 3 MW of installed wind power capacity. By comparison, the United States possessed 1680 MW of installed wind power capacity, Denmark hosted 436 MW, and Germany had 183 MW in the same year.[32] But it is not just a late start that has handicapped the wind power development program. As Figure 9.7 further illustrates, the pace of wind power development has also been flaccid. As a comparative, in 1992, China only possessed 6.14 MW of installed wind power capacity. Although the two nations initiated their respective wind power programs around the same time, China's installed wind power capacity exceeds Japan's by over 28 times.

Between 1981 and 1991, the Japanese government allocated an annual average of approximately US$4 million to wind power research (see Figure 9.4). To put this into perspective, the funds allocated for geothermal R&D in 1981 alone were twice as much as wind power researchers received during the entire decade in question.

In 1992, the government initiated sponsorship of a three-year program for testing prototype wind turbines.[33] Over a three-year period (1992–1994), wind power R&D funding increased to just under US$10 million per year. This financed the development of demonstration turbines, launching Japan's commercial wind power era. However, it is apparent in hindsight that the government was not sold on the results of the research, because

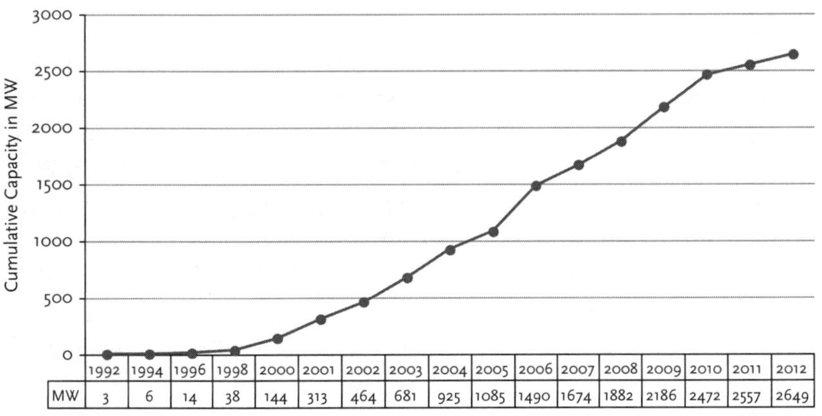

	1992	1994	1996	1998	2000	2001	2002	2003	2004	2005	2006	2007	2008	2009	2010	2011	2012
MW	3	6	14	38	144	313	464	681	925	1085	1490	1674	1882	2186	2472	2557	2649

Figure 9.7. Wind Power Development in Japan
Source: Japan Wind Power Association website, 2013.

by 1995 funding for wind power development was halved and remained at around US$5 million for the rest of the decade (see Figure 9.4).

In 1995, a new law was passed to permit independent power providers (IPPs) to sell power in the electricity wholesale market. Although there was still considerable resistance to wind power from the 10 regional utilities, this change enabled wind power developers to initiate trial projects. By 2000, installed wind power capacity had increased to 144 MW.[34]

In March 2000 the government, facing pressure to decarbonize its electricity grid in order to live up to its Kyoto Protocol emission reduction pledge, took another step toward liberalizing the nation's transmission and distribution (T&D) network. A new law was passed that allowed power producers and suppliers to sell electricity directly to extra-high-voltage users (those requiring more than 2 MW of capacity).[35] This law expanded the pool of electricity consumers beyond the 10 regional utilities and allowed wind power developers to implement sales strategies to bypass obstructive utilities. This contributed to a tripling of installed wind power capacity to 464 MW by the end of 2002.

In 2003, the government introduced the nation's first renewable energy mandate, a renewable portfolio standard (RPS). Under the terms of the legislation, Japan's utilities were required to purchase a specified amount of energy from a pool of renewable technology options, which included small and medium-sized hydropower, geothermal power, solar PV, wind power, and biomass. Under the terms of the RPS, power providers were to receive a purchase price that was equal to the price paid by the end-consumer. Utilities were given the choice of either purchasing the power from IPP's, generating the power through self-investment or purchasing surplus credits from other regional utilities.[36] This latter option was underpinned by a banking mechanism that allowed utilities to sell or store credits for any surplus purchases of renewable energy that exceeded the annual targets.[37]

On the surface, the RPS was a promising step toward encouraging renewable energy expansion; in practice it was ineffective, because the RPS targets were not ambitious enough to nurture innovation and market commitment (next page, Table 9.1).[38] The 2003 RPS quota of 7.32 TWh represented only about 0.75% of the nation's annual electricity production.[39] With wind power providers having to compete with commercially viable geothermal and small hydro projects, the market opportunities arising from the RPS were limited. This was exacerbated by a clause which accorded special treatment to solar photovoltaic energy. For each kilowatt hour of solar PV electricity that was purchased, the purchasing utility would receive credit for 2 kWh of renewable energy.

Since its launch, the modest development trajectory of the RPS has been criticized by renewable energy advocates.[40] The RPS quota for 2014 was

Table 9.1 JAPAN'S RENEWABLE PORTFOLIO STANDARD TARGETS 2003–2014 (IN TWH)

2003	2004	2005	2006	2007	2008	2009	2010	2011	2012	2013	2014
7.32	7.66	8.00	8.34	8.67	9.27	10.33	12.20	13.15	14.10	15.05	16.00

Source: IEA. 2009. Global Renewable Energy: Policies and Measures: International Energy Agency.

only 119% higher than the RPS quota for 2003, representing a targeted annual market growth rate of 8%. Industry groups argued that these low quotas were necessary to avoid adversely impacting the competitiveness of Japanese industry. However, one early study into the financial impact of the RPS repudiated this contention. The study found that the RPS program would increase the cost of electricity produced by ¥0.1/kWh (approximately US$0.001 per kWh).[41] If the findings of this study are valid, given that the cost of natural gas has tripled in Japan between 2002 and 2012,[42] an RPS of ten times the magnitude would have been viable without adversely impacting industrial competitiveness. In short, it was apparent from the onset that the RPS quota was too low and the targeted pace of development was too slow to entice the level of competition necessary to build a market.

In 2004, the T&D network was further liberalized. Power producers and suppliers were permitted to sell electricity directly to consumers that required more than 500 kW of power capacity. The following year, the capacity limit was lowered to 50 kW. This basically gave wind power providers access to approximately 60% of total electricity demand.[43] However, given the stochastic nature of wind power flows, in order to fully tap into these new market opportunities, wind power providers would have to either store the energy captured or combine wind systems with peak-load technologies in order to attenuate the power fluctuations. Storage adds to the cost, while twinning wind systems with peak load technologies gives rise to added complexities. Neither case represents a trouble-free solution.

Despite criticisms of the RPS, this policy somewhat quickened the pace of wind power development. By the end of 2011 installed wind power capacity had grown from 464 MW, prior to the initiation of the RPS, to over 2557 MW. At a capacity factor of 25%, this amounts to an expansion of about 4.6 TWh in annual wind power generation capacity. Assuming that all the wind power added during this period benefitted from the RPS, this suggests that wind power accounted for approximately 79% of the RPS quota between 2003 and 2011. The comparative competitive success of wind power under this program is largely because geothermal project developers have faced well-organized opposition from Japan's onsen industry and mini-hydro projects are hindered by limited potential and siting permit barriers. Wind power was the economically logical alternative to succeed under the RPS, despite the double incentives given to solar PV projects.

As 2011 began, there were indications that bolder renewable energy expansion targets would be initiated by the governing Democratic Party of Japan, led by Naoto Kan.[44] However, amidst party scandals, bloated fiscal deficits, and economic recession, Kan's administration was preoccupied and slow in formulating a new renewable energy strategy that would deliver on its campaign pledge to reduce Japanese GHG emissions to 70% of 1990 levels by 2020.[45] Incentive to speed up efforts was renewed on March 11, 2011, when the Fukushima disaster took place.

With the fate of Japan's nuclear power program under intense public scrutiny, the DPJ finally released a new renewable energy development program that would go into effect on July 1, 2012. A series of feed in tariffs (FIT) were announced with rates that were some of the highest in the world. The FIT surcharge would be covered by passing the additional cost on to the end-consumer. Wind power systems of 20 kW or more would receive ¥22 (approximately US$0.22) per kWh (after consumption tax), and wind power systems of less than 20 kW would receive ¥55 (approximately US$0.55) per kWh (after consumption tax). The program mandated that rates for the FIT be reviewed on April 1 of every year. Under the new scheme, electric utilities are obligated to enter into contracts with any wind power provider that meets government certification standards.[46]

Early indications are that solar energy technologies are clearly benefitting from the new FIT system.[47] However, continued opposition from utilities and project siting challenges have so far hindered the pace of wind power expansion. In regard to wind power, the government appears to see offshore wind as possessing the greatest realizable potential. Recently, Environment Minister Nobuteru Ishihara was quoted as anticipating an offshore development target of 1 GW by 2020, which he lauds as being 40 times greater than existing offshore capacity.[48] However, given that in the previous seven-year period (2006–2012), onshore wind power capacity grew by 1.159 GW, one wonders if the government truly expects the new FIT to usher in meaningful change.

9.4 UNDERSTANDING THE GENERAL FORCES FOR CHANGE

9.4.1 Sociocultural Landscape

As was detailed in the introduction, the Japanese public has been heavily influenced by well-organized government campaigns to garner public acceptance of a nuclear-centric energy policy. This has resulted in a unique phenomenon in Japan where public perception of energy policy priorities is remarkably congruent with existing government energy policy. A recent study into energy policy perspectives in 10 nations revealed that Japan exhibited far and away the highest level of public accord when it came to prioritizing national energy goals. The study revealed that the Japanese public viewed the highest energy priority to be securing adequate supplies of oil, gas, coal, and uranium. The respondents highlighted the need to reduce GHG emissions as the second priority and the need to enhance research and development on new and innovative energy technologies as the third priority.[49]

In regard to nuclear power, prior to the Fukushima disaster the general public appeared to be acquiescent to the notion that nuclear power development was a necessary evil for Japanese industry to maintain international competitiveness. For example, a 2007 Japanese public opinion poll in Japan indicated that 53% were in favor of maintaining status quo and 13% were in favor of expanding the nuclear power program. Interestingly, in April 2011, another poll conducted by the Asahi Newspaper reported that 51% of the respondents wanted to maintain status quo and 5% supported an expansion of nuclear power.[50] To summarize, prior to Fukushima 66% of respondents were in favor of maintaining or bolstering nuclear power capacity; immediately after the world's second-largest nuclear power disaster, there were still 56% in favor of maintaining or bolstering nuclear power capacity.

The 2011 Asahi poll did report that 89% of the respondents expressed some unease (33%) or much unease (56%) in regard to the Fukushima incident.[51] Directly after the Fukushima disaster, in other words, it was clear that sensitivities toward nuclear power had become amplified and further information might sway public opinion, one way or another. This has happened. Since the Fukushima disaster, the industry has been subject to much more media scrutiny, which in turn has enhanced awareness of the shortcomings of nuclear power governance, leading to a dip in public support for nuclear power. Criticisms over a lax safety culture within TEPCO and the inadequacy of government oversight have featured prominently in analyses of the Fukushima disaster. Consequently, a November 2011 poll by NHK revealed that 70% of respondents wanted nuclear power to be either eliminated or significantly reduced.[52] Additionally, those who are opposed to nuclear power have become more assertive. This is exemplified by the Occupy Kasumigaseki anti-nuclear protests in front of the prime minister's official residence in Tokyo. At its peak, the Friday protests attracted over 200,000 people.[53]

Generally, support for renewable energy is consistent with sociocultural beliefs. The majority of Japanese adhere to Shinto ideology, which is an indigenous Japanese religion that at its core emphasizes the symbiosis between humans and nature. This connection is symbolized by the reverence with which Japanese celebrate seasonal shifts through cultural festivals such as cherry blossom viewing (called "hanami" in Japanese) in the spring, the star festival ("tanabata") in the summer and observing the changing of the leaves ("koyo") in the fall. Moreover, there is a strong environmental ethic in Japan that has been escalating since the 1970s.[54] Consequently, there now tends to be a higher willingness to pay for environmentally friendly products and technologies in Japan.

When it comes to the topic of what technology might serve to replace nuclear power, the political and public darling tends to be solar power, which according to one survey is supported by 71% of the general public.[55] The Shinto premise of harmonizing with nature partially explains why Japanese would prefer more expensive solar PV technology to wind power. Japanese society tends to be very conservative. The aesthetics of a community are highly valued and reverence for traditional landscapes is high.[56] Therefore, despite the low-carbon appeal of wind turbines, the aesthetic invasiveness of this technology tends to engender a degree of NIMBY opposition.

In summary, sociocultural forces in Japan are pro-renewable, but not necessarily pro-wind. There is currently a degree of public support for curtailing the size of the nuclear power program in Japan, but this is more out of concern over the credibility of the nuclear power safety regime than it is out of concern over the technology itself. The general impression that one gets when discussing the matter of nuclear power with policymakers in Japan is that a public majority will be willing to support the continued presence of nuclear power provided they receive assurances that governance will be improved.

9.4.2 Economic Landscape

The good news is, without question, Japan is still an economic force. It is the third-largest economy in the world and the Japanese economy is still highly competitive, ranking tenth in the World Economic Forum's Global Competitiveness Report for 2012–2013. On a per capita basis its citizens are also comparatively affluent, ranking thirty-eighth in the world with a per capita GDP (PPP) of US$36,000 in 2012. The nation also enjoys enormous reserves of foreign exchange and gold, amounting to over US$1.35 trillion at the end of 2012.[57] This suggests that Japan enjoys a high degree of economic security, thanks to investment income.

The bad news is Japan's economic landscape is deteriorating. At one time it was the second-largest economy in the world. Today, China has surpassed

Japan, relegating it to third position. Industrial competitiveness is deteriorating. South Korean, Taiwanese, and Chinese firms have begun to challenge Japanese leadership in technology sectors that have been traditionally dominated by Japanese firms such as Sony, Toshiba, Panasonic, Toyota and Honda. Unemployment, which ranged between 2 to 3% during the economic heydays of the 1980s, now hovers between 4 to 5%; as of 2010, 16% of the population was living below the poverty line.[58]

Although adequately assessing the economic health of a given nation requires a level of analysis that is beyond the scope of this chapter, Figure 9.8 highlights an influential and troublesome trend. National affluence has grown, thanks to years of trade surplus. Between 2000 and 2010 Japan's trade balance plateaued, despite a 75% increase in global GDP. This suggests that the nation's industries have been challenged in overseas markets, and affluent Japanese citizens have been purchasing increasingly higher volumes of imported goods. In 2011, for the first time since 1980, Japan posted an annual trade deficit, a development that can be thought of as a transfer of wealth from domestic parties to other nations. Although this trade deficit was caused primarily by elevated natural gas imports following the Fukushima disaster, it is a trend that cannot be allowed to continue if the nation wishes to maintain industrial competitiveness.

The greatest economic concern is that Japanese industry now faces intense international competition. Mechanization has cascaded downward to other emerging economies, allowing nations such as South Korea, Taiwan, and even China to successfully compete head-on with Japan. Moreover, international pressure during the 1990s forced Japan to liberalize its domestic

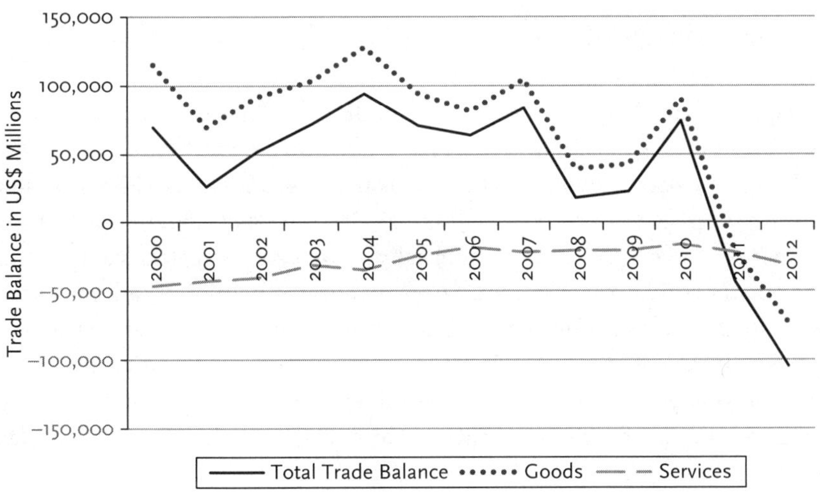

Figure 9.8. Japan's Annual Trade Balance 2000–2012

Source of data: Balance of Payments Statistics (Ministry of Finance and Bank of Japan), 2013.

markets, giving rise to an increase in foreign market competition and undermining the domestic cash cow conditions that enabled Japanese firms to reduce prices overseas.

Unfortunately, even though conditions have changed, the industrial development policy embraced by the Ministry of Economy, Trade, and Industry (METI) has not evolved to keep pace. The focus on implementing policies that will help industry to reduce the cost of factors of production is still a dominant pillar in METI industrial policy when actually what is needed is a transformation of industry both technologically and operationally.

The custom of setting policy to exploit current realities rather than future trends has had a major impact on the evolution of Japan's energy sector. METI—and by proxy the Japanese government—supports energy technologies that are currently the most cost effective. In the past that meant that coal, liquid natural gas (LNG), nuclear power, and hydropower were the preferred energy sources. Consequently, the nation has nurtured the development of energy infrastructure that is designed to optimize contributions from these technologies. Although wind power has now achieved a degree of commercialization that in most countries makes it preferable to nuclear power and LNG, the Japanese government does not perceive this to be the case for Japan because the conditions under which electricity technology costs are compared are biased toward the incumbent technologies. This will be discussed further in section 9.5.3.

9.4.3 Technological Landscape

Technologically, Japan's nuclear regime is as well entrenched as any nuclear regime in the world. It has historically enjoyed strong political support, benefited significantly from government funding, and has been insulated from public and media scrutiny.[59]

There is a strong sense that the nuclear power regime is biding its time, waiting for public opposition to diminish. In order to facilitate this the government has been very quick to publically reorganize its nuclear safety regime and bolster regulatory oversight.[60] Concurrently, it has quietly allowed a couple of nuclear power construction projects and uranium reprocessing plans to proceed. It is widely accepted in all circles that the 0% nuclear power option that was one of the three alternatives under study by the government is not really viewed as a feasible alternative. Moreover, even the 15 to 25% nuclear power option has been heavily criticized by industry supporters as unrealistic and uneconomical. Japanese Prime Minister Shinzo Abe is supportive of a full return to a nuclear powered energy policy.[61]

Meanwhile, wind power developers have faced significant siting challenges and persistent resistance from utilities. The costs for siting wind turbines are so expensive in this densely populated nation that wind power generation costs become inflated. Moreover, there have been reports of wind power developers being forced to store electricity prior to selling it into the power grid in order to mitigate power fluctuations.[62] These factors, along with political opponents that tend to exaggerate the true cost of wind power generation,[63] have resulted in wind power costs in Japan that do not conflate with international experience. The government estimates the cost of wind power in Japan to be ¥10–14 (US$0.10–0.14) per kilowatt hour,[64] which is over double the cost in most other nations.[65]

There is also no indication that the government is going to fully integrate the grid, which is a critical technical prerequisite for optimizing load management. It continues to support the concept of private monopolies controlling regional grids while liberalizing wholesale energy production to allow renewable energy developers better access to the grids. The trouble is that without a system for effectively integrating the regional grids, the potential for renewable energy will be constrained by transmission and distribution limitations.[66] This is exacerbated by the fact that Japan's electricity system is built around two separate platforms with different power frequencies. The Kansai (west) grid operates on 60 Hz and the Kanto (east) grid operates on 50 Hz, necessitating the installation of frequency conversion facilities.[67]

On a positive note, a new wind power technology is currently under development in Japan: the wind lens.[68] The wind lens utilizes a diffuser shroud to amplify the amount of wind funneled into the turbine. The result is a significant improvement in wind capture, estimated to be 300% greater than that of existing turbines. If this type of development continues to progress to a commercial stage, is possible that the enhanced economics of wind power will make it hard to avoid supporting wind power.

9.5 INFLUENCES ON GOVERNMENT POLICY

9.5.1 Sociocultural → Political

In many respects, insufficient critical engagement on the part of the general public toward activities of the state has been a silent enabler of nuclear power development. During the pioneering era of nuclear power in the 1950s and 1960s, government efforts to accentuate the benefits of nuclear power and downplay the risks went virtually uncontested by Japanese NGOs, media, and opposing political groups. Even during the 1990s and early 2000s, when a series of safety breaches occurred in various nuclear power plants around the nation, public alarm and media scrutiny was fleeting.[69]

Today, despite elevated public and media scrutiny of the nuclear power industry, political opposition from influential actors has crumbled as alarm over the economy casts a pall over costly initiatives, such as facilitating a transition away from nuclear power. In the 2012 Japanese election, DPJ candidate and incumbent Prime Minister Yoshihiko Noda announced a campaign pledge to make Japan nuclear free by 2030, an initiative requiring investment of nearly US$500 billion in renewable energy expansion projects over two decades.[70] Meanwhile, LDP candidate Shinzo Abe emphasized the importance of facilitating economic recovery above all else. The result was a landslide victory for the LDP. What is clear from the 2012 election is that the general public perceives Japan's economic woes to be top priority and anything that gets in the way of economic recovery (i.e., higher energy costs) needs to be heavily scrutinized. Since its election, the LDP is stealthily moving forward with plans to slowly restart the nuclear power program. Reactors which pass the new safety standards were slated to start operation again beginning in the fall of 2013,[71] but this timeline has been extended to the fall of 2014 following reports that the radioactive water leaks at Fukushima were far worse than initially anticipated.

9.5.2 Economic → Political

In addition to technological entrenchment, there is currently an enormous amount of stranded investment in Japan's nuclear power sector. Much of this investment lies dormant in Japan's idled nuclear power reactors. Although this is a cost currently borne by Japan's private utilities, which own the reactors, the government already has a controlling stake in the nation's largest utility (TEPCO); therefore, a prolonged shutdown of the nation's reactors will likely require further capital infusion by the government, at least to shore up TEPCO. If the government phases out nuclear power, another stranded investment will be the much maligned Rokkasho Nuclear Waste Reprocessing Plant that is scheduled to start operation in October 2014,[72] and which cost US$20 billion and took 20 years to complete. All of the nuclear power infrastructure—the plants, the waste reprocessing facilities, R&D facilities, and nuclear waste storage facilities—represent financial and ideological commitments that make it difficult for Japan's pro-nuclear politicians to turn away from.

Providentially, the scrutiny over nuclear power sired by the Fukushima disaster has also engendered a reevaluation of wind power as an option for bolstering domestic energy security. During the process, wind power potential estimates are inexplicably on the rise.[73] Furthermore, the government's perspective on the economics of nuclear power is being challenged,[74] and international wind power cost estimates are being introduced in energy policy circles to encourage greater scrutiny over why wind power cost estimates have been so high in Japan (to be discussed in section 9.7). Regrettably, renewable energy policy is still in the hands of the LDP, an ardent proponent of nuclear power.

9.5.3 Technological → Political

By far the greatest technological influence on political behavior stems from the technological lock that the nuclear power industry has engendered in Japan. Year after year, US$5–US$10 billion have been funneled into nuclear power research.[75] Since the beginning of the nuclear power program, the end goal was and still is the achievement of a closed loop nuclear fuel cycle, featuring fast breeder reactors and advanced waste reprocessing technologies. To date there has been little to indicate that the finish line is anywhere in sight. The nation's only fast breeder reactor has been offline for all but nine months since its completion in 1991.

The financial commitment that Japan has made to nuclear power has sired a powerful regime that continues to influence government policy. Moreover, the program is the offspring of Japan's LDP, which has governed Japan for all but four years since 1955. The amount of political capital expended and the concomitant ideological entrenchment suggests that it is improbable that LDP politicians will suddenly change direction and abandon Japan's nuclear power program.

In terms of wind power development, competing against a heavily supported nuclear power industry is not the only significant hurdle. Another key barrier has been resistance to wind power from the regional utilities. As outlined earlier, the regional utilities have developed an infrastructure that is conducive to managing large nuclear or fossil fuel-fired power plants and have exhibited resistance to the notion of connecting numerous wind power projects to the grid. Simply put, for Japan's utilities, wind power is a nuisance.[76]

Going forward, there is growing support for (or at least acceptance of) offshore wind power development. There are currently a number of projects in the works, with the future of offshore wind dependant on how the actual economics play out for these projects. Needless to say, if nuclear power is phased out, the economic appeal of offshore wind power development will certainly improve. If nuclear power makes a comeback, it will be an uphill battle for wind power developers.

9.6 POLITICAL INFLUENCES ON POLICY

9.6.1 National Political Structure

Japan's is a parliamentary, representative democracy, wherein the prime minister serves as head of government and leads a cabinet that directs the day-to-day operations of government. The prime minister is elected by Diet (Japan's Parliament) members, and therefore, must retain the confidence of the Diet's House of Representatives in order to remain in office.

The prime minister's cabinet directs ministerial policy. In other words, if the prime minister and his cabinet decided that Japan should endorse 50% nuclear power, then the only thing that could stand in the way of successfully implementing such a policy would be if a majority of members in the House of Representatives issued a no-confidence vote, forcing the ruling party to either nominate another candidate for prime minister or call a national election. In short, under this system, majority governments enjoy a considerable amount of autonomy.

9.6.2 Governing Party Ideology

The current prime minister is Shinzo Abe of the Liberal Democratic Party (LDP), which regained power in the December 2012 elections, winning 61% of the seats in the House of Representatives. As mentioned earlier, the LDP has governed Japan for all but four years since 1955. The LDP's grip on power in Japan has been so dominant that some pundits have referred to the Japanese political system as a "one and a half party" democracy.[77]

Prime Minister Abe is considered to be a conservative and a nationalist. This is his second go around as prime minister. In 2006, he was chief cabinet secretary to then-Prime Minister Junichiro Koizumi, and when Koizumi stepped down, the LDP members elected Abe as party leader. Abe's initial tenure only lasted one year. Marred by internal scandal and waning popularity, Abe's one year stint as prime minister was the first of a series of one-year prime ministerial tenures. In 2012, he campaigned on a platform that prioritized policies to spur economic growth. Since his election the stock market has responded favorably. However, Abe's foreign policy has been antagonistic toward China, which is now the nation's number one trading partner, and this is expected to adversely affect exports to China and the capacity of Japanese firms in China to operate effectively.

Within a week of taking office, Abe reversed his campaign pledge to phase out nuclear power, announcing that he supports the build-up of newer, safer nuclear power reactors. Although he has so far allowed the FIT of 2012 to continue, it is widely expected that Abe's return to power represents a boon to Japan's nuclear power industry.[78]

9.6.3 Fiscal Health

Economically, Japan is a disaster in progress. It has an aging populous; by 2050, it is estimated that one-third of all Japanese will be over 65 years of age. This trend is fueling a number of problems. First, the cost of social programs is increasing while the ratio of workers to dependants is declining. Consequently, government budgets have been squeezed to try and make up for the financial shortfall through existing tax revenues. Second, Japanese industry is becoming top-heavy. It is now laden with aging employees who exhibit comparatively low levels of productivity on a cost per employee basis. Declining organizational productivity, in conjunction with increased international competition, means that Japanese industry is far less profitable than it used to be. Since the late 1980s annual tax revenue has fallen by 30%, and the Japanese government has been forced to make up for the fiscal shortfalls through the issuance of public debt instruments. Outstanding Japanese government bonds now exceed US$9 trillion.[79]

Japan's gross debt to GDP ratio for 2013 is projected to reach 235%.[80] To put this into perspective, when the Greek economy began to collapse, its gross debt to GDP ratio was estimated at 150–170%. A high gross debt to GDP ratio means that the government must make substantial interest payments to service the debt, and this reduces the amount of funds available for program financing. In Japan's case, its high debt compounds the fiscal challenges described in the previous paragraph.

In order to counter progressive annual fiscal deficits, Japanese policymakers began studying the desirability of increasing the consumption tax from the current level of 5% to 10%. Many economic studies, including an influential review by the IMF, concluded that an increase to 10% would not succeed in restoring fiscal health.[81] Therefore, prior to the Fukushima disaster, economists in Japan were debating further initiatives to improve fiscal governance.

Aside from the social and environmental chaos associated with the Fukushima disaster, the clean-up costs and the subsequent decision to implement a moratorium on nuclear power plant operation has exacerbated an already dire fiscal predicament. Cleanup costs are conservatively estimated to be in the neighborhood of US$257 billion,[82] and this estimate was made prior to the July 2013 revelation that radioactive water leaks were far more serious than initially thought.[83] Moreover, the crippling blow to TEPCO's Fukushima nuclear complex along with the government mandated shutdown of Japan's nuclear reactors pushed TEPCO into insolvency. In 2012, the government was forced to inject US$12.4 billion into TEPCO and assume voting control over its board.[84] In 2012 alone, the government issued a record US$693 billion in government bonds, equating to 12% of nominal GDP.[85] The Fukushima disaster has placed financial demands on the government at the worst possible time.

In order to stave off the economic losses to utilities stemming from generating power with more costly natural gas-fired power plants, the government has permitted the added cost to be passed through to the end-consumers, and this has further exacerbated the government's fiscal dilemma. Many industrial power consumers were forced to endure electricity price increases of 15% or more in 2012.[86] For energy-intensive industries, this has undermined industrial profitability and might over the short term strangle corporate tax revenues during a period when the government actually needs more, not less, tax revenue.

9.6.4 Policy Regime

A prominent characteristic of LDP policy through the years has been consistency. The Japanese consensus building practice of "nemawashi" (binding the root) has permitted the LDP to carry out a number of cabinet reshuffles, replacing prime ministers on a regular basis without administrative chaos. In terms of energy policy, the pro-nuclear actors that were influencing policymaking five ago are virtually the same actors that influence policymaking today, suggesting that emergent government policy can be essentially extrapolated from government policy established five years ago. There is considerable political will to restart the nuclear power program; in doing so this provides a foundation for relatively inexpensive electricity, while concurrently reducing GHG emissions.

The only thing that has changed is that the government is now aware that the general public would currently not be keen to see the nation return to its pre-Fukushima plan of expanding nuclear power capacity to satisfy 40% of the nation's electricity needs. Consequently, the government has been forced to revisit its lackluster support for renewable energy. One of the key hurdles to supporting renewable energy in the past has been political, industrial, and general public concerns over higher energy costs. However, the electricity price increases that have ensued since the Fukushima disaster have somewhat acclimatized the end-consumer (including industry) to higher electricity costs. Therefore, the government has capitalized on this higher willingness to pay by structuring the FIT program to allow the renewable energy premium to be passed on to the end-consumer. In many ways this is a no-lose strategy, because the government does not have to directly subsidize the FIT and when the end-consumer tires of higher energy costs, the LDP will be ready to reintroduce the notion of nuclear expansion.

9.7 THE CULMINATION OF INFLUENCES

The account of Matsutaro Shoriki's campaign for building support for nuclear power reflects the modus operandi of Japan's LDP. The LDP has a long history of actively influencing the direction of economic development in the nation and embracing tactics to engender public support when support is waning.[87] In the current manifestation of this tactic, the LDP has structured its policies around the one theme that the vast majority of Japanese are concerned with—the health of the economy. Anything that gets in the way of the quest for economic recovery (i.e., higher energy costs) becomes a target for government action.

In regard to energy policy, the LDP and its industrial supporters are positioning renewable power as a technology that necessitates a choice between economic prosperity and enhanced environmental governance. Comments from the Abe administration are increasingly reverting to the old standard—nuclear power is cheaper, and therefore, a necessary evil. The success that the LDP has had with this economic recovery above all else rhetoric has forced opposing politicians to temper their support for an energy transition. Perhaps of greater concern, the economic recovery above all else mantra seems to also be permeating special interest groups and media watchdog groups. Groups that, in most countries, would be relied upon to provide a critical opposition to risky government initiatives have been comparatively moot in response to Abe's support for nuclear power expansion.

With that said, there is still a vociferous segment of grassroots opposition that has emerged post-Fukushima that is managing to air its views publicly through public demonstrations, social media and fringe media channels. Two years after the Fukushima disaster these small anti-nuclear groups have demonstrated a degree of resilience, although the number of protesters have somewhat declined. At the same time, there has been an infusion of public engagement and an elevation in public alarm since the

July 2013 admission by TEPCO that the equivalent of at least one Olympic sized swimming pool worth of polluted water has been leaking every eight days from the crippled reactors.[88] This new adverse development is putting pressure on the government to at least demonstrate efforts to diversify the energy mix and, for this reason, it is likely that the LDP will continue to support the FIT despite the added costs passed along to the consumer.

Economically, the government's staunch support for nuclear power continues to perpetuate the dissemination of distorted economic cost estimates that bear little resemblance to the economic cost estimates in other nations.

The government has consistently promoted statistics that suggest nuclear power is the cheapest form of utility-scale electricity generation (see Table 9.2). However, an investigation into the capacity factors that the government uses for estimating generation costs suggests a bias in favor of nuclear power. As Table 9.2 indicates, the Japanese government estimates the cost of nuclear power based on a capacity factor of 70 to 85%. This is a dubious claim in itself given the amount of time that Japanese nuclear power plants are shut down just for safety inspections. Moreover, these estimates do not include the billions of dollars that the government funneled into R&D for nuclear power. Nor does it include the processing and storage of nuclear waste.[89] To put these latter costs in perspective, Benjamin Sovacool, in a study based on the US nuclear reactor industry, estimated that waste and decommissioning costs alone add US$0.22 to US$0.49 / kWh to the cost of generating power through nuclear plants. In total, Sovacool estimated the cost of generating nuclear power from a new 1000 MWe nuclear plant to range between US$0.41 and US$0.80/kWh.[90] This is a far cry from the Japanese government's estimate of US$0.048–0.062/kWh.

Table 9.2 COMPARATIVE ELECTRICITY GENERATION COSTS IN JAPAN

Power source	Generation Cost (¥ per KWh)	Generation Cost (US¢ per KWh)	Capacity Factor (%)
Hydroelectric	¥8.2–13.3	US¢8.2–13.3	45
Oil-fired	¥10.0–17.3	US¢10.0–17.3	30–80
LNG-fired	¥5.8–7.1	US¢5.8–7.1	60–80
Coal-fired	¥5.0–6.5	US¢5.0–6.5	70–80
Nuclear	¥4.8–6.2	US¢4.8–6.2	70–85
Photovoltaic	¥46.0	US¢46.0	12
Wind	¥10.0–14.0	US¢10.0–14.0	20

Source: ANRE. 2008. *FY2007 Annual Energy Report*. Japan: Ministry of Economy, Trade and Industry.

Conversely, the government's cost estimate for wind power reflects a bias in the opposite direction. It bases wind power cost estimates on a capacity factor of 20%. Even back in 2007 when the estimate was generated, wind power capacity factors of 25 to 30% were the norm. If one bases an estimate on a 30% capacity factor, the wind power cost estimate in Table 9.2 would fall by 25% to ¥7.5–10 (US$0.075–0.10) per kWh. Suddenly, wind power would become a commercially competitive technology that does not produce radioactive waste that is costly (and dangerous) to manage.

The amount of sunken investment in Japan's nuclear power industry compels the LDP to continue to support development of the industry. Moreover, the belief that nuclear power is the most inexpensive energy technology makes continued support for the technology a no-brainer.

In the fall of 2014 the government will start up its Rokkasho nuclear waste reprocessing plant, which represents the next step toward its vision of a closed nuclear cycle. In order to justify the US$20 billion that has already been expended on this project, the nation will to have to commission new mixed-oxide fueled (MOX) nuclear power plants, which are capable of utilizing this processed waste. This portends an escalation in investment of MOX plants and fast breeder technology.

Continued support for nuclear power is a risky proposition should any of the existing plants suffer technological problems. Therefore, the government is working to ensure that the nuclear safety regime is improved and that the utilities which operate these plants exhibit a higher standard of due diligence in regard to ensuring safety.

On the other hand, political awareness that Japan's energy mix must be diversified and that renewable energy must play a greater role is fueling pressure on utilities to show a higher level of support for renewable energy. Currently there is an elevated level of interest in expanding geothermal and offshore wind power capacity, and these two technologies in particular can expect enhanced government support going forward.

9.8 WHAT TO EXPECT GOING FORWARD

It should be clear from the contents of this chapter that the future of wind power in Japan hinges on the future of nuclear power. A return to a nuclear power expansion agenda likely relegates wind power in Japan to a marginal, supplemental technology. A phase-out of nuclear power forces the government to embrace other commercially attractive low-carbon technologies, suggesting that wind power and geothermal power would have a significant role to play in Japan's energy mix.

The current body of evidence suggests that a nuclear power renaissance can be expected. Ideologically, the nuclear power regime is too strong and too politically connected to be ousted by a disaster brought on by a freak act of nature. Such an event is too easily dismissed as a once in a lifetime event. Moreover, the public appears to be buying into the LDP's mantra that tightening up nuclear power governance will allow the nation to enjoy cheap energy at minimal future risk.

A key factor in favor of nuclear power is the economic impact of a transition away from nuclear power. The stranded public investment, which must be amortized over time, along with the threat of insolvency to the private regional utilities, make it extremely hard for the government to support a phase-out of nuclear power over the short term. Given the dire financial challenges currently facing the Japanese government, government initiatives to finance a transition to renewable energy are politically unwise, as the ousted DPJ came to discover. This implies that any sort of transition needs to be led by private renewable energy developers supported by the private utilities, who already have a vested interest in nuclear power. Viewing this from the utilities' perspective, this situation is akin to expecting tobacco companies to voluntarily transition into the bubblegum business.

An additional economic and political barriers to a transition away from nuclear power is what to do with the nation's new waste reprocessing facilities. The government's US$20 billion investment has sired facilities that are capable of producing up to eight tons of weapons grade plutonium per year. This gives rise to a political dilemma if Japan discontinues its nuclear power program. Potential foes such as North Korea and perhaps China can justifiably question why a nation with no nuclear power program continues to process quantities of weapons grade plutonium that could conceivably build thousands of atomic bombs. Moreover, abandoning the Rokkasho investment on the verge of operational start-up represents bad politics. Opposition parties would be quick to attack the wisdom of the LDP's investment in this project if the nuclear power program were to be discontinued and the reprocessing plant mothballed.

Finally, there is one other reason that undermines the rationale for phasing out nuclear power—a phase-out would not insulate Japan from the threat of nuclear disaster. Across the East China Sea, China is moving ahead with a nuclear power program which, if it fully materializes, will be unprecedented in scale and scope. To the west and southwest, both South Korea and Taiwan boast active nuclear power programs. If a severe nuclear power mishap were to occur in any of these nations, the chances are high that Japan would be adversely affected. Japanese policymakers can be forgiven for questioning the logic of incurring the massive costs associated with a nuclear power phase-out for the sake of domestic safety when neighboring nations are continuing with nuclear power expansion programs.

So while the government waits for public opposition to nuclear power to wane, it continues on with preparations for a return to nuclear power. Concomitantly, it continues to support a renewable energy FIT program that is designed to engender animosity. The program, which boasts some of the highest FIT premiums in the world, passes along these costs to end-consumers, engendering a high degree of consumer dissonance. It also does not provide any mechanisms for coordinated, strategic development of wind power projects meaning that community opposition and utility opposition remains. Last but not least, the government has done nothing to reengineer grid management to allow renewable energy to flourish. As the situation now stands, for-profit utilities are being forced to accept stochastic wind power flows into their respective grids without any economic inducement. In the end, the government will be able to say that it made an effort; but there is a difference between making an effort and making an effort that is designed to succeed.

The true state of affairs at Fukushima may wind up being the decisive factor regarding the future of Japan's nuclear power program and the prospects for an enhanced role for renewable energy. As of September 2013, the extent of the radiation leaks was still unclear, but there is already public alarm that the radioactive material that has been leaking from the plant is far worse than the government and TEPCO have reported. This new crisis has eroded public confidence that TEPCO is capable of managing the clean-up, and increased skepticism over the government's sincerity in terms of transparently overseeing it. If the pollution winds up being even greater than currently purported or the price tag for cleaning up the Fukushima site rises significantly, the ability of the government to continue down the path of a nuclear renaissance will be severely inhibited. A nuclear renaissance necessitates an apathetic Japanese public and that is hard to manufacture when the Fukushima crisis continues to worsen.

NOTES

1. Economic policy speech by Mr. Shinzo Abe, Prime Minister of Japan, Wednesday, June 19, 2013, www.kantei.go.jp/foreign/96_abe/statement/201306/19guildhall_e.html.
2. Valentine, Scott Victor. 2013. "Wind Power Policy in Complex Adaptive Markets." *Renewable and Sustainable Energy Reviews* 19 (0): 1–10.
3. Japan Atomic Energy Commission (JAEC). 2008. White Paper on Nuclear Energy 2007. Tokyo: Japan Atomic Energy Commission.
4. Amari, Akira. 2006. "Japan: A New National Energy Strategy." *The OECD Observer*, 258/259: 6. http://connection.ebscohost.com/c/articles/24430390/new-national-energy-strategy.

5. A record of this speech in its entirety is available at www.americanrhetoric.com/speeches/dwightdeisenhoweratomsforpeace.html.

6. Reported in Kuznick, Peter. 2011. "Japan's Nuclear History in Perspective: Eisenhower and Atoms for War and Peace." *Bulletin of the Atomic Scientists*, April 13. http://thebulletin.org/japans-nuclear-history-perspective-eisenhower-and-atoms-war-and-peace.

7. Sovacool, Benjamin K., and Scott Victor Valentine. 2012. *The National Politics of Nuclear Power: Economics, Security and Governance*. Milton Park, UK: Routledge.

8. For further information on Japan's nuclear weapons program see Maas, Ad, and Hans Hooijmaijers (eds.). 2009. *Scientific Research In World War II: What Scientists Did in the War*. New York: Routledge.

9. Kuznick, Peter. 2011. "Japan's Nuclear History in Perspective: Eisenhower and Atoms for War and Peace." *Bulletin of the Atomic Scientists*, April 13. http://thebulletin.org/japans-nuclear-history-perspective-eisenhower-and-atoms-war-and-peace.

10. Aldrich, Daniel P. 2008. *Site Fights: Divisive Facilities and Civil Society in Japan and the West*. Ithaca, NY: Cornell University Press.

11. Pickett, Susan E. 2002. "Japan's Nuclear Energy Policy: From Firm Commitment to Difficult Dilemma Addressing Growing Stocks of Plutonium, Program Delays, Domestic Opposition and International Pressure." *Energy Policy* 30 (15): 1337–1355.

12. Kuznick, Peter. 2011. "Japan's Nuclear History in Perspective: Eisenhower and Atoms for War and Peace," *Bulletin of the Atomic Scientists*, April 13. http://thebulletin.org/japans-nuclear-history-perspective-eisenhower-and-atoms-war-and-peace.

13. Sovacool, Benjamin K., and Scott Victor Valentine. 2012. *The National Politics of Nuclear Power: Economics, Security and Governance*. Milton Park, UK: Routledge.

14. Ishikawa, M., M. Kawasaki, and M. Yokota. 1990. "JPDR Decommissioning Program—Plan and Experience." *Nuclear Engineering and Design* 122 (1): 357–364.

15. Ministry of Economy, Trade and Industry (ANRE). 2008. *FY2007 Annual Energy Report*. Tokyo: Ministry of Economy, Trade and Industry.

16. Cirincione, Joseph, and Jon Wolfsthal. 2004. "Producing Plutonium at Rokkasho-mura." Proliferation Analysis, October 12, http://carnegieendowment.org/2004/10/12/producing-plutonium-at-rokkasho-mura/20kp.

17. Aldrich, Daniel P. 2008. *Site Fights: Divisive Facilities and Civil Society in Japan and the West*. Ithaca, NY: Cornell University Press.

18. Sovacool, Benjamin K., and Scott Victor Valentine. 2012. *The National Politics of Nuclear Power: Economics, Security and Governance*. Milton Park, UK: Routledge.

19. More on this story can be found at www.thebulletin.org/web-edition/op-eds/reform-the-japanese-power-system-nationalize-tepco.

20. More on this story at Nakamoto, Michiyo. "Tepco Faces Revolt Over Price Rise." *The Financial Times*, April 4, 2012, www.ft.com/intl/cms/s/0/bf0affbc-7d5f-11e1-bfa5-00144feab49a.html#axzz2T8f3fVXw.

21. Pew Research Global Attitudes Project: www.pewglobal.org/2012/06/05/japanese-wary-of-nuclear-energy/.

22. World Nuclear Association: www.world-nuclear.org/info/Country-Profiles/Countries-G-N/Japan/.

23. Tabuchi, Hiroko. 2012. " Japan's New Leader Endorses Nuclear Plants." *New York Times*, December 30. www.nytimes.com/2012/12/31/world/asia/japans-new-prime-minister-backs-more-nuclear-plants.html?_r=0.

24. Amari, Akira. 2006. "Japan: A New National Energy Strategy." *The OECD Observer*, 258/259: 6. http://connection.ebscohost.com/c/articles/24430390/new-national-energy-strategy.

25. Ibid.

26. Ibid.

27. The industrial position is elaborated on in a policy note by Keidanren (Japan's association of industry) titled "Keidanren's Views on the "Options for Energy and the Environment," July 27, 2012, www.keidanren.or.jp. For an analysis, see Valentine, Scott Victor. 2011. "Japanese Wind Energy Development Policy: Grand Plan or Group Think?" *Energy Policy* 39 (11): 6842–6854.

28. Ushiyama, Izumi. 1999. "Renewable Energy Strategy and Japan." *Renewable Energy* 16 (1–4): 1174–1179.

29. Global Wind Energy Council. 2012. *Global Wind Statistics 2012*. www.gwec.net/wp-content/uploads/2013/02/GWEC-PRstats-2012_english.pdf.

30. Valentine, Scott Victor. 2011. "Japanese Wind Energy Development Policy: Grand Plan or Group Think?" *Energy Policy* 39 (11): 6842–6854.

31. CIA World Fact Book, Japan: www.cia.gov/library/publications/the-world-factbook/geos/ja.html.

32. Statistics for these nations are included in their respective chapters in this book.

33. Inoue, Yoshinori, and Kumiko Miyazaki. 2008. "Technological Innovation and Diffusion of Wind Power in Japan." *Technological Forecasting and Social Change* 75 (8): 1303–1323.

34. Federation of Electric Power Companies of Japan (FEPC). 2013. *Electricity Review Japan: 2013*. Tokyo: The Federation of Electric Power Companies of Japan.

35. Ibid.

36. International Energy Agency (IEA). 2009. *Global Renewable Energy: Policies and Measures*. Paris: International Energy Agency.

37. Japan Ministry of Economy, Trade and Industry (ANRE). 2008. *FY 2007 Implementation Status of the Law on Special Measures Concerning New Energy Use by Electric Utilities*. Tokyo: Agency for Natural Resources and Energy, Japan Ministry of Economy, Trade and Industry.

38. Englander, Dave. 2008. "Japan's Wind-Power Problem." *Greentech Media*, April 23. www.greentechmedia.com/articles/read/japans-wind-power-problem-828.

39. International Energy Agency (IEA). 2009. *Global Renewable Energy: Policies and Measures*. Paris: International Energy Agency.

40. Valentine, Scott Victor. 2011. "Japanese Wind Energy Development Policy: Grand Plan or Group Think?" *Energy Policy* 39 (11): 6842–6854.

41. Nishio, Kenichiro, and Hiroshi Asano. 2003. *The Amount of Renewable Energy and Additional Costs under the Renewable Portfolio Standards in Japan*. Tokyo: Central Research Institute of Electric Power Industry Report. http://criepi.denken.or.jp/jp/kenkikaku/report/detail/Y02014.html.

42. Statistics available at the Center for Climate and Energy Solutions. See www.c2es.org/publications/looming-natural-gas-transition-united-states.

43. Federation of Electric Power Companies of Japan (FEPC). 2013. *Electricity Review Japan: 2013*. Tokyo: The Federation of Electric Power Companies of Japan.

44. Source: Democratic Party of Japan (DPJ), www.dpj.or.jp/policy/rinen_seisaka/sei-saku.html.

45. Valentine, Scott, Benjamin K. Sovacool, and Masahiro Matsuura. 2011. "Empowered? Evaluating Japan's National Energy Strategy Under the DPJ Administration." *Energy Policy* 39 (3): 1865–1876.

46. Japan Ministry of Economy, Trade and Industry (METI). 2013. Feed-in Tariff Scheme in Japan. Tokyo: Japan Ministry of Economy, Trade and Industry.

47. For further reading see Watanabe, Chisaki. 2013. "Feed-In Tariffs Ready to Make Japan World No. 2 Solar Market after China." *Japan Times*, April 10.

www.japantimes.co.jp/news/2013/04/10/national/feed-in-tariffs-ready-to-make-japan-world-no-2-solar-market-after-china/.

48. *Wall Street Journal*. 2013. "Japan Wants Big Offshore Wind Power Expansion," March 9. www.marketwatch.com/story/japan-wants-big-offshore-wind-power-expansion-2013-03-09.

49. Valentine, Scott, Benjamin K. Sovacool, and Masahiro Matsuura. 2011. "Empowered? Evaluating Japan's National Energy Strategy Under the DPJ Administration." *Energy Policy* 39 (3): 1865–1876.

50. A discussion on the capricious nature of public opinion in Japan is provided in Penney, Matthew. 2012. "Nuclear Power and Shifts in Japanese Public Opinion." *Asia Pacific Journal*, February 13, www.japanfocus.org/events/view/130.

51. Ibid.

52. Ibid.

53. One of the organizers is the Metropolitan Coalition Against Nukes. More on their activities can be found at http://coalitionagainstnukes.jp/en/.

54. Tsuru, Shigeto. 2000. *The Political Economy of the Environment: The Case of Japan*. London: The Athlone Press.

55. Penney, Matthew. 2012. "Nuclear Power and Shifts in Japanese Public Opinion." *Asia Pacific Journal*, February 13. www.japanfocus.org/events/view/130.

56. Maruyama, Yasushi, Makoto Nishikido, and Tetsunari Iida. 2007. "The Rise of Community Wind Power in Japan: Enhanced Acceptance Through Social Innovation." *Energy Policy* 35 (5): 2761–2769.

57. All statistics taken from the CIA's World Fact Book, www.cia.gov/library/publications/the-world-factbook/index.html.

58. CIA's World Fact Book: www.cia.gov/library/publications/the-world-factbook/index.html.

59. Valentine, Scott Victor, and Benjamin K. Sovacool. 2010. "The Socio-Political Economy of Nuclear Power Development in Japan and South Korea." *Energy Policy* 38 (12): 7971–7979.

60. Federation of Electric Power Companies of Japan (FEPC). 2013. *Electricity Review Japan: 2013*. Tokyo: The Federation of Electric Power Companies of Japan.

61. See more on this story at http://rt.com/news/japan-nuclear-energy-return-854/.

62. Englander, Dave. 2008. "Japan's Wind-Power Problem." *Greentech Media*, April 23. www.greentechmedia.com/articles/read/japans-wind-power-problem-828

63. Valentine, Scott Victor. 2011. "Japanese Wind Energy Development Policy: Grand Plan or Group Think?" *Energy Policy* 39 (11): 6842–6854.

64. Japan Ministry of Economy, Trade and Industry (METI). 2010. *2010 Annual Report on Energy*. Tokyo: Japan Ministry of Economy, Trade and Industry.

65. Refer back to Chapter 2 for a discussion on wind power costs.

66. Valentine, Scott Victor. 2011. "Japanese Wind Energy Development Policy: Grand Plan or Group Think?" *Energy Policy* 39 (11): 6842–6854.

67. Federation of Electric Power Companies of Japan (FEPC). 2013. *Electricity Review Japan: 2013*. Tokyo: The Federation of Electric Power Companies of Japan.

68. This is being developed at Kyushu University. The website for the project is www.riam.kyushu-u.ac.jp/windeng/en_aboutus_detail04.html.

69. Sovacool, Benjamin K., and Scott Victor Valentine. 2012. *The National Politics of Nuclear Power: Economics, Security and Governance*. Milton Park, UK: Routledge.

70. Williams, Carol J. 2012. "In Wake of Fukushima Disaster, Japan to End Nuclear Power by 2030s." *Los Angeles Times*, September 13. http://latimesblogs.latimes.com/world_now/2012/09/in-wake-of-fukushima-disaster-japan-to-end-nuclear-power-by-2030s.html.

71. *Japan Times*. 2013. "Japan May Restart Nuclear Reactors from Fall: Motegi," April 24. www.japantimes.co.jp/news/2013/04/24/national/japan-may-restart-nuclear-reactors-from-fall-motegi/.

72. Federation of Electric Power Companies of Japan (FEPC). 2013. *Electricity Review Japan: 2013*. Tokyo: The Federation of Electric Power Companies of Japan.

73. A presentation made by the Japan Wind Power Association highlights a government sponsored wind power study in 2000 which estimated potential to be 6.4 GW. In 2010, a new study (Study of Potential for the introduction of Renewable Energy for FY 2010) estimated potential to be in excess of 250 GW. The slides to the presentation are at http://jwpa.jp/pdf/roadmap_v3_2.pdf.

74. There is an excellent critique of nuclear power economics provided by the Citizen's Nuclear Information Center at www.cnic.jp/english/newsletter/nit113/nit113articles/nit113cost.html.

75. International Energy Agency (IEA). 2004. *Renewable Energy: Market and Policy Trends in IEA Countries*, edited by International Energy Agency, Paris.

76. Valentine, Scott Victor. 2011. "Japanese Wind Energy Development Policy: Grand Plan or Group Think?" *Energy Policy* 39 (11): 6842–6854.

77. Hall, John Whitney. 1990. *Japan: From Prehistory to Modern Times*. Japan: Charles E. Tuttle Publishers.

78. Tabuchi, Hiroko. 2012. "Japan's New Leader Endorses Nuclear Plants," *New York Times*, December 30. www.nytimes.com/2012/12/31/world/asia/japans-new-prime-minister-backs-more-nuclear-plants.html?_r=0

79. Koike, Yuriko. 2012. "Japan's Fiscal Crisis Comes of Age," August 13. Project Syndicate. www.project-syndicate.org/commentary/japan-s-fiscal-crisis-comes-of-age-by-yuriko-koike. It bears noting that Yuriko Koike is Japan's former defense minister and national security adviser. She was also chairwoman of Japan's Liberal Democrat Party and currently is a member of the National Diet.

80. Harner, Stephen. 2012. "IMF Raises Alarms over Japan's Dangerous Fiscal Course." Forbes, November 16. www.forbes.com/sites/stephenharner/2012/10/16/imf-raises-alarms-over-japans-dangerous-fiscal-course/.

81. Harner, Stephen. 2012. "IMF Raises Alarms over Japan's Dangerous Fiscal Course." Forbes, November 16. www.forbes.com/sites/stephenharner/2012/10/16/imf-raises-alarms-over-japans-dangerous-fiscal-course/.

82. Saoshiro, Shinichi. 2011. "Japan Sees Atomic Power Cost Up by at Least 50 pct by 2030—Nikkei." *Reuters*, www.reuters.com/article/2011/12/06/japan-nuclear-cost-idUSL3E7N60MR20111206.

83. For more on this story see Kigar, Patrick J. 2013. "Fukushima's Radioactive Water Leak: What You Should Know." *National Geographic News*, August 7. http://news.nationalgeographic.com/news/energy/2013/08/130807-fukushima-radioactive-water-leak/.

84. Obe, Mitsuru. 2012. "Tepco Gets Bailout, but Cedes Power." *Wall Street Journal*, April 27. http://online.wsj.com/article/SB1000142405270230481130457736955 0496643114.html.

85. Koike, Yuriko. 2012. "Japan's Fiscal Crisis Comes of Age," August 13. Project Syndicate. www.project-syndicate.org/commentary/japan-s-fiscal-crisis-comes-of-age-by-yuriko-koike.

86. Ohira, Kaname, and Mari Fujisaki. 2012. "Taxpayers, Electricity Users Finance TEPCO Bailout." *Asahi Shinbun*, July 31. http://ajw.asahi.com/article/0311disaster/fukushima/AJ201207310068.

87. Sovacool, Benjamin K., and Scott Victor Valentine. 2012. *The National Politics of Nuclear Power: Economics, Security and Governance*. Milton Park, UK: Routledge.

88. For more on this story see Kigar, Patrick J. 2013. "Fukushima's Radioactive Water Leak: What You Should Know," *National Geographic News*, August 7. http://news.nationalgeographic.com/news/energy/2013/08/130807-fukushima-radioactive-water-leak/.

89. The readers are again referred to the critique of nuclear power economics provided by the Citizen's Nuclear Information Center at www.cnic.jp/english/newsletter/nit113/nit113articles/nit113cost.html.

90. Sovacool, Benjamin, K. 2011. *Contesting the Future of Nuclear Power*. Singapore: World Scientific Publishing.

CHAPTER 10

Strategic Control Over Wind Power Development Policy

If you're not confused, you're not paying attention.

—Tom Peters, author of *Thriving on Chaos*

10.1 INTRODUCTION

Chapter 3 introduced a three-step framework that could be applied to case study analysis in order to extract insights for refining wind power development policy. The first step of the framework entailed the analysis of a sufficient number of national case studies to identify prominent commonalities that influence wind power development. In this book Germany, Denmark, China, the United States, Japan, and Canada were chosen as nations for analysis. Germany and Denmark—two nations that have laudable and sustained successes in wind power development—were selected in order to provide insight into successful wind power development policies. China and the United States, which have both experienced boom and bust periods of wind power development, were picked to provide insight into factors that cause such oscillations in development. Japan and Canada, which are two nations that have underperformed in regard to wind power development, were selected to provide insight into barriers to wind power diffusion. Although only six nations were included in this study, additional wind power policy analysis undertaken by the author in Australia and Taiwan provide general confirmation of the external validity of the findings that will be summarized in this chapter.[1]

In this chapter the social, technological, economic, and political (STEP) factors that emerged as influential for either supporting or impeding wind power development in the six case study nations will be summarized. The intention of this compendium is to provide policymakers and interested stakeholders with greater clarity regarding the factors that must be strategically managed in order to enhance the scale scope and pace of wind power diffusion.

The factors introduced in this chapter should not be misconstrued as constituting a best practice list for optimizing wind power policy success. As was pointed out in the introductory chapter, energy policy is designed and implemented within a contextually unique environment that involves a seamless web of dynamically evolving forces. Consequently, the notion that it might be feasible to construct a universally applicable manual of best policy practice is a fool's errand.

The premise that is encapsulated in this compendium is that the STEP factors that are enumerated below have been empirically shown to be of relevance to wind power development policy in more than one nation. Therefore, in spite of the caveat that contextual differences will cause these factors to be more or less pertinent to the development of wind power policy in different nations, a prudent policymaker will be able to make use of the insights provided in this chapter to guide analysis of what needs to be done to optimize energy policy within his or her national context. The insights are not intended to enumerate a list of best practice; rather, the insights that are presented in this chapter are intended to help guide the analytical process to optimize wind power development.

10.2 INFLUENTIAL SOCIAL FACTORS

As should be apparent from reading the case studies, social factors are gateway forces in that they influence the scale and pace of wind power development in any given nation. When social conditions are right wind farms begin to spring up around the country,—yet, if social factors are not effectively managed, they can restrict progress. Analysis of the case studies presented in this book suggests that at least nine social factors influence the fortunes of wind power development. In this section, each factor will be critically examined in order to extract policy lessons that are manifest in these insights.

10.2.1 Social Factor 1: Community Perspectives

Research suggests that social support or resistance to community wind power projects plays a major role in enabling wind power diffusion. Of particular salience for policymakers is the understanding that community opposition or support stems from a medley of concerns or interests, which may or may not be based on factual evidence.

Research done on this topic in United States and Australia has documented numerous disparate sources of NIMBY opposition[2]—some of which are valid, and some of which are based on misperceptions.[3] For example, progress on the Cape Wind project in Nantucket Sound in the United States has faced stiff resistance from various stakeholder factions. There have been concerns expressed by environmental groups, fishermen, business groups that rely on tourism, householders who are concerned about the erosion of land, and residents who rue the debasement of seaside vistas. Although there is very little empirical evidence to support a claim that wind power development causes long-term economic damage to a host community, when it comes to the success of wind power projects, reversing negative perceptions can be as intractable as resolving actual problems.[4]

Consequently, the lesson to policymakers should be clear—the fundamental drivers of public sentiments must be clearly defined in order to devise strategies to minimize NIMBY resistance.[5] Denmark and Germany have exhibited exemplary practice in managing community perception of wind power by pursuing transparent, inclusive siting policies that strive to engage stakeholders, correct misperceptions, clarify concerns, and identify solutions through participative problem solving.

10.2.2 Social Factor 2: Information Asymmetry

As the previous paragraph implied, wind power development can be derailed by misperceptions. Research indicates that criticism attributed to wind power systems such as turbine noise, shadow flicker, threat to avian mortality, degradation of land value, and even aesthetic impairment is frequently overstated, because critics typically fail to acknowledge solutions which have been put forth to mitigate such problems. Similarly, support for wind power development that is based on inaccurate perceptions can also lead to community dissonance. Therefore, overselling the benefits of a wind power project should also be avoided.

The lesson for policymakers and project developers is that correcting misperceptions is a critical aspect of optimizing wind power diffusion. Project planners are well advised to communicate with stakeholders frequently, accurately, and comprehensively. Stakeholder engagement is not an activity that either policymakers or project developers typically excel at; however, if one wished to minimize community resistance to a given project and reduce any negative backlash directed broadly at wind power development policy, a strategy to overcome misinformation should be a precursor to the site planning process.

10.2.3 Social Factor 3: Civic Activism

Depending on the focus of public engagement, civic activism can be either a positive or negative force for wind power development. Therefore, it is useful to try and understand the dynamics that influence civic activism.

Generally speaking, there is evidence that civic activism tends to become more institutionalized as a society becomes more affluent, and this engenders higher levels of effectiveness.[6] However, there are enough exceptions to the rule (i.e., Japanese public acquiescence over nuclear power and Canadian public acquiescence over the nation's high carbon footprint) to suggest that forces other than affluence also influence civic activism.

Insights from the case studies suggest that forces arising from social norms, political structure, bureaucratic management of environmental problems, national economic structure, and settlement patterns all conspire to shape the scale and scope of civic activism in a given nation. For example, on one end of the spectrum there is a cultural norm that inspires citizenry in Germany to embrace civic activism and proactively participate in numerous special interest groups. On the other end of the spectrum, citizenry in nations that exhibit a more autocratic approach to governance (such as China) tend to be far less willing to engage in public activism, and when they do engage, they tend to do so in a stealthy manner. In other nations—such as Japan, Canada, and Denmark—citizens tend to cede governance to bureaucrats and oversight to NGOs until there is evidence that the bureaucrats have failed in their fiduciary duties in a particular thematic area. Then activism tends to heat up. This suggests that in most nations, civic activism is not a static notion—it can be conspicuous in one area and nonexistent in another.

The main lesson that policymakers should draw from these observations is that wind power development policy must be malleable enough to respond to emergent social concerns. As is the case in Germany, where citizens from states with high concentrations of wind power capacity have begun to protest wind power development, policymakers must be sensitive to the impact that wind power diffusion is having on public opinion.

Similarly, policymakers should also be on the lookout for emergent opportunities to enhance public support for wind power. For example, in all six of the case study nations, public alarm over the environment threats posed by climate change and widespread aversion to the risks associated with nuclear power have opened a window of opportunity for wind power development in these nations—a window of opportunity that can be exploited by tapping into these emergent public sentiments.[7]

10.2.4 Social Factor 4: Habitat Patterns

Insights from the case studies appear to suggest that support for wind power development tends to be inversely related to population density but support is contingent upon an economic benefit being conveyed to community members. For example, in the United States wind power is strongly supported in many rural farming communities, because farmers see wind power as an added revenue source. Similarly, in Denmark and Germany rural communities are generally supportive of wind power development, provided that there are clear local economic benefits. In the absence of local benefit, opposition tends to rise.[8]

Research also indicates that community sentiments toward local vistas play an influential role in wind farm acceptance. For example, in rural communities in Japan ensuring community benefits from wind systems does not guarantee community acceptance, because small Japanese villages exhibit a high level of sensitivity toward projects that might alter traditional vistas.[9] Similarly, in the United States offshore wind power development faces stiff opposition in many rural maritime communities, due to concerns over aesthetic impairment to community vistas.

The main lesson for policymakers and project developers in regard to habitat management is that wind power projects are best planned in remote locations; however, this is not in itself sufficient to ensure community acceptance. Projects need to benefit the local community and need to be planned in an inclusive manner to allow all community stakeholders the chance to voice concerns. Although project developers may be able to avoid such a time-consuming, contentious activity by simply contracting with a private farmer to lease land, in the long run, failure to seek collaborative siting of wind farms can result in greater siting restrictions in the future. This also highlights the desirability of working with local governments and municipal stakeholders to identify prospective sites which are favorably predisposed to hosting wind power projects. In short, cobbling together wind power development strategies based on wind power potential studies alone is a recipe for civic opposition.

10.2.5 Social Factor 5: Affluence

The case studies introduced in this book provide evidence that the level of affluence of a community or nation can influence wind power development prospects in at least three ways. First, as highlighted earlier, increased affluence tends to engender enhanced levels of public activism in regard to environmental governance. This was apparent in regard to Denmark, Germany, the United States, and to a lesser extent in Canada and Japan. In fact, there are even indications that public activism is on the rise in China, where pollution has become such a bane to communities that members of the general public are beginning to rally behind calls for tougher pollution control laws.[10]

Second, affluence fuels a higher willingness to pay for enhanced environmental governance. Fundamental support for this observation is found in the Kyoto Protocol. The Kyoto Protocol contains a declaration made by Annex I nations acknowledging a responsibility to make the first financial commitments to greenhouse gas emission abatement.[11] Meanwhile, leaders of non-Annex I nations, such as Hu Jin Tao, have been quoted as vociferously resisting legally binding commitments under the pretext that developing economies cannot afford the higher energy costs that such commitments would engender.[12]

A final way in which affluence can influence wind power development is antithetical to the first two observations—increased affluence can give rise to diluted levels of civic activism in certain thematic areas. In impoverished communities or nations, environmental problems are visible threats that incite civic activism when they become severe enough. Conversely, in affluent communities or nations the most invasive environmental problems have been addressed so the problems are no longer in the public eye. Exigency to abate greenhouse gas emissions (GHG) is more apparent to nations such as China, where air pollution is a visible problem, than it is in a nation such as Canada where GHG emissions are an invisible environmental threat.

The lesson imparted by these observations is that affluence tends to foster positive perceptions of wind power as a cleaner form of energy generation, but affluence will not necessarily engender proactive public advocacy for wind power development. In order to mobilize public support for wind power development, impassive public attitudes must be reversed by shedding light on the "invisible" threat posed by excessive GHG emissions.

10.2.6 Social Factor 6: Uncertainty and Change

Evidence from the case studies appears to suggest that two phenomenon that relate to decision making under conditions of uncertainty impede the transition from conventional energy to renewable energy technologies. First, the general public tends to discount the risks associated with incumbent technologies. The enormous costs associated with nuclear power in Japan and coal-fired power in China and Canada have been put forward as key justifications for facilitating a transition away from conventional energy sources. However, clearly these costs were significantly underestimated when the technologies were becoming entrenched in these nations; otherwise, the investments would not have been made in the first place.

Second, the general public tends to add a risk premium to new technologies. As an example, the blossoming wind power program in Canada's province of Ontario has suddenly encountered severe public resistance stemming from fears that the noise of turbines can have adverse psychological repercussions for humans and animals. These fears continue to be perpetuated despite studies released by the Canadian government that such fears are unfounded.[13]

These two observations are broadly supported by research from Nobel laureate Daniel Kahneman and colleagues who contend that an endowment effect exists which causes individuals to place a higher value on things that they already possess.[14] This can partially explain why the majority of citizens in nations such as Japan and France continue to support nuclear power in spite of the global alarm caused by the Fukushima disaster. It does not, however, explain why a nation like Germany—in contrast to Japan and France—would turn its back on nuclear power.

Shedding light on this anomaly requires deeper understanding of other forces that influence risk assessment. In addition to the endowment effect, risk assessment is influenced by factors such as affluence, the perceived availability of substitute technologies, education, media practice, and social norms. Affluence tends to make societies more risk-averse, because there's more to lose. The perceived lack of substitute technologies makes groups more risk-tolerant, as exemplified by Japan's continued support for nuclear power. Education can either enhance risk tolerance or enhanced risk aversion toward a given technology, depending on the nature of knowledge being imparted. The manner in which the media reports on issues influences risk perception. Liberalized media markets exhibit a greater diversity of opinion, thereby better informing the general public as to the risks of each technology. Social norms also influence a society's perspectives on risk. As Geert Hofstede postulates, some societies are simply more risk tolerant than others.[15]

The key insight for policymakers in regard to analyzing public support for incumbent and competing energy technologies is that a host of complex factors conflate to influence risk perception. For example, in risk-averse Japan, prior to the Fukushima disaster nuclear power risk tolerance was cultivated through a conservative media that rarely contested government policy, three decades of disaster-free nuclear power operation, and government assurances over the safety of the technology. In contrast, in Germany, after the Fukushima disaster, the nation's liberal media tapped into already prevalent social concern regarding nuclear power and catalyzed a complete about face in regard to nuclear power policy, despite the technological lock engendered by preestablished nuclear power facilities.

These insights suggest that policymakers can more effectively catalyze a transition away from conventional technologies by comprehensively communicating the risks inherent with incumbent technologies and the benefits (and low-risk profile) of wind power as a substitute technology.

10.2.7 Social Factor 7: Vocational Influences

Evidence extracted from the case studies suggests that there are two vocational elements that can influence wind power development policy. The first element is the level of employment provided by each energy technology sector in a given nation or community. There can be an elevated degree of public opposition to wind power technology in nations (or communities) where the extraction of fossil fuel resources has historically been a significant employer. Examples of this were documented in the case studies pertaining to Canada, the United States, and China. The common conduits for public opposition in this regard are labor unions, which can gain the ear of policymakers and, in extreme cases, organize civic protest. This is because the ascendance of an energy technology such as wind power tends to come at the expense of the incumbent technologies, potentially resulting in transitional job losses.

The second vocational element that can influence the effectiveness of wind power development policy is the general level of vocational aptitude in a given society. As Chapter 1 outlined, on a per kilowatt hour basis wind power employs far more people than conventional technologies do. However, occupational competence is necessary in order to take advantage of the job creation benefits attributed to wind power. In nations such as Denmark and Germany, technological competence is embedded in small workshops staffed by experienced craftspeople who are well-suited to carrying out the types of activities necessary to support an emergent wind power manufacturing sector.

The lessons that policymakers can draw from this are twofold. First, policy designed to enhance wind power diffusion policy in nations where conventional energy has been a key employer must anticipate and manage transitional losses to avoid fueling public opposition. Second, in order to optimize the employment benefits associated with wind power development, policy must be designed to support the acquisition of skills that are necessary to fully harness the job creation opportunities embedded in the wind power technology supply chain.

10.2.8 Social Factor 8: Agricultural Sector Structure

There is considerable evidence that the structure of a nation's agricultural sector can influence support for wind power development. Wind power projects engender much higher levels of public acceptance when local farmers can invest in wind power projects, either individually or as part of a community cooperative. This largely explains rural support for wind power development in Denmark, Germany, and more recently, in the United States. In fact, research indicates that the mere opportunity to invest in a community wind power project can engender support for a project, even from community members who do not decide to invest.[16] Conversely, evidence from both Germany and Denmark indicates that wind power development policy that centers on the proliferation of large-scale, corporately owned wind farms tends to foster community opposition.

The obvious lesson for policymakers and project developers is to ensure that mechanisms exist to enable individuals or community groups to financially benefit from the projects and to ensure that local contractors enjoy a preferred standing in regard to providing supply chain support for wind projects that are constructed in the community.

10.2.9 Social Factor 9: Environmental Impact of Energy Mix

The case studies reveal useful insights into the relationship between the environmental impact of a nation's chosen energy mix and public attitudes toward energy. The main tenet is that when externalities associated with energy technologies are more visible, public opposition toward the offending technology is enhanced. In regard to wind systems, this can be a double-edged sword. On the one hand, in communities that host nuclear power plants or where the particulate matter emitted from coal-fired power plants is a visible blight, one can expect enhanced support for wind power development provided that the development actually helps attenuate the visible environmental threat. On the other hand, the sudden appearance of wind turbines can be seen as a defilement of the aesthetics of a community. In many industrialized nations, conventional electricity plants are built with technologies which minimize visible pollution and are tucked away in remote locations. Consequently, in many industrialized nations the emissions and risks associated with these advanced conventional electricity generation systems are concealed from public scrutiny. Under such circumstances, a proliferation of wind turbines can be seen as a negative development.

The lesson for policymakers and wind power developers, particularly in industrialized nations, is to ensure that the public is informed of the array of environmental risks associated with each energy technology. Nevertheless, this might not be enough to foster support for wind power in communities where the desire to preserve vistas is high. As the German case study illustrated, NIMBY opposition is now emerging in some German states where wind power capacity is nearing 40% of total electricity generation capacity. This is in spite of the high levels of environmental awareness that exists among the German populace. There is evidence that even pro-wind communities have limits beyond which wind projects become an esthetic concern. Dispersing developments to ensure that communities are not overwhelmed by the sight of wind farms at every turn should be an integral part of site planning.

10.3 INFLUENTIAL TECHNOLOGICAL FACTORS

Technological factors tend to frame the scale and pace of wind power development. The comparative state of technology defines what level of development is both acceptable and feasible over a given time frame. The research presented in this book indicates that there are at least seven technological factors that influence the fortunes of wind power. In this section, each technological factor will be critically examined from a policy perspective.

10.3.1 Technological Factor 1: Grid Resilience

Grid resilience is an important issue when it comes to wind power development, because the capacity of grids to accommodate stochastic electricity streams generated by wind power becomes a technological bottleneck at higher levels of installed capacity.

Electricity grids are not created equal. Each nation possesses an electricity transmission and distribution network that exhibits unique characteristics, which engender conditions of either higher or lower resilience. For example, as Chapter 7 described, the United States grid is a patchwork of regional, municipal, and custom-purpose electricity grids, some of which are publicly owned and some of which are privately owned. A federal body

(FERC) regulates the transmission of electricity across this network. As Chapter 6 explained, electricity grids in China are 100% state-owned, and the two main state-owned utilities in China enjoy a monopoly in their zones of coverage. However, these utilities have to gain approval from the central government for electricity price increases and to secure funds for grid expansion. As Chapter 8 illustrated, the electricity grids in Canada are largely owned and operated by the provinces, although a few provinces have privatized some of the transmission and distribution services. Provinces negotiate with other provinces when it comes to sharing power flows.

Insights gleaned from the case studies suggest that there are at least four features of electricity grids that influence the capacity to accommodate high contributions from wind power. The first feature is the nature of ownership. Public ownership over transmission and distribution (T&D) networks tends to enhance grid resilience when compared to private T&D models. This is because the profit motives of private T&D firms deter such firms from making grid fortification investments that may provide public benefits, but do not add to the bottom line. The second feature that influences the capacity of the grid to accommodate enhanced levels of wind power is coverage. Generally speaking, the larger the grid, the higher the resilience. This is because larger grids have higher levels of load balancing slack capacity built into the system and this enables higher contributions from wind power without adding backup power. The third feature is system sophistication. On the one hand, some national grids are governed by sophisticated load balancing systems, which are technologically able to accommodate stochastic electricity flows more effectively. On the other hand, other national grids—like the US network—are aging, poorly coordinated, and susceptible to system failure. Integrating high levels of wind power into such systems further reduces network stability. The fourth feature that enhances the resilience of electricity grids is access to neighboring grids. As was evident in the Danish and German case studies, the EU grid provides a safety buffer and permits high levels of installed wind power capacity in these nations. During periods of wind power surplus the EU grid acts as a clearinghouse for selling off surplus capacity, and during periods of wind power deficiency the EU grid can help balance the supply load at a fraction of the cost of domestic backup systems.

These insights underscore three key lessons for policymakers who are interested in structuring a national grid to accommodate high levels of installed wind power capacity. First, the T&D infrastructure should be unified and preferably state-owned. This ensures that load balancing spare capacity is aggregated and available for balancing flows when needed. Second, expanding T&D capacity between neighboring grids provides added grid security when accommodating higher levels of wind power. Third, policymakers must ensure sufficient investment in both generating capacity and T&D infrastructure. In many nations, investment in T&D infrastructure tends to lag behind power generation capacity expansion and this inevitably leads to the problems experienced in the United States and China where delays in connecting wind projects to the grid result in stranded generating capacity or where unacceptably high levels of power are lost during the T&D process.

10.3.2 Technological Factor 2: Energy Mix

As should be apparent from the case studies, energy mix plays an instrumental role in a nation's capacity to accommodate higher levels of wind power without adding generating capacity. As the Canadian case study suggested, 30 to 40% wind power contribution is viable given that 62% of the nation's electricity is currently produced by hydropower, which is the most responsive of all utility-scale electricity generation technologies. Conversely, in Japan, prior to the Fukushima disaster dominant contributions from nuclear power and coal-fired power hindered Japan from incorporating high levels of wind power capacity without adding backup capacity or storage. One of the few promising developments to come out of the Fukushima nuclear disaster is that the Japanese energy mix has changed—natural gas-fired power systems have increased. As a result, Japan's electricity network is now able to accommodate much higher contributions from wind power due to the responsiveness of natural gas systems.

The lesson for policymakers is that in order to accommodate higher amounts of wind power capacity, a nation's power mix needs to be strategically managed in order to minimize costs. The starting point should be maximizing hydropower capacity. The next step should be to replace base-load technologies (coal-fired power plants and nuclear power plants) with peak-load technologies (natural gas and geothermal power). This will not only provide much-needed flexibility for balancing stochastic power flows, it will also enhance economic stability because the capricious nature of coal and oil prices are emergent threats to energy-intensive industry.

10.3.3 Technological Factor 3: Technological Regimes

The economic and political might underpinning dominant energy technologies can engender technological lock—a degree of entrenchment that is highly impervious to competitive market entry efforts. Insights from the case studies suggest that this is particularly evident in nations that have strong nuclear power programs. For example, in Japan, regional utilities have invested heavily in nuclear power. The financial and operational commitment made to nuclear power engenders a high degree of resistance to change and impedes independent wind power providers from gaining a competitive foothold. Similarly, in China, the coal industry is still a source of employment and tax revenues in many provinces. This provides economic and political rationale for maintaining coal-fired power plants, despite widespread knowledge that coal-fired power is a major contributor to the pollution problems that plague the nation.

Technological lock should be assessed separately in an analysis of a nation's energy mix because some nations exhibit energy profiles that are dominated by one or two technologies but the political and economic might of the firms that support the technologies is not sufficient to engender technological lock. For example, in Germany, nuclear power firms have been on the defensive for so many decades that the ability to preserve market share through political pressure is limited. In Denmark, the nation boasts reserves in natural gas; however, the firms within this industry are not strong enough to engender a degree of technological lock in regard to natural gas-fired power.

Accordingly, when assessing the threat of technological lock, policymakers need to look at the individual and collective capacity of the dominant technologies to influence market developments. If there are powerful stakeholders that are capable of insulating markets from competitive challenge, policymakers must be prepared to initiate policy to open doors for competitive activity. This may mean advocating market push policies (renewable portfolio standards) rather than market pull strategies (feed-in tariffs).

10.3.4 Technological Factor 4: Distance to the Grid

In each of the case studies presented in this book, the economic and technological challenges of transmitting power from wind farms to demand centers were highlighted as key challenges for policymakers. In the case of China, it was noted that some operational wind farms have been unable to get the electricity generated to the grid. In Germany, policymakers face the costly challenge of trying to harness wind power potential from the north and deliver it to demand centers in the south. In Denmark, the nation is now grappling with substantial costs for connecting offshore wind farms to the electricity grid. In all cases, there is a shared challenge—many of the most attractive wind sites are in remote locations and must be transmitted back to demand centers, thereby amplifying the cost of wind power.

The applied insight is that strategic network planning can yield significant savings in terms of connecting wind farms with the national power grid. The Danish case illustrates how strategic planning at the national level goes beyond identifying wind power potential. Effective strategic planning looks to earmark sites that tie effectively together and avoids areas where community opposition is more likely.

10.3.5 Technological Factor 5: Nationwide Potential

Estimates of realizable wind power potential influence the level of political enthusiasm for supporting wind power. As the Danish and German case studies illustrated, a key rationale for robust support for wind power stems from comprehensive national wind power potential studies which indicate that, if fully realized, wind power could make a major contribution to the national energy supply. On the other hand, as the Japanese case study exemplified, estimates of relatively low wind power potential (in comparison to national electricity demand) tend to fuel political apathy toward wind power development. However, as the Japanese case study further highlighted, wind power potential estimates tend to exhibit political bias. Depending on the vested interests of the parties undertaking a survey, wind power potential estimates can vary significantly. In a recent case study done on wind power in Taiwan, it was discovered that competing wind power potential estimates put forth by the national utility and a prominent wind power developer deviated by 360%, suggesting that the nature of assumptions that go into realizable wind power potential estimates can significantly alter the results.[17]

The main lesson that policymakers should glean from this evidence is that whenever wind power potential studies are commissioned, policymakers should be sure to ensure that the estimate's assumptions are clearly explicated and vetted for accuracy. Critical elements that influence wind power potential estimates include assumptions relating to: i) the size of wind turbines being installed, ii) capacity factor estimates, iii) the spacing of wind turbines within wind farms, and iv) land use assumptions.

10.3.6 Technological Factor 6: Electricity Sector Structure

The manner in which the electricity sector is structured directly influences the ability of wind power developers to sell wind power into the grid. Under extreme situations, when utilities are monopolies, the prospects of wind power development depend significantly on the perspective held by the monopoly. For example, the Japanese case study demonstrated how the nation's regional utilities enjoy monopoly positions and have been obstructive of policy efforts to establish even minor market footholds for wind power. In the Canadian case study, provincial monopolies which also control generation facilities in many provinces has ceded the fate of wind power development to the whims of bureaucrats working within the utilities. Similarly, in the Chinese case study, government control over the electricity sector resulted in limited prospects for wind power until about 2001, when national leaders finally began to get serious about expanding wind power development—catalyzing a market boom.

This evidence suggests that two market reforms are most likely to enhance wind power development. First, the electricity grid should be nationalized as a public service but laws should be created to allow any form of electricity generation to be sold into the system, provided that the electricity flows meet certain production standards. However, development of production standards must be technologically neutral. For example, some nations have bidding systems that require generators to contractually agree to deliver a fixed output of electricity at a future time. This disadvantages renewable energy technologies because renewable energy power flows are more stochastic and cannot be controlled in the same way that coal fired-power plants can control energy output.

Second, the electricity generation function should be fully liberalized in order to encourage enhanced competition. The government body that oversees operation of the nation's electricity grid should not possess electricity generation infrastructure because this gives rise to conflicts of interest and engenders technological lock. Rather, the national grid operator should focus on developing smart grid infrastructure to effectively balance flows from various technologies and from numerous generators. By doing so, grid resilience is assured and national energy security is bolstered by diversifying and decentralizing supply.

10.3.7 Technological Factor 7: Technological Network Links

Research indicates that political decisions made regarding the nature of links between industry, academia and policymakers influence whether collaborative links benefit or disadvantage wind power developers. At one extreme, government policy that is biased toward energy technologies other than wind power can potentially create a well-insulated regime that hinders market entry for wind power developers. Perhaps the best example of this is in Japan, where government support for nuclear power has been so strong that renewable energy technology firms have found it difficult to make market inroads, despite Japan being a world leader in a number of renewable energy platforms.[18] At the other extreme, government policy that is biased toward wind power can foster links between academia and industry which is instrumental in maximizing the benefits of a transition to wind power. Perhaps the best example of this is wind power development in Denmark, where government efforts to nurture stakeholder collaboration spawned the development of a wind turbine manufacturing sector, fostered strategies for attenuating community siting concerns and engendered investment schemes to broaden ownership of wind farms.[19]

It appears from research that the structure of a nation's energy market plays a substantial role in determining which energy technology networks receive the most government support. In markets where conventional energy firms are large and profitable, the network of advocates in support of preserving the status quo tends to enjoy political and economic clout. A prominent example of this is in the United States, where America's fossil fuel lobbyists have successfully managed to ensure that a lion's share of public energy R&D funding are consistently channeled to fossil fuel energy projects, such as clean coal or carbon capture and sequestration (CCS). This is also true in Japan of its nuclear power industry, which has annually enjoyed billions of dollars of public R&D support.

The lesson for policymakers is that successful diffusion of any new energy technology—including wind power—is enhanced by government efforts to nurture a diverse network of actors, who work together to address and resolve emergent problems. In Denmark's case, it has been contended that government sponsored initiatives to accurately map wind power potential, enlist the support of municipal government planners, encourage the transfer of technical knowledge between academics and entrepreneurs, and encourage stakeholder collaboration to resolve emergent problems were the principal factors underpinning the success of wind power development in the nation.[20]

10.4 INFLUENTIAL ECONOMIC FACTORS

The case studies indicate that economic factors have by far the greatest sway over the success of a nation's wind power development efforts. This should not be surprising in a world that is driven largely by neo-classical economic theory. The research presented in this book indicates that there are at least seven economic factors that influence the effectiveness of wind power development policy. In this section, each economic factor will be introduced and critically examined from an applied policy perspective.

10.4.1 Economic Factor 1: Internalization of Externalities

The term "externalities" refers to costs or benefits associated with the use of a particular energy technology that are not incorporated into the end-cost of electricity generated by the technology in question. Every technology produces hidden costs and benefits that somehow must be factored into the end-price if one is to equitably compare the true cost of each technology. For example, in some communities wind turbines are perceived to defile the community's aesthetics, and as such pose a cost in terms of hindering personal enjoyment of a community's vistas. On the other hand, wind power systems mitigate greenhouse gas emissions when compared to most other conventional technologies, and as such sire a benefit that should be factored into cost comparisons between wind power and other technologies. As another example, coal-fired power plants emit a number of atmospheric pollutants which adversely impact individuals and communities, yet these costs are rarely incorporated into the overall price of coal-fired electricity. These two examples highlight an important insight regarding externalities and electricity technologies—typically wind power provides as many external benefits as external costs, whereas for most of the conventional energy technologies, external costs outweigh external benefits. The implication is that when externalities are internalized into the various energy technologies, the economic viability of wind power is enhanced.[21]

The obvious lesson to draw from these findings is that an effort should be made to internalize all externalities associated with each technological platform in order to ensure economic comparisons are comprehensive and equitable. The problem is that many externalities are hard, if not impossible, to quantify in a noncontentious manner. For example, particulate matter (PM) emitted from coal-fired power plants in China is considered to be the chief culprit behind 1.2 million premature deaths caused by air pollution in China in 2010.[22] How does one put a price on 1.2 million human lives in order to calculate the unaccounted for mortality costs caused by coal-fired power in this regard? As another example, CO_2 emitted from fossil fuel-fired power plants is contributing to the accumulation of greenhouse gases in the atmosphere. However, in order to estimate the contribution that one tonne of CO_2 is making to climate change, one must first estimate the total cost of climate change, which is currently a contentious exercise.

10.4.2 Economic Factor 2: Subsidies to Energy Technologies

Conventional electricity technologies in any given nation become entrenched partly through government support. Consider coal-fired power is a case in point. In most nations, governments have provided infrastructure for transporting coal to the plants (i.e., rail links), subsidized coal procurement, and provided T&D infrastructure to get the electricity from the plant to the electricity grid. Nowadays, it is not unusual to see government research funds allocated to clean coal or CCS research. Consequently the actual cost of adopting coal-fired power technology is underreported, creating a false economy. The nuclear power industry is even worse than the coal-fired power industry in terms of hidden government subsidies. If the government wasn't responsible for nuclear power R&D, nuclear storage and underwriting the threat of nuclear disaster, it is likely that nuclear power plants would not be built in any nation because the costs would render the technology to be commercially unviable.[23]

Comparatively, wind power has not received the same level of support as fossil fuel power and nuclear power technologies. This suggests that while many of the conventional technologies have now matured (where cost declines associated with technological innovation slow down considerably), wind power is entering the early stages of commercialization; as such, wind power should experience comparatively greater cost reductions (on a percentage basis). However, wind power manufacturers are still playing a costly game of catch-up to try and match the cost profiles of some of these conventional technologies that have enjoyed decades of government subsidization. In order to accelerate the transition from early commercialization to technological maturity, government support for wind power should be at least as robust as government support for conventional technologies. Remarkably, at this juncture in time, this is still not the case in many nations. As this book documented, in Japan, Canada, the United States, and China conventional technology firms still receive far more government support than wind power technology firms receive.

The lesson for policymakers is clear: in order to expedite a transition away from carbon-intensive energy technologies and nuclear power, policymakers must eliminate conventional fuel subsidies and ramp up financial support for wind power and other renewable energy technologies in order to even the competitive playing field.

10.4.3 Economic Factor 3: Entrenched Investments

Conventional electricity generation plants are expensive capital assets that represent sunken costs once they are built. Therefore, once a plant is built, there is strong incentive to extend the life of the plant for as long as possible in order to avoid undertaking new capital costs. Moreover, some governments prefer to amortize capital investments of this type by issuing bonds which are then retired over time. The decision to decommission a power plant before it has been fully amortized results in a continuing debt obligation for an asset that is no longer of use. This places strain on a government's fiscal budget. Take for example Japan's current dilemma. It has been suggested that phasing out Japan's nuclear power program will force four of Japan's 10 utilities into insolvency.[24]

Research suggests that if a nation's energy mix is dominated by aging assets (as is the case in the United States), there is greater propensity for a rapid uptake of wind power because developmental opportunities stem from both replacing aging assets and supply expansion initiatives. On the other hand, in nations such as China, where new coal-fired power plants have been constructed at a pace of one per week for the past decade,[25] the opportunities for wind power development stem more from supply expansion initiatives than replacement of aging assets.

The sticky influence of sunken investments can be amplified when a public or private monopoly plays a role in both grid operations and power generation. Under such an industry structure, even if wind power holds the greatest economic appeal, the transitional pace will be curtailed until the monopoly decides to begin replacement of aging assets. The case studies provide evidence that this is the case in Japan and certain provinces in Canada. In nations where the power generation market has been liberalized, competition plays a more influential role in determining which power generation assets sell power into the grid. Under liberalized market conditions, the pace of wind power development may be less influenced by the age of existing infrastructure because the grid operator has no financial stake in extending the lives of older assets.

These insights tell us that financial conditions in support of a transition away from carbon-intensive technologies differ based on industry structure and the age of generating assets. China and many of the other rapidly developing economies that have recently been adding huge blocks of coal-fired electricity generation capacity will be hard-pressed to expedite a transition away from carbon-intensive electricity generation over the next 20 to 30 years. There is strong financial incentive to allow the existing assets to operate until the end of their useful lives. On the other hand, the United States is currently facing enormous costs for upgrading its aging electricity generation and transmission infrastructure. Therefore, the exigent nature of America's refurbishment challenge actually means that the United States could transition away from carbon-intensive technologies at a much faster pace.

The lesson for policymakers is that bold strategies designed to expedite a transition away from carbon-intensive technologies are rarely achievable in practice. Sunken investment in electricity generation infrastructure is like a hog in a python—it has to work its way through the digestive track before the python can contemplate further activities. Policy decisions that are being made today typically will not have much effect for at least one decade from now. Therefore, the importance of discouraging investment in conventional energy cannot be overstated. It is easy to put off making hard decisions for decades and simply allow conventional technologies to serve as interim measures, but this engenders a degree of technological entrenchment that is hard to break.

It is for this reason that many wind power advocates point to the decoupling of T&D activities from electricity generation as a necessary structural reform for optimizing wind power diffusion. Management of the grid should be considered to be a public good because users should not be excluded from it and assigning property rights to the grid diminishes resilience and economies of scale. As such, nations that wish to balance grid resilience, affordability, system reliability, and unfettered access can benefit from nationalizing the grid service function. On the other hand, electricity generation should be considered as a private good because assigning property rights to electricity generation can improve system affordability. Private goods are best left to competitive markets.[26]

10.4.4 Economic Factor 4: Competitive Health of Firms in the Energy Sector

Insights from the case studies suggest that wind power development in a given nation is influenced by both the financial health of competing technology firms and the financial health of domestic wind power firms. In markets where there are private conventional energy firms that enjoy a strong market foothold, the threat that wind power development poses to corporate profitability encourages defensive responses that can derail wind power development. Private, well-funded firms tend to react to stiff competition by adopting marginal cost pricing strategies which make it difficult for new competitors to gain market share.[27] The case studies provide evidence of this being undertaken by the nuclear power industry in Japan and the coal and oil industries in the United States and Canada. Conversely, publicly owned conventional energy firms are guided by a broader array of strategic objectives and rarely take competition to this level.

Two other strategies that private conventional energy firms have been known to adopt to deter competition include funding lobby groups to influence political support and funding special interest groups to provide research-based advocacy for preservation of the status quo. Naturally, none of this is very effective if the conventional energy firms do not possess the financial might to substantially impact market dynamics.

The existence of a domestic wind power manufacturing sector also influences wind power development prospects in a given nation, because a competitive advantage of wind power is a superior job creation profile when compared to conventional energy firms. If a nation lacks wind power system manufacturing activity, the appeal of the technology and the benefit to the nation is lessened.

The takeaway from this is that the evolution of energy systems occurs in a dynamic, competitive environment where a key goal is typically to make it difficult for competitors to gain market share because elevated market share engenders economies of scale that enhance profitability. Therefore, policymakers should be aware that conventional energy firms will attempt to paint their technologies in a favorable light through biased research. Wind power advocates must be prepared to counter such claims.

10.4.5 Economic Factor 5: Community Economic Benefits

The case studies clearly suggest that communities which benefit economically from wind power development are most likely to support further capacity expansion. Conversely, NIMBY opposition is most likely to arise in communities that do not stand to gain economically from wind power development.

Given these insights, the challenge for wind power policy advocates is to craft proposals to ensure that communities benefit from wind power development in as many ways as possible. The case studies have indicated that initiatives could include: i) designing investment schemes to benefit community investors, ii) integrating local content requirements into wind power projects, iii) delegating the siting process to municipalities, and iv) ensuring that electricity stays in the community in which it is generated.

Evidence from the case studies also highlights the likelihood of anxiety to change in communities where conventional energy has historically been a source of community employment and investment. If wind power developments displace technologies that have financially benefited a community, policymakers must be sensitive to the fact that there will be transitional losses associated with a shift away from conventional energy. Mechanisms must be developed in order to attenuate these losses and mitigate the opposition that arises from disenfranchised community members. This can also apply to the perceived danger of economic loss in mining communities when coal-fired power plants are replaced by wind power systems. Given the global nature of natural resource markets, such perceptions typically lack substance; but nevertheless, the concerns are valid in the eyes of the people that hold such perceptions. In public policy, rectifying misperceptions can easily be as difficult as responding to actual problems associated with change.

10.4.6 Economic Factor 6: National Industry Composition

Research indicates that there are at least two ways that industry composition influences the prospects of wind power development. First, in nations where conventional energy resources are extensive or downstream energy businesses have proliferated, there tends to be heightened resistance to transitioning away from conventional energy technologies because of the potential risk to industries involved in the conventional energy supply chain.

A second way that industry composition can influence the prospects of wind power development relates to energy intensity. Some nations exhibit higher industrial energy usage profiles than others. As a result, there is an elevated concern that higher energy costs will undermine national competitiveness. In nations such as Japan, where industry associations possess considerable political clout, these concerns can translate to political pressures to avoid change.

The general lesson is that energy-intensive firms will be particularly sensitive to transitioning into wind power. Such mega manufacturing entities typically also possess substantial financial might, which translates into political influence. Therefore, policymakers that hope to facilitate high levels of wind power capacity development need to consult early and often with industry groups to try to allay concerns over unsubstantiated economic threats stemming from a structural change to electricity provision and begin to craft solutions to minimize real threats associated with such a transition.

10.4.7 Economic Factor 7: Economic Growth Rate

Research indicates that there is a strong correlation between economic growth and electricity demand.[28] A nation with a high economic growth rate will likely be faced with the challenge of expanding its electricity supply. This can be advantageous to wind power development firms. When an electricity network needs to be expanded, it provides amplified opportunities for wind power development. On the other hand, the need to expand electricity supply poses economic costs on a host nation both in terms of financing the electricity generation infrastructure and augmenting the grid. The financial demands of this can lead to nations making investment decisions that are not optimal in the long run. China is a case in point. Between 1980 and 2000, China invested heavily in coal-fired power in order to capitalize on readily available technology and cheap domestic reserves but now finds itself in a bind over how to finance a transition away from this overdependence on coal-fired power.

The correlation between economic growth and electricity demand does not necessarily mean that the fortunes of wind power development will be undermined in nations plagued by low economic growth. This is because, in nations where the electricity supply does not need to be expanded, installed generation capacity still needs to be updated as older facilities become obsolete. Moreover, if the grid does not need to be expanded in order to accommodate new supply sources, government resources can be freed up for subsidizing replacement of conventional energy by wind power systems.

These mixed signals associated with economic growth patterns highlights an important lesson for policymakers. In addition to emergent demand needs, the evolving nature of a nation's electricity network is influenced by the structure of the sector, the relative power of the players involved, the nature of past and current subsidies, and the age of existing generating assets. Both high and low economic growth conditions can engender conditions in support of wind power expansion. The trick lies in cobbling together policy that best taps into the emerging dynamics that impact the electricity sector.

10.5 INFLUENTIAL POLITICAL FACTORS

Political factors tend to frame the environment within which wind power developers must compete. In the early days of wind power, without government subsidies, there was little chance of substantive development due to the adverse economic profile of wind power. As wind power costs continue to fall and the costs of conventional energy generation increase a degree of economic convergence is taking place; however, government subsidies are still important. Nevertheless, politics has a much greater role to play in enabling wind power development than merely closing the economic divide. The research presented in this book indicates that there are at least nine political factors that influence the effectiveness of wind power development policy. In this section, each political factor will be introduced and critically examined in order to identify policy lessons that are manifest in these insights.

10.5.1 Political Factor 1: Political Ideologies

Research suggests that greater agreement over energy strategy among competing political parties can engender either boom or bust periods for wind power development. For example, in Japan, a remarkably homogeneous take on what constitutes energy security, shared by political parties and the public, has been largely responsible for the phlegmatic performance of wind power development in the nation. Conversely, in Denmark and Germany, staunch support for wind power across the political spectrum has provided wind power developers with a high degree of confidence that support for wind power development progress will continue relatively unabated, thereby, justifying investment in manufacturing activities.

In the more common situation where political ideologies over energy strategy conflict, the impact on wind power development depends on how the political interface between parties is handled. In the event that the ruling party is not a staunch advocate of wind power, a pro-wind opposition party can put pressure on the ruling party to provide concessions, thus ensuring that wind power firms can stay afloat until a change of governing factions. In the event that the ruling party is in favor of wind power, an anti-wind opposition party can place pressure on the government to ensure that wind power development is carried out in a more effective manner in terms of economic efficiency, siting strategy, and public acceptance.

This highlights the importance of networking and bridge building across political platforms. Maintaining open channels of communication and actively nurturing common ground ensures that wind power development will always have at least a modicum of support on the public policy agenda. For policymakers from ruling parties that are pro-wind, the critical lesson to take home from this research is that while in power, reaching out to wind power opponents can be an effective strategy to building the political goodwill necessary to ensure that a change in governing coalitions will not catalyze the wild fluctuations in government support for wind power that can unsettled developers.

10.5.2 Political Factor 2: Lobbyist Environment

Comparatively speaking, some nations are characterized by elevated levels of political participation by well-coordinated special interest groups. At one extreme lies the United States, where special interest groups exert considerable influence on the development of public policy. At the other extreme lies China, where the special interest groups that do exist are relatively ineffective in terms of legally influencing public policy.

Generally speaking, research indicates that the prevalence of special interest lobby groups tends to adversely impact the fortunes of wind power development. This is because special-interest groups require funding to be effective, and the more funding a special interest group has, the more effective it can generally be. Given that conventional energy concerns typically hold dominant market positions, conventional energy special interests groups tend to enjoy higher levels of funding, and therefore hold greater political sway.

With that said, this trend appears to be undergoing change. In Denmark and Germany, the success of wind power firms has enabled the wind power industry to muster strong special-interest support. Meanwhile, public opposition to nuclear power and political opposition to coal-fired power have dampened the effectiveness of conventional energy industry groups. Consequently, in both nations, wind power advocates have exerted considerable influence over the evolution of energy policy. In the United States, an emergent wind power manufacturing industry is beginning to generate the critical financial mass necessary to chip away at the political fortress that has been insulating conventional energy technologies for decades. The job creation benefits of wind power are attracting political support from both parties in the United States.

The admonition for wind power advocates is that better coordination needs to be mustered between wind power firms and alliances between alternative energy technologies need to be improved in order to strip away the economic and political cloak that has insulated conventional energy technologies from competition. It would be far more effective for firms from various alternative energy technology sectors to band together to muster a unified response to conventional energy lobbyists rather than to attempt to compete simultaneously with conventional technology firms and each other.

10.5.3 Political Factor 3: Fiscal Health

Regardless of how wind power development is initiated through government policy (i.e., through FIT or renewable portfolio standards), there are some financial demands associated with wind power development that governments will be hard-pressed to avoid (or to pass on to end-consumers). For example, in Denmark, the success of the wind power program was largely predicated on time-consuming and costly market preparation activities. The nation commissioned mapping exercises to ensure that wind power development was carried out in a strategic and effective manner and established institutions (such as the Risø Laboratories) for linking industry, academia, and government stakeholders. In Germany, the government is now trying to get to grips with the challenge of providing grid connection to wind power developers that wish to exploit offshore energy potential. Even if the government decides not to directly subsidize wind power development, these grid connection costs will have to be incurred before being recovered from electricity rate premiums. The ability and willingness of a nation to undertake such costs depends significantly on its fiscal health.

Nations that are faced with fiscal deficits or high levels of public debt find it more difficult to directly subsidize the development of wind power. Such nations need to creatively cultivate solutions that pass along the costs to end-consumers in an expedient manner. In the United States, numerous utilities have adopted voluntary green purchase programs to allow consumers who are willing to pay more for green electricity to do so. In the process, this subsidizes development. In the Canadian province of Ontario, despite facing a paralytic fiscal deficit, the government is managing to subsidize wind power development by passing rate hikes through to the end-consumer, thereby generating the funds necessary to provide a healthy feed-in tariff. In Japan, the government is grappling with the largest level of public debt among all industrialized nations and has historically been reluctant to pass along electricity rate hikes to end-consumers, who are already paying some of the highest rates in the world. However, since the Fukushima disaster, the high cost of natural gas-fired power has led to further electricity rate increases. This has desensitized the Japanese public from adverse reactions to rate hikes and has made it possible for the government to devise an aggressive FIT for supporting wind power which can then be passed along to the end-consumer.

The lesson for policymakers is that a transition to wind power does indeed come at a financial cost, which the government must somehow be able to bear. However, a number of policy tools are available to governments to permit creative and strategic deployment of policies designed to allay the fiscal pain.

10.5.4 Political Factor 4: Nationalization of Utilities

The case study research indicates that nationalization of utilities is not necessarily desirable in terms of catalyzing enhanced wind power capacity expansion. To a certain extent, this finding is counterintuitive. One would assume that public monopolies, which are guided by more than profit-seeking motives, should be more open to wind power development. Moreover, one might argue that public monopolies permit wind power policy to be more effectively implemented thanks to the centralized nature of grid and electricity generation management. However, the reason why nationalization of utilities is not necessarily desirable—and indeed may in fact restrain wind power development—is threefold. First, due to limited profits, national utilities typically lack the funds necessary to build infrastructure for enhanced wind power development. Second, in comparison to firms which operate in a competitive environment, monopolies tend to be less efficient and, as such, are apt to incur higher costs when developing wind farms.[29] Third, centralization of the electricity generation function tends to engender myopic strategy that favors familiar technologies. Simply put, if the senior management of the public utility does not understand the benefits of wind power, wind power will not be embraced regardless of the benefits.

The main lesson has been summarized earlier—the optimal strategy for supporting wind power development is to nationalize the grid but privatize electricity generation. The goal should be to establish a grid that is capable of handling a variety of electricity generation technologies and then allow free-market competition (with external costs internalized) to dictate which technologies thrive. The key caveat is to ensure that all external costs are internalized. This means that an effort needs to be made to quantify environmental costs, health costs, potential safety risks, and direct and indirect economic benefits over the entire lifecycle of each energy technology. Admittedly this process requires a degree of subjective extrapolation that can render the process contentious; however, making the attempt to do so is clearly preferable to the alternative of doing nothing, an alternative which sadly prevails in every energy market in the world.

10.5.5 Political Factor 5: National Interests and International Pressures

The research presented in this book strongly suggests that national interests trump international pressures when it comes to greenhouse gas abatement. Nations where policymakers have managed to link national interest to a transition away from carbon-intensive energy technologies engender far greater stakeholder support. In Germany, Denmark, China, and to a lesser extent the United States, a promising wind power manufacturing sector underpins government support for wind power diffusion. Conversely, in Japan, the government does not believe that wind power conveys commercial advantage, and as such the nation continues to support nuclear power, a technology that the government perceives to be in the nation's commercial interests. Similarly, in Canada, the government views the continued exploitation of fossil fuel resources to be in the national interest. The evolution of wind power in China and the United States provides perhaps the strongest support for the contention that national interests trump international pressures to mitigate greenhouse gas emissions. In both nations, wind power diffusion was negligible until the cost of wind power started to drop, the cost of fossil fuel energy began to rise and promising wind power manufacturing industries began to emerge. Then national interests and international GHG abatement pressures converged, and wind power development took off in both nations.

Another factor that influences whether or not nations succumb to international pressure to abate CO_2 emissions is the perceived level of economic damage caused by such a transition. Globally, the mantra of many developing nations is that committing to binding emission reduction targets will adversely affect economic development and this is unacceptable. Many of the industrialized nations have echoed similar concerns. The industrialized nations that have been most vociferous sport the worst GHG emission reduction track records. Indeed, the challenge of meeting GHG reduction targets is viewed as so daunting in both Japan and Canada that these two nations have declared an intention to withdraw from the Kyoto Protocol framework.

The main lesson for wind power advocates is that wind power development policy will not be successful if it is promoted solely as a way to contribute to goals established by the international community. This is not sufficient motivation for many of the stakeholders that influence domestic energy policy. Rather, wind power development must be promoted as a commercially prudent investment in an era of inflated fossil fuel prices, an avenue for job creation and a vehicle for enhancing national energy security by reducing dependence on imported fossil fuel supplies. National interests trump international pressures.

10.5.6 Political Factor 6: Historic Trends in Public Energy R&D

All of the case studies provide evidence that the technological platforms to which governments have directed R&D investments in the past catalyze political lock in favor of the technologies that have been favored. Research suggests that there are sound reasons for this. Government funding for energy technology R&D tends to follow an ideological undercurrent driven by powerful advocacy coalitions that possess core ideologies which align with a specific technology.

Changing the core values of long standing regimes is not easy and will likely not materialize from within the regime.[30] Politically, policy reversals tend to send negative messages to voters that the policymakers lack strategic foresight. Operationally, a well-funded R&D community tends to garner financial and political power, which helps ensure a priority stake over future R&D provisions. For example, despite staunch public opposition to nuclear power, up until the Fukushima disaster, technocrats within Germany's political system managed to keep the nation's nuclear power sector afloat. This is also the case in Japan and in Canada's province of Ontario, where despite public opposition to nuclear power, a political commitment to nuclear power persists. In the United States and China, public support for the coal industry is being perpetuated in the form of public funding of CCS technology research.

Generally speaking, the tendency for R&D funding to be path dependent tends to work against wind power development because conventional energy technologies have historically received the lion's share of government R&D. However, there is evidence that once the trend has been broken by funds shifting from one technology to another, the new technological recipient tends to enjoy a similar level of entrenched support. For example, in Denmark, strong support for wind power development has nurtured a government R&D climate that favors further wind power R&D and resists funding technologies such as solar power which have less of a commercial allure. In Germany, the Fukushima crisis was the decisive cut that severed government support for nuclear power. Consequently, support for wind power R&D has become amplified and wind power has become the new technology wunderkind for government assistance.

Interestingly, even when political regimes change, energy regimes tend to endure. One explanation for this is that electricity networks are costly assets and financing a significant departure from status quo can impair a government's efficacy in other areas which need financing. Consequently, the adage "if it ain't broke, don't fix it" tends to guide energy policy. Impetus for change arises only under circumstances of obvious need (i.e., Germany's decision to phase out nuclear power after the Fukushima disaster) or when investment requirements for facilitating a technological transition are roughly on par with the investment requirements for maintaining current technological platforms. Therefore, in most nations, seeing small progressive shifts of government R&D into promising technologies is more likely than seeing massive paradigm shifts in what is funded.

In attempting to craft effective wind power development policy, a review of public R&D initiatives provides an indication of how much resistance there will be to facilitating a transition to wind power. In nations such as the United States, China, Canada, and Germany, where coal-fired power has historically enjoyed high levels of government support, it may be necessary to concede support to CCS research in order to appease coal advocates and avert damaging pushback from conventional technology stakeholders.

10.5.7 Political Factor 7: Inclusiveness of the Energy Policy Process

The case studies suggest that centralized policy regimes that are dominated by a few actors tend to be prone to policy capture, whereby powerful special interest groups have an increased ability to influence policy decisions and powerful policymakers have a heightened ability to direct policy without contest. Japan's energy regime exemplifies an extreme manifestation of policy capture by special interest groups. A pro-nuclear power coalition has managed to gain dominant influence over Japanese energy policy. China exemplifies an extreme manifestation of policy capture by a handful of influential energy policymakers. In China's case the judgment of a few influential policymakers has significantly influenced the fortunes of wind power.

The Chinese case study suggests that policy capture can be either beneficial or disadvantageous for wind power development; however, in most nations the economic power of the conventional energy sector tends to foster policy regimes that are favorably disposed toward supporting conventional energy interests. Even in cases where insular policy regimes are pro-wind, there is always the danger of political change resulting in a reversal of fortunes. For example, a government decision to drop support for wind power and redirect all investment to nuclear power development is not inconceivable. Under such a scenario, the wind power boom in China could disintegrate virtually overnight.

These insights suggest that more inclusive policy regimes that incorporate participation from a diversity of stakeholders engender conditions that are conducive to progressive wind power development. This is because even if powerful stakeholder factions support conventional energy, the employment and environmental benefits attributed to wind power tend to provide justification for at least some support for wind energy. As the US case study demonstrated, extreme policy shifts, such as the feast and famine conditions caused by the erratic renewal process of the production tax credits, causes far more damage to wind power development than modest yet consistently progressive development strategies do.

10.5.8 Political Factor 8: Central and Subnational Government Coordination

Evidence from the case studies suggests that the level of coordination between different levels of government influences the sustainability of wind power development programs. For example, in China, poor coordination between the central government planners, provincial authorities and regional utilities have led to wind farms being constructed in areas that lack T&D infrastructure. Consequently, the power that is generated from the wind turbines does not always reach the electricity grid. In Japan, the recent wind power FITs that have been announced are among the highest in the world and this has elevated the scale and number of wind projects under development. However, lack of coordination between the central government, prefectural authorities and municipal planners has forced wind power developers into "cold call" development patterns where developers must try and negotiate suitable sites in the face of community (and sometimes municipal) resistance. Contrary to this, successful diffusion of wind power in Germany and Denmark has largely been attributed to amplified levels of coordination between central, regional, and community level policy planners and high levels of collaboration between citizens, industry, academia, and government officials to address emergent problems as wind power capacity increases.

The lesson for policymakers and wind power developers is to ensure that wind power projects are designed through an inclusive process that is bottom-up, to ensure that the concerns and needs of stakeholders in targeted host communities will be adequately addressed. Even in situations where the central government is incapable of unifying wind power policy, subnational actors can optimize success by fostering collaborative planning to the greatest extent possible. This is particularly true for nations such as Canada or the United States, where wind power development is driven by provincial or state level policy strategy.

10.5.9 Political Factor 9: Perceived Risk to National Security

One final political factor that influences the fortunes of wind power development in a given nation relates to risk. Research indicates that the political instability that characterizes many fossil fuel exporting nations is viewed by many fossil fuel dependent nations as a threat to national security.[31] In all six case studies the inception of wind power development harkens back to the 1970s, when the two oil crises catalyzed fossil fuel price inflation. In response, all of the case study nations began to explore ways to diversify energy supplies and one of the strategies was to attenuate risk through technological diversification. Similarly in recent times, concerns over oil sponsored terrorism and the capricious nature of fossil fuel prices have prompted all six case study nations to elevate support for wind power development.

Wind power advocates that can effectively promote wind power development as an essential cog in a program for improving national security stand a far greater chance of engendering broader support.

10.6 CONCLUDING THOUGHTS

The nature of the complex adaptive environment within which energy policy is designed and implemented precludes the development of rigid, prescriptive strategies for enhancing wind power development. Simply put, energy policy is carried out in nations that are unique from sociocultural, technological, economic, and political perspectives. A specific policy strategy that has benefited one nation (e.g., Denmark) cannot be imported directly into another nation with the expectation that similar results will ensue. However, the case studies that were presented in this book provide evidence that there are common factors which can catalyze or impede wind development in different nations. An astute policymaker should be aware of these common factors and use this knowledge as a staring point to devise initiatives that can enhance wind power development. The insights provided in this chapter catalogue these common factors. They constitute a foundation from which policymakers can begin to customize policy to suit the unique national context in which the policy will be applied.

This chapter has put forth the contention that successful wind power development policy necessitates that strategic attention be given to the management of nine social factors, seven technological factors, seven economic factors, and nine political factors that have influenced the efficacy of wind power development policy in the six case study nations. Policymakers that give attention to these factors will undoubtedly cobble together more effective wind power development strategies. However, as the framework proposed in Chapter 3 suggested, wind power development strategies can be further improved by understanding how these factors interrelate and by highlighting any dominate influences that warrant special attention. Chapter 11 will take up these issues and conclude our investigation.

NOTES

1. For more on this supporting research the reader is directed to Valentine, Scott Victor. 2010. "A STEP Toward Understanding Wind Power Development Policy Barriers in Advanced Economies." *Renewable and Sustainable Energy Reviews* 14 (9): 2796–2807; Valentine, Scott Victor. 2010. "Braking Wind in Australia: A Critical Evaluation of the Renewable Energy Target." *Energy Policy* 38 (7): 3668–3675; and Valentine, Scott Victor. 2010. "Disputed Wind Directions: Reinvigorating Wind Power Development in Taiwan." *Energy for Sustainable Development* 14 (1): 22–34.
2. Firestone, Jeremy, and Willett Kempton. 2007. "Public Opinion About Large Offshore Wind Power: Underlying Factors." *Energy Policy* 35 (3):1584–1598.
3. Gross, Catherine. 2007. "Community Perspectives of Wind Energy in Australia: The Application of a Justice and Community Fairness Framework to Increase Social Acceptance." *Energy Policy* 35 (5): 2727–2736.

4. Valentine, Scott Victor. 2011. "Sheltering Wind Power Projects from Tempestuous Community Concerns." *Energy for Sustainable Development* 15 (1): 109–114.

5. Valentine, Scott Victor. 2010. "A STEP Toward Understanding Wind Power Development Policy Barriers in Advanced Economies." *Renewable and Sustainable Energy Reviews* 14 (9): 2796–2807.

6. Tsuru, Shigeto. 2000. *The Political Economy of the Environment: The Case of Japan.* London: The Athlone Press.

7. The development and subsequent exploitation of these windows of opportunity is eloquently described in Kingdom, John W. 1984. *Agendas, Alternatives and Public Policies.* Boston: Little, Brown Publishers.

8. Valentine, Scott Victor. 2013. "Wind Power Policy in Complex Adaptive Markets." *Renewable and Sustainable Energy Reviews* 19 (1): 1–10. http://dx.doi.org/10.1016/j.rser.2012.11.018.

9. Valentine, Scott Victor. 2011. "Japanese Wind Energy Development Policy: Grand Plan or Group Think?" *Energy Policy* 39 (11): 6842–6854. doi: 10.1016/j.enpol.2009.10.016.

10. As an example see Spegele, Brian. 2013. "Behind Chinese Protests, Growing Dismay at Pollution." *Wall Street Journal*, May 17,. http://online.wsj.com/article/SB10001424127887323398204578488913567354812.html.

11. United Nations. 1998. *Kyoto Protocol to the United Nations Framework Convention on Climate Change.* New York: United Nations.

12. Valentine, Scott Victor. 2013. "Enhancing Climate Change Mitigation Efforts through Sino-American Collaboration." *Chinese Journal of International Politics* 6 (2): 159–182. 10.1093/cjip/pos021.

13. For more on this story see Artuso, Antonella. 2012. "No More Wind Turbines Until Study Released, MPP Demands." *North Bay Nugget*, July 21.

14. Kahneman, Daniel, Jack L. Knetsch, and Richard H. Thaler. 1990. "Experimental Tests of the Endowment Effect and the Coase Theorem." *Journal of Political Economy* 98 (6): 1325–1348. 10.2307/2937761.

15. For more on this see Geert Hofstede's website, http://geert-hofstede.com/index.php.

16. Wizelius, Tore. 2007. *Developing Wind Power Projects: Theory and Practice.* Oxford: Earthscan.

17. Valentine, Scott Victor. 2010. "Disputed Wind Directions: Reinvigorating Wind Power Development in Taiwan." *Energy for Sustainable Development* 14 (1): 22–34.

18. Valentine, Scott Victor. 2011. "Japanese Wind Energy Development Policy: Grand Plan or Group Think?" *Energy Policy* 39 (11): 6842–6854. 10.1016/j.enpol.2009.10.016.

19. Kamp, L. M. 2004. "Wind Turbine Development 1973–2000: A Critique of the Differences in Policies Between the Netherlands and Denmark." *Wind Engineering* 28 (4): 341–354.

20. Ibid.

21. Sovacool, Benjamin K. 2008. *The Dirty Energy Dilemma: What's Blocking Clean Power in the United States.* New York: Praeger Publishers.

22. Wong, Edward. 2013. "Air Pollution Linked to 1.2 Million Premature Deaths in China." *New York Times*, April 1. www.nytimes.com/2013/04/02/world/asia/air-pollution-linked-to-1-2-million-deaths-in-china.html?_r=0.

23. Sovacool, Benjamin K. 2011. *Contesting the Future of Nuclear Power.* Singapore: World Scientific Publishing.

24. More on this story which is based on data from Japan's Agency for Natural Resources and Energy is available on the New York Times website at www.nytimes.com/

interactive/2012/08/30/business/estimating-the-cost-of-a-nonnuclear-japan.
html?ref=energy-environment.

25. Sovacool, Benjamin K., and Scott Victor Valentine. 2012. *The National Politics of Nuclear Power: Economics, Security and Governance*. Milton Park, UK: Routledge.

26. For a discussion on public and private goods, please see Mankiw, N. Gregory. 1997. *Principles of Economics*. New York: Harcourt Publishing.

27. For insights on marginal cost strategies, see Doyle, Peter. 1998. *Marketing Management and Strategy. 2 ed.* Harlow: Prentice Hall Publishing.

28. Tester, Jefferson W., Elisabeth M. Drake, Michael J. Driscoll, Michael W. Golay, and William A. Peters. 2005. *Sustainable Energy: Choosing Among Options*. Cambridge, MA: MIT.

29. Valentine, Scott Victor. 2010. "Disputed Wind Directions: Reinvigorating Wind Power Development in Taiwan." *Energy for Sustainable Development* 14 (1): 22–34.

30. This is the basis for advocacy coalition theory. For more on this see Sabatier, Paul A. 1988. "An Advocacy Coalition Framework of Policy Change and the Role of Policy-Oriented Learning Therein." *Policy Sciences* 21 (2–3): 129–168.

31. For more on this discussion see Campbell, Kurt M., and Jonathon Price (eds.). 2008. *The Global Politics of Energy*. Washington: The Aspen Institute.

CHAPTER 11

Applied Policymaking

Knowledge has to be improved, challenged, and increased constantly, or it vanishes.

—Peter Drucker

11.1 POLICY FORMULATION

Chapter 10 summarized nine social factors, seven technological factors, seven economic factors, and nine political factors that have influenced the fortunes of wind power development in the six case study nations covered in this book. The premise underpinning the previous chapter is that successful wind power development policy depends on strategic management of forces of change within four contextual areas depicted in Figure 11.1.

There are three basic tenets underpinning this model. First, the environment in which wind power policy is formulated and implemented can be better understood by comprehensive analysis of conditions within four contextual areas: the sociocultural context, the economic context, the technological context, and the political context. Within each of these four areas there are dominant forces (variables) that have proven to be influential in hindering or helping wind power development. The trouble is that for each nation, the relative importance of each influential variable differs because energy policy in each nation is influenced by a unique conflation of sociocultural, technological, economic, and political conditions.[1] For example, a high degree of information asymmetry is evident in both Japan and China. Citizens of both nations lack adequate information about the pros and cons of energy technologies to make informed decisions. In Japan, information asymmetry helps explain why there is so little support for wind power and why the government has been able to continue its advocacy of nuclear

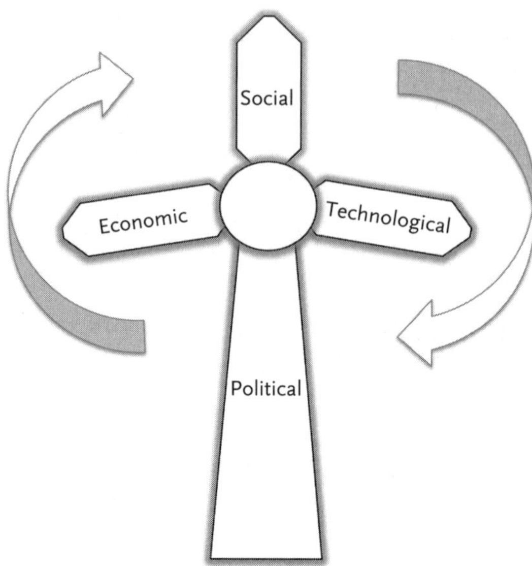

Figure 11.1. The *Political SET* Model

power. In China citizens are also kept largely in the dark about energy sector developments, but this is not a problem for wind power development because the government is committed to supporting wind power whether the public consents or not. In short, information asymmetry is a barrier to wind power development in Japan, but in China, it is not.

Second, the analysis of STEP forces is complicated because variables *within* each of these four contextual areas interact in unpredictable ways due to the complexity of variable interrelations. Cause-and-effect links are extensive which means that numerous positive and negative feedbacks catalyze chaotic systemic evolution. For example Canada possesses a wealth of hydropower capacity that suggests a high degree of grid resilience. Furthermore, Canada boasts comparatively broad grid coverage so wind power developers are afforded fairly easy access to the grid. In comparison, the Chinese grid is less resilient, and wind power developers have far greater challenges in connecting to regional grids. Normally this should suggest that technological conditions in Canada are more conducive to supporting wind power development. However, the current reality is that conditions are more favorable in China because of another technological factor—electricity sector structure. Canadian utilities are structured to serve provinces. Consequently, there is no incentive for hydropower rich provinces to share peaking capacity in order to enhance wind power development in other provinces. On the other hand, Chinese utilities are structured to serve wider regions, and the state-owned utilities face a far easier task in terms of sharing flows between regions.

Third, in addition to the interplay of forces *within* each of the four contextual areas (STEP), there is interplay of forces *between* each of the four contextual areas. For example, influential sociocultural variables influence economic, political, and technological variables and positive and negative systemic feedbacks that radiate from initial cause and effect relationships proceed to catalyze conditions of evolutionary change throughout the entire STEP environment. To complicate things, the direction of causality is often two-way and not necessarily positive. To illustrate, in China, political will to abate pollution has engendered support for wind power development and lead to policies which have fueled the emergence of a wind turbine manufacturing industry. In the process, the economic and job creation benefits associated with the wind turbine manufacturing industry have begun to reinforce government support for wind power. In short, these two variables have had a mutually reinforcing effect on each other. As a contrasting illustration, high levels of public environmentalism in the Canadian province of Ontario has fostered public support for wind power development, and this has nurtured provincial government support for wind power. However, overly aggressive wind power development policy has engendered NIMBY opposition, undermining the climate of support for wind power. In short, these two variables have had negative feedback effects on each other.

It should be clear from this analysis that too little is known about the extensive interactions between the variables within the energy policy STEP environment to create a prescriptive framework for guiding energy policy. The STEP environment which affects energy policy is a complex adaptive system. The variables which impact wind power development evolve in response to influences from other influential variables within the same system catalyzing continuous change. The interplay between these influential variables is so complex and contextually bound that attempts to quantify relative influence is highly prone to error.[2] By extension, attempts to identify universally valid truths are futile. It is for this reason that the Political SET model should be considered to be an analytical tool, rather than a predictive framework.

The merit of the Political SET framework as an analytical tool is predicated on the premise that studying the contextual STEP forces which have most influenced wind power development in other markets can serve as a fundamental starting point for examining STEP forces in other countries. Indeed, the commonalities uncovered in the six case studies in regard to forces which inhibit or catalyze wind power development serve as evidence

Table 11.1 KEY STEP CONDITIONS INFLUENCING WIND POWER DEVELOPMENT			
Sociocultural Context	**Technological Context**	**Economic Context**	**Political Context**
Community perspective	Grid resilience	Internalization of externalities	Political ideologies
Information asymmetry	Energy mix	Subsidies to energy technologies	Lobbyist environment
Civic activism	Technological regimes	Entrenched investments	Fiscal health
Habitat patterns	Distance to the grid	Competitive health of energy sector firms	Nationalization of utilities
Affluence	Nationwide potential	Community economic benefits	National interests and international pressures
Uncertainty and change	Electricity sector structure	National industry composition	Trends in public energy R&D
Vocational influences	Technological network links	Economic growth rate	Inclusiveness of the energy policy process
Agricultural sector structure			Central and subnational government coordination
Environmental impact of energy mix			Perceived risk to national security

that utilizing the Political SET framework to guide analysis and commencing such an analysis by studying national conditions related to the 32 STEP variables outlined in Chapter 10 can serve as a comprehensive starting point for policymakers who are intent on optimizing wind power development. These 32 STEP variables, which are summarized in Table 11.1, do not provide prescriptive insight; rather, they provide analysts with guidance on what to think about when formulating wind power development policy.

11.2 UNDERSTANDING SYSTEM DYNAMICS

In employing the Political SET framework as an analytical tool, there are five further observations extracted from the case studies that can help analysts better understand the dynamics of the Political SET environment.

11.2.1 Economics Is Still King

Although the prospects of wind power diffusion in a given nation depend on how the STEP variables outlined in Chapter 10 conflate and evolve, there is one variable that stands out as exerting the dominant influence on the fortunes of wind power—comparative cost. Simply put, wind power systems cannot compete with coal-fired technologies or nuclear power when the negative externalities associated with these two latter technologies are not comprehensively internalized into the cost of electricity generation. Even in recent times, where the cost of coal has risen dramatically and wind power costs have continued to diminish, the cost gap between wind power and coal-fired power has not yet been bridged.

There is no nation that adequately internalizes these external costs. However, policymakers in nations that have been most successful in catalyzing wind power diffusion have made attempts to internalize some of the costs and benefits—through carbon taxes, feed-in tariffs, or other government subsidies—and to rectify historical financial support imbalances by ramping up renewable energy R&D support.

Nevertheless, all wind power support policies suffer from two shortcomings. First, any subsidies that are provided to wind power developers do not come even remotely close to the cost savings associated with mitigating the economic impact of elevated levels of climate change.[3] If the subsidies fully reflected the costs that will be averted, wind power would be the preferred economic option.[4] Second, wind power subsidies give energy consumers the wrong impression. By subsidizing wind power, the message sent to energy consumers is that wind power is commercially unviable and needs government support. The truth is that the current economics underpinning coal-fired power and nuclear power are distorted and inaccurate. These technologies are actually far more expensive than the wholesale price indicates, but the external costs (health costs, waste storage costs, climate change costs, etc.) are currently shouldered by other stakeholders. It may very well be that a better approach to fostering public support for wind power development would be to force power plants that employ conventional energy technology to fully internalize external costs so that the public understands that wind power technology should not be viewed as a charity case, but rather as a preferred economic solution.

The influence of economics on wind power diffusion should not be misconstrued as intimating that economics is the only factor influencing diffusion. As Chapter 10 summarized, there are other sociocultural, technological, and political forces that play enabling roles. In short, favorable economic conditions are necessary but not sufficient conditions for wind power diffusion.

11.2.2 Chain Reactions Can Be Partially Managed

This study should have hopefully highlighted the futility inherent in trying to establish a best practice manual for wind power policy formulation. Each nation is characterized by unique sociocultural, technological, economic, and political conditions. These conditions then come together in an elaborate dance to form a causal pattern of dynamic change that is far too complex to comprehensively model. Therefore, the best that we can hope for is to highlight conditions that have proven to be influential in catalyzing wind power diffusion and try and manage influential variables in a way to engender positive feedbacks that dampen resistance and drive change.

This should not be misconstrued to suggest that applying the Political SET framework is organized guesswork. The approach that is being advocated in this book is a common practice in applied chaos theory. As chaos theorist Eric Beinhocker contends, under conditions of uncertainty—which is characteristic of complex adaptive systems—the most resilient strategy is to enhance data collection and analysis to better predict likely trends and then prepare organizationally to react and alter course expediently when anticipated trends do not emerge in the manner predicted.[5] Thinking back to the Danish case study, this is precisely what Danish energy policymakers have been doing to drive wind power diffusion—apply best available knowledge, and monitor and react expediently to change.

Applied back to the Political SET framework, policymakers can use the insights from Chapter 10 to begin to craft strategy for optimizing wind power diffusion; however, they must be cognizant of the fact that each nation is shaped by a different conflation of sociocultural, technological, economic, and political conditions that will catalyze varied responses to initial policy initiatives. Under such circumstances policymakers need to be fully aware of market developments and must be prepared to adjust policy to meet emergent challenges.

11.2.3 Favorable Conditions Still Need to Be Managed

A recent avenue of wind power policy focuses on trying to explicate best practice.[6] The premise behind this is that policies that have worked in a number of nations can be replicated in other nations. Although the application of battlefield tested policies can engender positive feedback within a nation's energy policy STEP environment, there is no guarantee that best practice applied in a rigid manner will always engender positive feedback. As the case studies on wind power in Denmark and Germany exemplify, although the Danish and

German governments did many things right when cobbling together policy to support wind power development, the success of the programs engendered public dissatisfaction as large wind farms began to impair the aesthetics of certain communities. Failure to respond to this emergent problem would have stopped the wind power programs in these nations dead in their tracks had these two nations not adjusted policy to address these concerns.

In brief, wind power policy is not a static activity where policymakers simply have to apply a prescribed set of best practice principles and sit back while the program successfully unfolds.[7] Policy must evolve in response to the ever-evolving nature of the STEP environment that impacts wind power development. This does not mean that policymakers should always be adjusting policy. One aspect of effective wind power development policy is consistency, because consistency attenuates market risk. Market momentum can be maintained if market players can anticipate the direction and the magnitude of government support for wind power development. Consequently, policymakers must seek balance by responding to emergent challenges to wind power development through policy adjustment while being cognizant of the fact that extreme policy adjustments could have adverse consequences on market momentum.

11.2.4 First Mover Advantages Exist

The future of wind power development is still obscure. The energy sector is one of the most lucrative industrial sectors in the world. Trillions of US dollars are at stake and there are financial motives across the spectrum of technological platforms to invest in technological R&D in order to generate carbon-free electricity at the lowest cost. A groundbreaking technological development in any electricity generation platform is not outside the realm of possibility. Nuclear power technologies have been improving both in terms of operational safety and waste reduction. Carbon capture and sequestration research breakthroughs could allow the coal-fired power sector to maintain short-term dominance in electricity generation. Geothermal power and advanced solar thermal electricity generation technologies are already commercially viable in certain contexts. What transpires in all these areas significantly impacts the fortunes of wind power. Moreover, the potential of wind power is inextricably linked to innovations in a complementary technological area—energy storage. If the stochastic flows of wind power systems can be attenuated at low cost, the future of wind power will be positively affected.

Yet despite the uncertainties of technological evolution intrinsic to the global energy sector, one has reason for optimism over the prospects of wind power. In the absence of a technological breakthrough that will benefit a competing technology, wind power is currently the most versatile utility-scale renewable energy technology that is capable of competing economically with fossil fuel energy technologies and nuclear power. Accordingly, over the short to medium terms, global installed wind power capacity can be expected to increase substantially, particularly as the threats stemming from climate change become more evident and imminent. This short- to medium-term window of opportunity bodes well for wind turbine manufacturers because enhanced market share improves economies of scale and production, thereby fostering cost reductions. Enhanced market share also enables wind turbine manufacturers to commit more funds to R&D, which further enhances the prospects of technological innovation and cost reduction.[8] Indications point to the wind power sector experiencing growing market share over the next 20 to 30 years, and this suggests that nations that can nurture leading wind power manufacturing firms stand to reap benefits including enhanced domestic energy security, enhanced employment in the wind power sector, and the cultivation of a high growth industrial sector.

There is evidence that there are first mover advantages associated with nurturing wind power manufacturing competency. It is not a coincidence that the top wind power markets—the United States, China, Germany, Denmark, Spain, and India—also boast domestic wind system manufacturing firms that are among the largest in the world. The market is still in a developmental stage and no wind system manufacturer has managed to garner enough market share to enjoy monopolistic advantage. Accordingly, there is still scope for nations to foster competitive domestic wind system manufacturers through domestic support strategies; however, as the market matures, market entry will become increasingly difficult for new entrants.

11.2.5 Public Opposition Trumps Economic Impediments at High Levels of Installed Capacity

Given the commercial viability of wind power systems and the imperative for expedience in facilitating a transition away from carbon-intensive electricity generation systems, it seems to be a foregone conclusion that wind power capacity expansion should be on the political agenda of most nations. If so, what level of installed capacity should nations aim for?

The answer to this question is nation-specific, but some generalizations are possible. As the technology review of wind power presented in this book suggested, conservative estimates suggest that up to 10% contribution from wind power to the electricity grid can be accommodated by tapping into existing slack grid capacity to balance supply fluctuations. Moreover, in most nations, 20% contribution from wind power is generally achievable without necessitating added back-up capacity through strategic site planning, strategic management of the electricity system's fuel mix, and improved interconnections with neighboring grids. Much higher levels of wind power integration are possible in nations that boast high wind power potential, high capacity in peak-load technologies (i.e., hydropower, natural gas-fired power), or the capacity to interconnect with neighboring grids. We are already seeing evidence of wind power contribution levels reaching up to 40% in some regions without destabilizing the grid or adversely affecting electricity prices. In short, most nations can achieve far greater levels of wind power integration than currently exist.

At higher capacity levels, impediments to expanding wind power capacity appear to stem more from social barriers than economic barriers. Economically, even if a degree of added generation capacity must be added to permit greater levels of wind power integration, the costs are estimated to be negligible—certainly far less than the unaccounted for external costs associated with fossil fuel electricity generation.[9] Yet overcoming social barriers to achieving higher levels of wind power saturation appear to be far less tractable than overcoming economic barriers. As both the German and Danish case studies suggest, there appears to be a level of wind power saturation beyond which communities begin to experience dissonance associated with the physical invasiveness of so many wind power turbines. As wind power capacity continues to expand in nations around the world, social resistance to wind power may wind up being the predominant variable constraining development. Nevertheless, as outlined in this book, by cobbling together technological, economic, and political strategies aimed at attenuating public resistance, the sociocultural boundaries that frame the upper limits of wind power capacity can likely be expanded.

11.3 POLICY IMPLEMENTATION AND MONITORING ESSENTIALS

To this point, this study has been clear in explicating the types of STEP variables which should be proactively managed during the formulation of wind power development policy. However, effective wind power development policy is dependent on far more than effective policy formulation. Once formulated, policy must be effectively implemented and monitoring strategies need to be developed to ensure that the implementation process is achieving the desired results. Therefore, it would be remiss to conclude this study without addressing implementation and monitoring. Insights gleaned from the case studies suggest that there are four attributes related to policy implementation and monitoring that are imperative for sustainable wind power development success.

11.3.1 Malleability

As often mentioned throughout this book, wind power policy should be viewed as a dynamic undertaking because policy initiatives catalyze feedback responses in other influential areas, which in turn initiate further feedback responses throughout the system. Sophisticated monitoring systems are needed to evaluate how the STEP environment responds to policy implementation. This should not be done on an ad hoc basis. Policymakers must ensure that structures, strategies, and actor competencies are aligned and positioned to respond effectively and expediently to the changing dynamics of the energy sector.

Structurally, monitoring units need to be set up within policymaking departments and funding must be made available to allow these units to track developments in the STEP environment. This need not be an expensive undertaking; however, this must be institutionalized to ensure that results from the monitoring process are fed back into policy reformulation. The greatest error during the implementation process is to fail to provide strategic structure to permit policy reformulation in a timely manner.

Strategically, as STEP conditions change in response to policy or other influences, policymakers need to be prepared and empowered to modify policy or introduce new initiatives to address emergent challenges. For example, in Denmark and Germany, policymakers have been quick to respond to public opposition stemming from high concentrations of wind farms in certain regions. Policymakers in both nations have addressed this challenge by shifting development patterns and reestablishing financial incentives in target communities. Similarly, in China, the government has responded quickly to information that wind farms were unable to transmit energy to the electricity grid by expediting grid reinforcement projects.

In order to effectively traverse the bridge between monitoring and policy reformulation, the individuals responsible for these processes must possess sufficient interpersonal aptitude. Resistance to policy reformulation should be expected because those who have initially formulated policy typically do so with a degree of confidence that the policies will indeed lead to the desired results. This suggests that combining the actors involved in policy formulation and monitoring to as great an extent as possible is desirable. When this is not possible, individuals who lead the monitoring process must be willing and able to communicate effectively with policymakers and key stakeholders to ensure that the policy formulation, monitoring, and reformulation cycle is as cohesive as possible.

The importance of policy reformulation was evident in many of the successful initiatives documented in the case studies. In Germany and Denmark, policymakers periodically adjusted subsidies as the wind power market developed in order to control the pace of development. In the United States and China, policymakers in regional jurisdictions cobbled together ad hoc policies to complement central government support and induce wind power capacity development. Danish policymakers created formal structures to improve communication between key stakeholders—sharing knowledge, collaborating in R&D, and cooperating in strategic development as new challenges emerged. Good examples of highly responsive reformulation of policy are the Danish wind turbine replacement schemes that were designed specifically to enhance installed capacity without inflaming community concerns over the proliferation of wind parks. Similarly, in Germany, the strategic shift to offshore wind power development is exemplary of effective STEP environmental monitoring and responsive policy reformulation designed to avert public dissonance with wind power.

In summary, wind power policy can be considered to be like a fencing match. Policies are formulated and implementation typically commences in a clearly directed manner. Policy implementers continue advancing the program until they face resistance or emergent problems. When this happens, responsive policymakers adopt defensive strategies aimed at effectively parrying (eliminating) resistance or emergent problems. Once the challenges are parried, the emphasis then turns back to proactive policy designed to once again thrust the program forward to achieve positive results. The thrust-and-parry dynamic is an essential element of effective policymaking in complex adaptive systems and has proven to be instrumental to success in wind power policy in Denmark, Germany, and China. Establishing an open and flexible environment requires the establishment of formal structures and procedures to link monitoring to policy reformulation. It also demands institutional flexibility to permit policy to be adjusted in a timely manner to ensure that emergent problems do not reach crisis stages before being addressed. Finally, it requires the employment of individuals who are comfortable with change and who understand the importance of communicating broadly in order to close the cycle of policy formulation, monitoring, and reformulation.

11.3.2 Transparency and Broad Participation

Successful wind power programs encourage participation from a broad spectrum of actors and exhibit a high degree of transparency throughout the policy cycle. During policy formulation, initiatives to encourage participation lead to the creation of better policy. Broader participation helps mitigate unanticipated secondary problems that arise when policymaking is undertaken in isolation from key stakeholders. To illustrate, in California's Altamont Pass, poor planning led to the establishment of wind farms in the 1980s that were responsible for a high avian mortality rate, engendering severe opposition from environmental groups. If the planning process had been more transparent and encouraged broader participation, alternative wind farm siting strategies may have been developed to avert this problem.

When policy is being formulated, transparency in communicating the policy to the market helps reduce market uncertainty (and risk). As the US case study illustrated, uncertainty over whether or not the production tax credit would be extended engendered boom and bust cycles that made it hard for wind power development firms to establish business growth strategies and plan R&D. Conversely, unwavering support for wind power in Denmark and Germany, even amidst regime change, engendered a degree of market confidence that encourages wind power firms to make longer term investments.

At the implementation stage, frequent communication with key stakeholders yields valuable feedback to improve the effectiveness of a given policy. This is most apparent in Denmark where frequent interaction with municipal governments and community members has allowed the government to reformulate strategies to attenuate community dissonance associated with higher concentrations of wind power systems.

These insights agree with academic studies on transparency management, which indicate that transparency and participation engender greater stakeholder acceptance, enhanced innovation, better strategic cohesion, and elevated responsiveness to change.[10]

11.3.3 Strategic Balance

Advocacy coalition theory hypothesizes that policy regimes become established through coalitions of stakeholders, which share core ideologies that allow a given policy direction to dominate. Advocacy coalition theory further postulates that these entrenched regimes will resist change that necessitates rejection of core ideologies.[11] There is evidence that these insights apply to energy policymaking. For example, in Japan, the energy policy regime is dominated by pro-nuclear advocates who have vested interests in fighting for the continued support for nuclear power. In the face of these entrenched interests, fostering change is not easy.

Fortunately, advocacy coalition theory also provides some insight into how change can be strategically carried out. Work of Paul Sabatier suggests that change can be effected by appealing to the secondary interests of regime stakeholders in order to cause dominant regimes to fracture and lose cohesiveness.[12]

This implies that entrenched conventional energy regimes can be undermined by encouraging shifts of allegiance within coalitions that support incumbent regimes. For example, natural gas-fired power is a peak-load technology that complements wind power. Accordingly, although natural gas-fired power plants can be considered to be competitors to wind power plants, strategically focusing on exploiting the synergies between these technologies can foster support for wind power from a technological sector that otherwise would be in opposition. Similarly, some stakeholders who support coal-fired power do so because of economic justifications. For some, the coal industry might be a regional employer. Consequently, policies that are designed to help coal workers transition into the wind power field can help attenuate resistance from those employed by the coal sector. Other stakeholders may harbor a belief that coal-fired power is the cheapest source of electricity provision, thereby justifying support. Enhancing knowledge related to the externalities associated with coal-fired power may help to mitigate opposition of this type.

The point is that policymakers or wind power advocates that wish to foster change need to understand that entrenched interests have both economic and political power to impair progress. Accordingly, strategies to either appease entrenched interests or alter entrenched positions need to be considered at the agenda and formulation stages and must be implemented as part of the wind power development strategy.

11.3.4 Evolutionary Mindset

Research on successful wind power markets suggests that policy effectiveness is predicated on the understanding that wind power diffusion is an evolutionary process that features delineable stages that require different types of policy support.

At the inception—the foundation stage—policy must be designed to establish the structures necessary to support commercial wind power development. Wind power potential must be measured, prospective sites must be strategically selected, and networks must be established to encourage R&D, support technology transfer, and nurture collaboration between stakeholders. Failure to provide such a foundation sows the seeds of emergent problems when the program is actually launched. Denmark's methodical approach to planning for success is exemplary of the types of initiatives that must be formulated. For example, the research networks and communities which formed around the Risø Laboratories were instrumental in helping new wind turbine manufacturers to improve designs in a cost effective manner.

Once a foundation has been established, wind power development strategies can shift focus to capacity expansion. During the capacity expansion stage, wind turbine manufacturers, wind farm developers, and stakeholders in communities that are targeted for wind power development face steep learning curves that merit strategic attention.

In wind turbine manufacturing, aspiring domestic firms struggle to refine the technology to compete with more experienced foreign wind turbine manufacturers. Therefore, manufacturers require R&D seed support and access to advanced technological knowledge and may in some cases need to have market protection policies put in place in order to help them gain a market foothold. Wind turbine manufacturing firms in Denmark and China have benefited significantly from such policies.

Wind farm developers at the capacity expansion stage are simultaneously learning how to interact effectively with community stakeholders and struggling to improve strategies for wind park development in order to improve profitability. Policies designed to encourage applied knowledge transfer in developing sites and mechanisms designed to enhance collaboration between developers and community stakeholders can go a long way to fostering better wind power projects. Both Denmark and Germany seeded the collaborative process by establishing policies that encouraged community members to financially participate in wind power projects.

Community stakeholders also face a steep learning curve at the capacity expansion stage. Many community members are introduced to the prospect of their community hosting a wind farm without the benefit of applied experience. Consequently, they tend to harbor misperceptions concerning the pros and cons of wind farms. Then, once a wind farm is established, misperceptions are dispelled and new concerns emerge. Policies that are designed to help community stakeholders understand the main advantages and disadvantages of wind power development help attenuate dissonance caused by a gap between expectations and reality. In the United States, Germany, and Denmark, this has been achieved through open public forums where interested stakeholders are given an opportunity to come and learn about proposed developments and air grievances prior to the inception of the project.

Eventually, the wind power development process enters a third stage—maturity. During the maturity stage, wind power projects become more physically invasive and utilities are presented with greater challenges in terms of balancing loads. Typically at the maturity stage, larger scale developments become the norm. Cooperative or individual investments are superseded by wind farms that are owned and operated by larger corporate entities.

These developments pose new policy challenges. One key challenge at the maturity stage is maintaining community support for wind power expansion. Policies that continue to engender community benefits associated with wind farm developments are important. Moreover, policies aimed at encouraging the development of larger wind farms either offshore or in remote locations take on greater relevance. Consequently, policymakers also begin to confront challenges associated with connecting these remote projects to the grid. Germany is experiencing such challenges at the present time. In short, at the maturity stage, policymakers are increasingly taxed with the challenge of expanding the boundaries of public acceptance.

Failure to recognize that wind power development is an evolutionary process that requires different policies at different phases can lead to unanticipated inefficiencies. For example, in the early days of wind power development in the United States, insufficient attention was given to nurturing domestic wind power manufacturing firms that were capable of designing reliable wind turbines. Consequently, wind power systems that were installed in response to a subsequent subsidy were prone to failure, undermining utility support for this new technology. Similarly, both Denmark and Germany have discovered that policies aimed at encouraging a transition to large-scale wind farm developments need to be complemented by policies that continue to reward communities for hosting large-scale wind farms. Initial failure to incentivize communities to host such developments necessitated more elaborate policy adjustments that to this day are still underway.

11.4 AVENUES OF FURTHER RESEARCH AND CONCLUSION

In assessing the utility of the Political SET framework, one cannot help but be reminded of a quotation from Plato's Republic: "The learning and knowledge that we have is, at the most, but little compared with that of which we are ignorant." Due to the structured yet malleable approach advocated here, the insights provided in this book will help policymakers to fashion better wind power development policy and manage the process in a way that ensures that results approximate aspirations. However, there are avenues of future research that would significantly enhance the applied effectiveness of the Political SET model. Prior to concluding the study, seven questions for guiding future research will be briefly introduced with the hope that this encourages other researchers to expand on this work.

First, are some variables more prominent than others? In this chapter it was suggested that economic forces still rule the roost in terms of influencing wind power diffusion. It was subsequently noted that public opposition to high concentrations of wind power development could wind up replacing economic concerns as the prime factor in limiting wind power diffusion in mature wind power markets. This suggests that influential STEP variables within the Political SET framework are not equal. Therefore, when seeking to describe the Political SET conditions influencing a given nation, it would be beneficial to try and prioritize the STEP variables to ensure that more influential variables are given particular attention.

Second, are some variables consistently more influential? It may very well be that some variables are universally influential for wind power diffusion. For example, we know that the economic differential between wind power costs and the cost of conventional energy generation was a key factor in influencing wind diffusion in all six nations and we can speculate that this is likely true for all nations. Similarly, community perspectives in regard to wind power were seen as critical factors underpinning wind power development in five of the six nations (China being the exception). This therefore raises the prospect that some variables may be more universally relevant than others. Attempts to highlight which variables are more transferable to other national contexts would help improve external validity.

Third, is it possible to assign weights of importance to variables? At its extreme, the quest to prioritize variables based on comparative influence justifies attempts to quantify comparative influence. Due to the complex adaptive nature of energy sectors, this book contends that such a process is not achievable. However, this does not mean that attempts to try and quantify relative influence would not contribute to improved understanding of the interplay between variables. Attempts to quantify comparative influence force researchers to explain anomalies, fostering greater depth of understanding.

Fourth, are combinations of variables highly correlated? For example, to what degree is affluence correlated to community willingness to host wind power development? To what degree does the existence of domestic natural resource stocks influence the structuring of the electricity sector? It may be possible that combinations of variables are so highly correlated that they should be considered as ineluctably bound, implying that changes to one variable will result in predictable changes to another variable. Improved understanding of the relationships between the 32 variables presented in Chapter 10 would help to improve predictability in the face of change.

Fifth, can a set of variables be identified which can be isolated to improve wind power policy success? This study concluded that there were 32 variables which were demonstrated through the six case studies to influence wind power diffusion. It is conceivable that relationships between some of these variables are more influential than others. For example, it appears intuitively obvious from the case studies that dominant ideologies of political groups, the extent to which externalities are internalized, electricity sector structure, national affluence and community perspectives are five variables that seem to conflate to create a virtuous circle in certain nations. Conditions favoring wind power are advanced if the dominant political groups are pro-wind, policies exist that attempt to level the competitive playing field, the electricity sector is structured in a way to allow the government to better influence development, citizens are able and willing to pay higher electricity prices as externalities become internalized, and community perspectives are aligned in such a way to permit the establishment of invasive wind turbines. This begs the question: if these five factors alone were tightly managed, would these five forces be strong enough unto themselves to overcome impediments to wind power development caused by adverse conditions in some of the other 27 variables?

Sixth, how will the repercussions of climate change-related disasters and progressive fossil fuel price inflation influence the energy policy STEP environment? As the costs of climate change become clearer and the dangers more pronounced, one can hypothesize that many of the sociocultural, economic, and political impediments to wind power will diminish. Similarly, if fossil fuel prices continue to increase in response to elevated demand and dwindling supplies, the vested interests that have successfully stymied a transition away from fossil fuel-fired power will be increasingly hard-pressed to avoid disintegration of this advocacy regime. It may very well be that further developments in regard to climate change and fossil fuel costs will completely alter the dynamics of the STEP environment. More research is needed to understand how progressive developments in these two areas will impact the dynamics of the STEP environment.

Seventh, how will public perception evolve in a nation full of wind turbines? Currently, wind power turbines are still a novelty in many nations. For some host communities, the presence of a wind power farm is a symbolic badge of environmental honor. However, in Denmark and Germany there are already indications that public dissatisfaction is on the rise in many regions with high concentrations of wind farms. If this is occurring now in nations where wind power only contributes about 20% to the national power grid, what will public perception be like if wind power expands to 30 or 40% in these and other nations? More research is needed concerning the evolution of public perception as wind power diffusion enhances.

Notwithstanding these valuable avenues for future research, the Political SET model allows analysts to extract a much higher degree of contextually infused knowledge in a systematic manner. The 32 variables described in Chapter 10 that have played empirically verified roles in influencing wind power development provide analysts with a concrete foundation from which to commence analysis. If policy is formulated by taking into consideration the unique national make-up of these 32 STEP variables the prospects for policy success will be elevated. Furthermore, if the insights described in this chapter are then used to guide policy implementation, monitoring, and reformulation, the effectiveness of wind power policy can be maintained throughout the stages of wind power diffusion. Ensuring that the structure of policies are customized to synergize with the national dynamics of the energy sector's STEP variables necessitates infusing the policy process with malleability, transparency, public participation, strategic balance, and evolutionary understanding. Plato might be right in observing that the knowledge that we currently have pales in comparison to the knowledge that we do not have. But on the other hand, a rudimentary map is far better than not possessing a map and all.

NOTES

1. Valentine, Scott Victor. 2010. "A STEP Toward Understanding Wind Power Development Policy Barriers in Advanced Economies." *Renewable and Sustainable Energy Reviews* 14 (9): 2796–2807. 10.1016/j.rser.2010.07.043.
2. A very useful book that describes the strategic challenges posed by complex adaptive systems is Brown, Shona L., and Kathleen M. Eisenhardt. 1998. *Competing on the Edge: Strategy as Structured Chaos.* Cambridge, MA: Harvard Business School Press.
3. The most recent energy outlook report from the International Energy Agency (IEA) is illustrative of alarm in mainstream energy circles that the current pace of transition away from carbon-intensive electricity generation is too slow to abate undesirable costs attributed to the worst impacts of climate change. International Energy Agency (IEA). 2013. *World Energy Outlook 2013.* Paris: International Energy Agency.
4. For a cost comparison see International Renewable Energy Agency (IRENA). 2012. *30 Years of Policies for Wind Energy: Lessons from 12 Wind Energy Markets.* Abu Dhabi: International Renewable Energy Agency.

5. For more on this see Beinhocker, Eric. 1999. "Robust Adaptive Strategies." *Sloan Management Review* 4 (3): 95–106.

6. For example see International Renewable Energy Agency (IRENA). 2012. *30 Years of Policies for Wind Energy: Lessons from 12 Wind Energy Markets.* Abu Dhabi: International Renewable Energy Agency; and García, Clara. 2011. "Grid-Connected Renewable Energy in China: Policies and Institutions Under Gradualism, Developmentalism, and Socialism." *Energy Policy* 39 (12): 8046–8050. 10.1016/j.enpol.2011.09.059.

7. Valentine, Scott Victor. 2013. "Wind Power Policy in Complex Adaptive Markets." *Renewable and Sustainable Energy Reviews* 19: 1–10. http://dx.doi.org/10.1016/j. rser.2012.11.018.

8. More on the benefits of size and market share can be found in Doyle, Peter. 1998. *Marketing Management and Strategy. 2 ed.* Harlow: Prentice Hall Publishing.

9. Valentine, Scott Victor. 2011. "Understanding the Variability of Wind Power Costs." *Renewable and Sustainable Energy Reviews* 15 (8): 3632–3639. 10.1016/j. rser.2011.06.002.

10. For example, see Buen, Jorund. 2006. "Danish and Norwegian wind Industry: The Relationship Between Policy Instruments, Innovation and Diffusion." *Energy Policy* 34 (18): 3887–3897. 10.1016/j.enpol.2005.09.003; and Haggett, Claire. 2011. "Understanding Public Responses to Offshore Wind Power." *Energy Policy* 39 (2): 503–510. 10.1016/j.enpol.2010.10.014; and Valentine, Scott Victor. 2011. "Sheltering Wind Power Projects from Tempestuous Community Concerns." *Energy for Sustainable Development* 15 (1): 109–114. 10.1016/j.esd.2010.11.002.

11. For more on Advocacy Coalition Frameworks, see: Sabatier, Paul A., and Hank C. Jenkins-Smith, eds. 1993. *Policy Change and Learning: An Advocacy Coalition Approach, Theoretical Lenses on Public Policy.* Boulder, CO: Westview Press.

12. Sabatier, Paul A. 1988. "An Advocacy Coalition Framework of Policy Change and the Role of Policy-Oriented Learning Therein." *Policy Sciences* 21 (2–3): 129–168.